VACUUM DESIGN OF SYNCHROTRON LIGHT SOURCES

CONFERENCE PROCEEDINGS NO. **236**

AMERICAN VACUUM SOCIETY SERIES 12

SERIES EDITOR: GERALD LUCOVSKY
NORTH CAROLINA STATE UNIVERSITY

VACUUM DESIGN OF SYNCHROTRON LIGHT SOURCES

ARGONNE, IL 1990

EDITORS:

YELDEZ G. AMER
SAMUEL D. BADER
ALAN R. KRAUSS
RALPH C. NIEMANN
ARGONNE NATIONAL LABORATORY

American Institute of Physics New York

Authorization to photocopy items for internal or personal use, beyond the free copying permitted under the 1978 U.S. Copyright Law (see statement below), is granted by the American Institute of Physics for users registered with the Copyright Clearance Center (CCC) Transactional Reporting Service, provided that the base fee of $2.00 per copy is paid directly to CCC, 27 Congress St., Salem, MA 01970. For those organizations that have been granted a photocopy license by CCC, a separate system of payment has been arranged. The fee code for users of the Transactional Reporting Service is: 0094-243X/87 $2.00.

© 1991 American Institute of Physics.

Individual readers of this volume and nonprofit libraries, acting for them, are permitted to make fair use of the material in it, such as copying an article for use in teaching or research. Permission is granted to quote from this volume in scientific work with the customary acknowledgment of the source. To reprint a figure, table, or other excerpt requires the consent of one of the original authors and notification to AIP. Republication or systematic or multiple reproduction of any material in this volume is permitted only under license from AIP. Address inquiries to Series Editor, AIP Conference Proceedings, AIP, 335 East 45th Street, New York, NY 10017-3483.

L.C. Catalog Card No. 91-55527
ISBN 0-88318-873-2
DOE CONF-901144

Printed in the United States of America.

Contents

Preface viii

Upgrade Design for Vacuum System of TRISTAN Accumulating Ring 1
    H. Ishimaru, T. Momose, K. Kanazawa, Y. Suetsugu, and H. Hisamatsu

LEP Vacuum System Start-Up and Operation Experience 18
    O. Gröbner and P. Strubin

Design of the Vacuum System TNK Dedicated Synchrotron Source for X-ray Lithography 30
    V. V. Anashin, A. N. Bulygin, N. G. Gavrilov, E. P. Kollerov, V. N. Korchuganov, A. I. Nikitin, V. N. Osipov, and E. M. Trakhtenberg

NSLS Vacuum System Operating Experience Conditioning and Desorption Yields 39
    Henry J. Halama

Vacuum Experience and Experiments at Super-ACO 52
    P. Marin

Development of the ESRF Vacuum System 71
    M. Renier, D. Schmied, and B. A. Trickett

APS Storage Ring Vacuum System 84
    R. C. Niemann, R. Benaroya, M. Choi, R. J. Dortwegt, G. A. Goeppner, J. Gonczy, C. Krieger, J. Howell, R. W. Nielsen, B. Roop, and R. B. Wehrle

Vacuum System for SPring-8 Storage Ring and Test Experience 102
    S. H. Be, S. Yokouchi, I. Nishidono, Y. Morimoto, K. Watanabe, S. Takahashi, S. R. In, H. Daibo, and Y. OiKawa

Design of the Crotch for SPring-8 110
    Y. Morimoto, T. Shirakura, K. Konishi, S. Takahashi, S. Yokouchi, and S. H. Be

Vacuum Performance of the LNLS Injector Linac—Present Status 118
    Paulo Alberto Paes Gomes

APS Storage Ring Vacuum Chamber Fabrication 124
    George A. Goeppner

The Vacuum System of IHI's Compact SOR "LUNA" 142
    M. Oishi, M. Uesaka, S. Mandai, T. Nishidono, T. Nakashizu, and Y. Hoshi

Vacuum Control and Interlock System for VUV Photochemistry Beamline of HESYRL 152
    X. Xu, W. Xu, and C. Yao

Vacuum System Design of SRS Indus-I 155
    S. S. Ramamurthi, M. G. Karmarkar, and R. J. Patel

Copper Washer Positioner for Conflat Flange 168
    Guihe Li

SRRC Vacuum System 173
    Y. C. Liu and J. R. Chen

Vacuum System Experience at the Daresbury SRS 183
    R. J. Reid, S. F. Hill, and P. A. Crank

ELETTRA Vacuum System 188
   M. Bernardini

Manufacture of the ALS Storage Ring Vacuum System 197
   Kurt Kennedy

Vacuum System Design for the PLS Storage Ring 202
   C. D. Park, K. H. Kil, C. K. Kim, and S. M. Chung

LSU Electron Storage Ring Vacuum System Design 210
   Donald E. Geiler

ALS Insertion Devices 219
   E. Hoyer, J. Chin, K. Halbach, W. V. Hassenzahl, D. Humphries, B. Kincaid, H. Lancaster, and D. Plate

Carbon and Other Contaminants in Vacuum Systems 235
   Victor Rehn

The Design of ESRF Front Ends 266
   Trevor Mairs and Jean Claude Biasci

Free-Electron Laser Sources of Extreme-Ultraviolet Radiation and their Vacuum Requirements 278
   Brian E. Newnam

Review of Vacuum Systems for X-Ray Lithography Light Sources 300
   J. C. Schuchman

Comparison of Synchrotron Radiation Induced Gas Desorption from Al, Stainless Steel, and Cu Chambers 313
   A. G. Mathewson, O. Gröbner, P. Strubin, P. Marin, and R. Souchet

The Search for Low Photodesorption Coatings 325
   C. L. Foerster and G. Korn

Investigation of Photodesorption in a Chamber of the Electron Storage Ring 332
   Masanori Kobayashi

Photoelectron Effect on Photodesorption in a Chamber Irradiated by Synchrotron Radiation 347
   T. Kobari, M. Matsumoto, T. Ikeguchi, S. Ueda, M. Kobayashi, and Y. Hori

Glow Discharge Cleaning Effects on Aluminum Alloy by TDS and SIMS Surface Analysis 355
   G. Y. Hsiung, J. R. Chen, and Y. C. Liu

The Effect of "Diversey" Cleaning on the Surface Composition of 304 Stainless Steel 361
   T. W. Rusch, R. J. Liptak, and W. J. Eberle

Performance Characteristics of Lumped NEG Pump 374
   S. R. In, S. Yokouchi, S. H. Be, and T. Maruyama

Vacuum Performance of Vacuum Chamber with a NEG Strip of 8 and 16 m 381
   S. Yokouchi, S. R. In, T. Nishidono, H. Daibo, and S. H. Be

Vacuum System Design for a 1.2 GeV Electron Storage Ring with Non-Evaporable Getter Pumping 389
   H. F. Dylla, D. M. Manos, J. C. Citrolo, P. H. LaMarche, S. Raftopoulas, M. Ulrickson, A. G. Mathewson, A. Poncet, and F. Mazza

Modern Residual Gas Analyser (RGA) 404
   S. P. Shannon and A. P. James

Vacuum Valves for Synchrotrons and Storage Rings.. 418
    John A. Freeman

Author Index.. 427

Note: The author who delivered the talk is designated by italics.

Preface

The Second Topical Conference on Vacuum Design of Synchrotron Light Sources was held at Argonne National Laboratory on November 13–15, 1990. The conference was cosponsored by Argonne, the American Vacuum Society, and The University of Chicago. There were 126 participants representing 13 countries. The purpose was to bring together vacuum specialists working on existing synchrotron light sources, designing the next generation of insertion-device machines, or contemplating novel concepts that undoubtedly will make the future very exciting. The successful utilization of synchrotron radiation world-wide has stimulated numerous advanced source projects, including a machine presently under construction at Argonne. Vacuum design, vacuum materials, and surface treatment are of critical importance to the success of the myriad of international efforts to use synchrotron radiation in multidisciplinary basic and applied research. The purpose of the present volume is to provide a legacy of the meeting and, hopefully, to contribute to the successful commissioning and operation of these invaluable new resources. The order of presentation of articles follows closely that established by the conference program.

We acknowledge all parties who contributed to the conference and to the existence of this volume. It is hoped that users of this volume will derive some of the same sense of satisfaction that we have felt in assembling it.

Yeldez G. Amer
Samuel D. Bader
Alan R. Krauss
Ralph C. Niemann

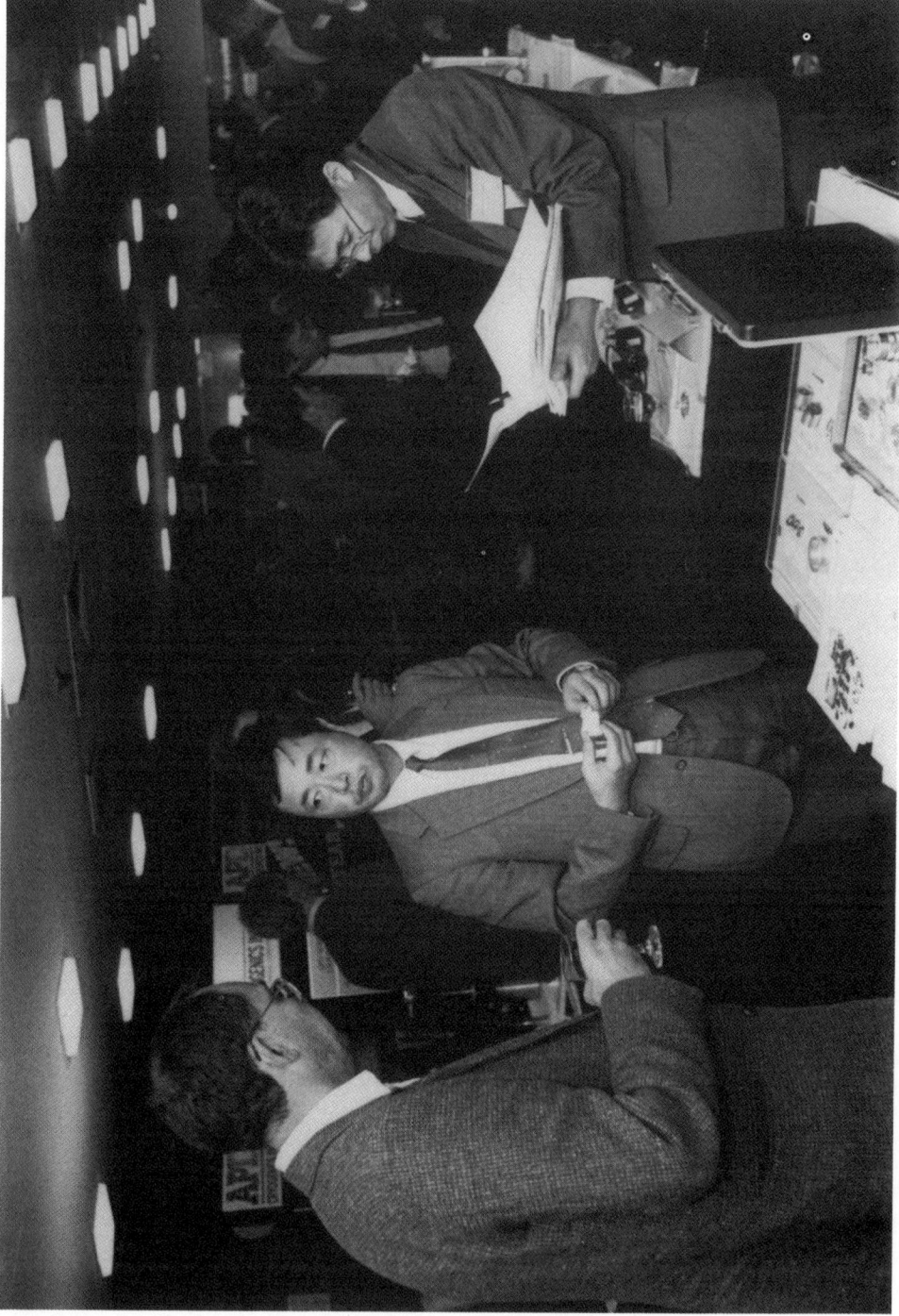

UPGRADE DESIGN FOR VACUUM SYSTEM OF TRISTAN ACCUMULATION RING

Hajime Ishimaru, Takashi Momose, Ken'ichi Kanazawa,
Yuhsuke Suetsugu, and Hiromi Hisamatsu
National Laboratory for High Energy Physics
1-1 Oho, Tsukuba, Ibaraki, 305, Japan

ABSTRACT

Present operational parameters of TRISTAN accumulation ring (TAR) are 6.5 GeV, 30 mA with 6 hours beam lifetime as a synchrotron radiation source and 8 GeV, 20 mA as an injector to the main ring. The accumulation ring has several modes of operation: as a synchrotron radiation source, an injector for the planned B(B-meson)-factory and an injector for the TRISTAN main ring. The required maximum beam current and lifetime depend on the application. In order to improve performance for use as a dedicated synchrotron light sourse, it is desirable to improve the beam current and lifetime. The design goal of the upgrade is 100 mA at 8 GeV. In order to achieve this goal, the effective pumping speed of the Distributed Ion Pump (DIP) was increased by a factor of 3. For the quadrupole magnet chambers, Non-Evaporable Getter (NEG) cartridges are more suitable. The effective average pumping speed along the ring as a whole will be three times higher than that of the present TAR. On the other hand, the present beam current can not be raised more than 30 mA at 6.5 GeV because absorbers of ceramic chambers for kicker magnets have poor cooling and radiation power shielding functions. The newly designed ceramic chambers are protected by new water cooled absorbers and offset structure. After installation, it is expected that the improved beam current and lifetime will be obtained quickly once the modification has been made. UHV performance of individual components must be confirmed by preassembling, prebaking, and preconditioning of the DIP, lumped Sputter-Ion Pump (SIP), Titanium Sublimation pump (TSP), and NEG before installation. All assembly is done in a clean room. The chamber material has been changed from A6063 alloy to high purity 99.99 % aluminum.

2 Upgrade Design for Vacuum System

I. INTRODUCTION

Design, research and development of TRISTAN accumulation ring (TAR) were begun around 1979. Construction was carried out during FY 1981 and 1982. In November 1983, beam current was achieved thereby setting the record for minimum time between construction and operation of any accelerator. Since completion of the TRISTAN main collider ring, TAR has served as an injector for the main ring. In addition, TAR also has been used for an internal target experiment, synchrotron radiation (SR) studies, and a B-meson-physics experiment. TAR presently provides 20 mA at 8 GeV as an injector to the collider, or 30 mA at 6.5 GeV when used as a storage ring for Synchrotron Radiation (SR) studies. The average pressure is 10^{-9} torr with the beam off and 10^{-8} torr with the beam on. Beam lifetime is about 6 hours. The average pressure rise is several times 10^{-10} torr/mA for time integrated beam current of about 200 Ah without any baking and discharge cleaning. As the number of SR and colliding beam experiments increases, it will be necessary to increase the beam current and beam lifetime.

To improve beam lifetime by one order of magnitude, the effective pumping speed must be increased by one order. The straight section between the bending (B) and quadruple (Q) magnets in TAR is the shortest in the existing electron storage rings. The main pump of TAR is a distributed ion pump (DIP) in a bending magnet (B-) chamber; however, there is no DIP in the quadrupole (Q-) magnet chamber. The latter fact limits beam lifetime. More information on gas composition and pressure distribution are necessary to investigate beam lifetime. On the other hand, the present beam current can not be raised more than 30 mA at 6.5 GeV because synchrotron radiation absorbers of ceramic chambers have poor cooling and radiation power shielding functions. Therefore new absorbers for the ceramic chambers must be designed. Since a sudden decrease in the lifetime due to a fine dust trapping[1] has been observed more than one in a day, assembly to decrease dust in the chamber must be taken into account.

II. BASIC DESIGN[2]

The final goal of the upgrade design is 100 mA at 8 GeV, with 10 hours beam lifetime. Installation will be done during a scheduled shutdown in the summer of 1991. After installation, it is anticipated that the desired beam current and lifetime will be obtained quickly. Failures in the past have been mainly due to compromise in design, therefore, the highest performance is based on our design. In ordering equipment, performance requirements must be added to the traditional structural and material requirements. UHV characteristics must be confirmed with preassembling, prebaking, precleaning using oxygen plasma[3] and preconditioning before installation. The highest performance is pursued utilizing extremely high vacuum (XHV) techniques[4]. No trouble of vacuum components can decrease pressure rise of the ring because the number of air exposures of the beam chamber during maintainance work can be decreased. As a result beam lifetime can be increased. Therefore, reliability of the components should be investigated thoroughly. To increase beam lifetime, the most effective distribution of pumps as well as gauges is required. As many of the synchrotron radiation ports as possible are installed in the bending magnet chambers. The chamber material is determined on the basis of past research and development. With regard to fabrication, surface treatment techniques established during the development of XHV technology are applied. Welding in a chamber with argon atmosphere, automatic welding for elliptical cross section, and electron-beam welding are positively adopted. All assembly is done in a clean room to suppress the influence of microdust.

III. R&D, AND DETAIL DESIGN
Beam Chambers

The chamber material has been changed from 6063-T6 to high purity (99.99 %) aluminum. Bulk material is used since it is difficult to obtain uniformity in a clad material[5]. The high purity aluminum is melted and degassed with an Ar/Cl_2 mixture using bubbling flushing to make an aluminum bullet. The high purity aluminum bullet is degassed in a vacuum furnace before extrusion. A special extrusion process using an Ar/O_2 mixture with water content in the ppm range is used in the chamber to keep the surface clean. Internal hydrogen gas content of the high purity aluminum chamber is less than 0.05 cc/100 gr Al[6]. The thermal outgassing rate[7] is 2×10^{-14} torr l/s cm^2 for the high purity aluminum chamber at 150°C after 24 hour baking. The EX-process and EL-process[4] are used for all machining processes. The cross section of the chamber is changed from racetrack type to elliptical one. The elliptical cross section improves the uniformity of contact forces at the RF shield installed in the bellows, that is, stress against chamber deformation is decreased. The gap between two water cooling pipes in a B-chamber is enlarged as much as possible to extract SR effectively in to the SR ports (Fig. 1). Pumping holes are changed from long rectangular holes (4 mm x 40 mm, 20 mm spacing, as shown by the double line) to circular holes (10 mm in diameter, 10 mm spacing) for easy fabrication in the Ar/O_2 atmosphere. Bending of B-chambers is also performed using a stretch forming process in an Ar/O_2 atmosphere. Seventeen SR ports (ten ports for electrons and seven ports for positrons) are installed in B-chambers (Fig. 2). A synchrotron radiation shield for undulators is installed in one of the SR ports. Power density at the shield is extremely high (about 10.7 Kwatts/m for 8 GeV, 100 mA beams). Absorbers with Be-Al-Cu structure are being considered for this high power density. Precise alignment must be available for contraction and expansion during baking. Q-chambers (Fig. 3) have additional pumping ports for the ST-707-C150 NEG cartridges. Beam position monitors (BPM) are electron-beam welded to the Q-chambers. The gap between the pickup electrode and the housing must have a width of 2 mm and the surface of the electrode and the housing is covered with TiN film to protect against unwanted multipactoring discharge due to bunched beam current. In addition the electrode material is changed from stainless steel to Ti. The pickup electrode with an SMA-type feedthrough is demontable if the feedthrough is damaged due to corrosion. Calibration of the feedthrough as well as the beam position monitor could be rendered unnecessary by fabricating the feedthrough and the beam position monitor more precisely than the traditional ones. Frames are used to fix the BPM against a Q-magnet. Electron beam welding is extensively used to minimize the welding stress. The beam chambers along the arc section are joined with welding without flanges. For a long straight section, an RF section and an interaction section, separate beam chamber unit are joined each other with flanges. An RF shield is installed between the flanges. Installation in the tunnel is done using automatic welding machines. The RF shield is equipped with an elliptic bellows (Fig. 4). The aluminum alloy bellows are coated with SiO_2[8] to protect bellows surface from radiation corrosion. Fig. 5 shows absorbers for bellows.

<u>Pumps</u>

For the DIP in B-chambers, pumping characteristics in the UHV range are critical. The simple electrode structure absence of connecting bolts, and use of thermal-expansion-free parts (Fig. 6) suppress abnormal discharge, short circuit, and insulation damage in the present DIP electrodes. Ceramic insulation is protected from contamination by a cup. For Q-chambers, NEG cartridges are adopted as a main pump. NEG modules and sputter ion pumps (SIP) are used in the straight sections. Titanium sublimation pumps (TSP) are used at the synchrotron radiation ports in the B-chambers. SIPs are installed as close to the beam as possible. The main pumps serve to evacuate CO effectively. High voltage conditioning[4] is applied to DIP and SIP. The roughing pump system with a port surrounding the beam chamber to improve conductance has a 300 l/s turbomolecular pump (TMP) with magnetic bearings. The pumping speed for CO in 10^{-10} torr range was measured precisely. Partial pressure (especially for CO) was measured during operation to give precise correspondence between beam lifetime and pressure. The Star-cell pump element is housed in a high purity aluminum shell. Pumping speed for CO and H_2 was about twice that obtained with an ordinary stainless steel housing in the $10^{-9} - 10^{-10}$ torr range[4]. Pumping speeds of the DIP and SIP were measured in the 10^{-10} torr range with background pressure of the order of 10^{-11} torr. The measurement is done to determine the effect of preconditioning and baking as well as the effect of magnetic field[9] (Fig. 7). The pump and pressure distribution for unit half cell are shown in Fig. 8. The pressure distribution is obtained by considering the gas load due to SR and adopting as many DIPs, SIPs, NEGs, and TSPs as possible. The predicted effective pumping speed is at least three times higher than the present TAR.

Vacuum Monitoring

As many gauges are installed as possible to get information for beam lifetime and to detect small leaks. Cold cathode gauges and B-A gauges are used to measure 10^{-11} torr. Many quadrupole mass filters are used. The gauges are placed close to the beam for precise monitoring.

Other Components

Gate valves, light angle valves, high voltage feedthroughs for DIPs and SIPs, and the absorbers for the ceramic chamber have had problems in the past and need to be made more reliable. Gate valves with dual flat-face seals should have lifetime and reliability of more than 10^4 operations[10], provided that no small dust particles destroy the sealing characteristics. Improved angle valves using Helicoflex seals have an operating life of 5,000 cycles with constant tightening torque, and have leak rates less than 10^{-11} torr l/s[11]. The Helicoflex was welded to the valve body. So far, several operation tests after transportation were carried out, however, early failures were inevitable. From now on, repeated tests amounting to several % of the life time should be applied in order to prevent early failures. The high voltage feedthroughs must be durable to the high voltage conditioning at 15 kV which is higher than the operation voltage of 5.5 kV. The feedthrough is exposed to convection flow to suppress the corrosion due to synchrotron radiation. The feedthroughs have to have mineral insulated cables to suppress radiation damage and corrosion. Ceramic chambers with internal metallic film (TiMo) but no flange have a wide aperture of 100 mm and offset of 10 mm to escape SR. Absorbers[12] for ceramic chambers are made more effective (Fig. 9) than the traditional ones by supplying cooling water. Scrapers, gaskets, an internal target, gauges, and other small components are prebaked and preconditioned in a clean vacuum furnace[13]. Super-dry nitrogen[14] (10 ppb in H_2O) is introduced to improve reevacuation characteristics. Elastomer O-rings in roughing pump systems have been changed to metal seals to suppress radiation damage.

IV. INSTALLATION

Installation will be made during a short scheduled shutdown in summer. A B- and a Q-chamber will be welded into a connected unit in the clean room of the assembly factory. Using metal sealed flanges at both ends of the unit, UHV performance will be confirmed by applying pretreatments. On DIP, preconditioning is applied using B-magnets. In a miniclean room in the tunnel, one can automatically produce a weld simply by removing the flanges. The chambers are evacuated to the order of 10^{-8} torr. Initial bake-out is started. After pressure in the 10^{-7} torr range is reached, DIP and SIP are started. After bake-out in the tunnel a more strict leak test[15] of 10^{-11} torr l/s is applied. To maintain effective CO pumping speed, NEG cartridges and strips are activated at low pressure. TSPs are also preconditioned using other chambers. Sufficient rebaking and reconditioning must be applied to the DIP and SIP to prevent the exchange desorption due to water vapor. At the initial operation with beam, parallel operation of the main pumps with the roughing pump system must be used.

V. SCHEDULE

Final design was completed by the end of FY 1990. Design and arrangement of magnets and chambers are done and controlled using CAD. Fabrication test of B- and Q-chambers, bellows, DIP and NEG have been performed. UHV characteristics of DIP and NEG were obtained. By the end of FY 1990 three combined units of a B- and a Q-chambers was completed. The units will be fully evaluated. Beam loading test on the unit will be done in TAR after installation in summer 1991. After confirming the characteristics and reliability of the units, whole chambers will be produced in FY 1991-1992. During the scheduled shutdown in summer 1993, whole chambers will be renewed.

VI. SUMMARY

Beam lifetime and current are mainly limited by the vacuum characteristics of TAR. In the future, the role as an SR ring or electron-positron colliding ring will increase rather than relative to the present role as the accumulation ring for the main colliding ring. To meet this status change, a redesign of TAR has been investigated. The new TAR design is expected to overcome problems encountered in the past. Based on previous research and development, as well as 10 year's operating experience at TRISTAN, we expect to be able to reliably produce an 8 GeV, 100 mA stored beam with a 10 hour lifetime.

ACKNOWLEDGEMENTS

The authors wish to thank Professors Y. Kimura, K. Takata, and S. Kurokawa for their continuous encouragements. Detailed design using CAD has been done by Ishikawajima Harima Heavy Industries Ltd.

REFERENCES

1. H. Saeki, T. Momose, and H. Ishimaru, to be published in Rev. Sci. Instrum.
2. H. Ishimaru, T. Momose, K. Kanazawa, Y. Suetsugu, and H. Hisamatsu, Proceedings of Acc. Sci. Technol., Osaka Univ., Dec. 1989, 121.
3. M. Saitoh, J. Uramoto, and H. Ishimaru, Proceedings of the 13th Symposium on Ion Sources and Ion-assisted Technol., Tokyo, June 1990, 45.
4. H. Ishimaru, J. Vac. Sci. Technol. A7 (3), May/June 1989, 2439.
5. H. Ishimaru, K. Narushima, and K. Kanazawa, American Institute of Physics, Conference Proceedings, No. 171, 1988, 219.
6. Y. Kato, T. Kitamura, E. Isoyama, M. Hasegawa, T. Momose, and H. Ishimaru, J. Vac. Sci. Technol. A6, Nov./Dec. 1988, 3111.
7. S. Saitoh, K. Shimura, T. Iwata, T. Momose, and H. Ishimaru, to be submitted to J. Vac. Sci. Technol.
8. M. Miyamoto, Y. Sumi, S. Komaki, K. Narushima, and H. Ishimaru, J. Vac. Sci. Technol. A4, (1986) 2515.
9. Y. Suetsugu, and M. Nakagawa, to be published in Vacuum.
10. H. Ishimaru, MRS BULLETIN, July 1990, 23.
11. K. Itoh, K. Waragai, H. Komuro, T. Ishigaki, and H. Ishimaru, J. Vac. Sci. Technol. A8, May/June, 1990, 2836.
12. T. Momose, O. Kon'no, H. Hirayama, and H. Ishimaru, KEK Preprint 90-59 July 1990, A.
13. K. Itoh, T. Ishigaki, A. Kamikawana, and H. Ishimaru, J. Vac. Sci. Technol. A7, 1989, 2435.
14. H. Ishimaru, K. Itoh, T. Ishigaki, and S. Furutate, 37th AVS, Toronto, Canada 1990.
15. F. Watanabe, and H. Ishimaru, J. Vac. Sci. Technol. A8, May/June 1990, 2795.

FIGURE CAPTIONS

Fig. 1 A cross section of a bending magnet chamber.
Fig. 2 A bending magnet chamber with a synchrotron light port and a pumping port for TMP, NEG cartridge and TSP.
Fig. 3 A cross section of a quadrupole magnet chamber with beam position monitor.

8 Upgrade Design for Vacuum System

Fig. 4　　Forming type aluminum alloy bellows with thick ends. An RF shield made of BeCu is installed inside of the bellows. The outer surface is coated with SiO_2.
Fig. 5　　Absorbers for bellows.
Fig. 6　　A distributed ion pump unit.
Fig. 7　　Magnetic field dependence of pumping speed of the distributed ion pump unit.
Fig. 8　　A distribution of pumps, and calculation of pressure and pumping speed distribution.
Fig. 9　　Aluminum alloy bellows and absorbers for a ceramic chamber.

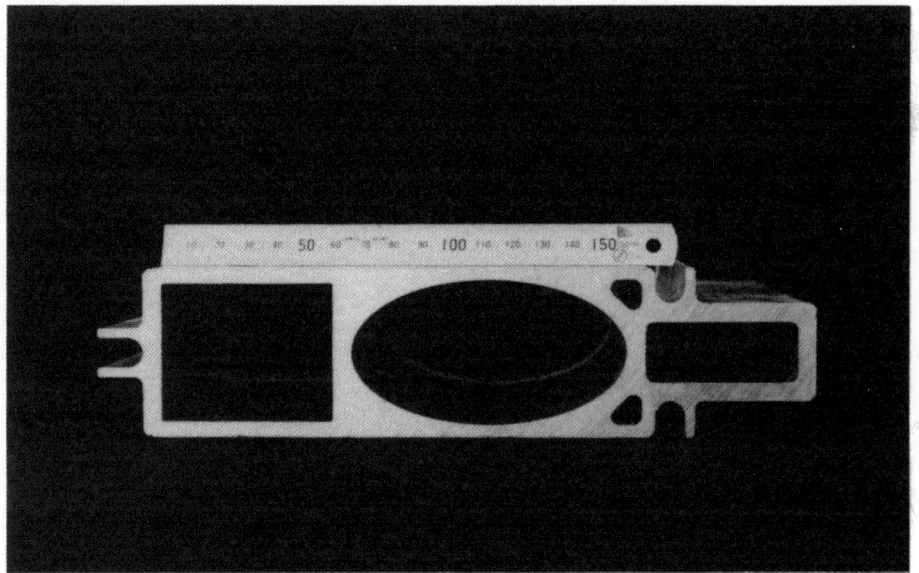

Fig.1 A cross section of a bending magnet chamber.

10 Upgrade Design for Vacuum System

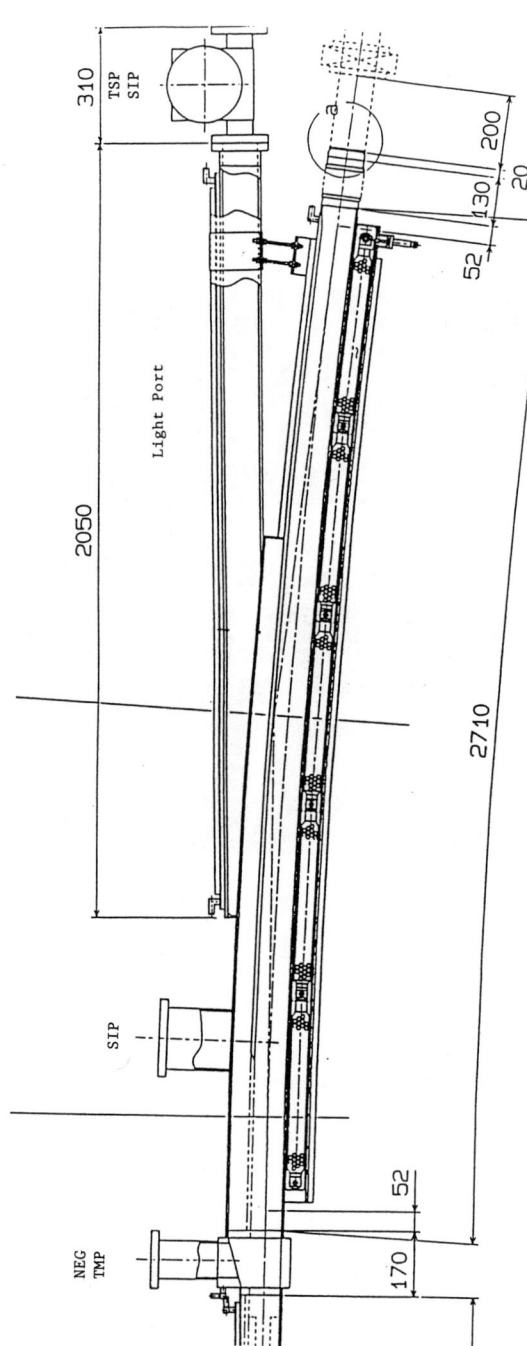

Fig. 2   A bending magnet chamber with a synchrotron light port and a pumping port for TMP, NEG cartridge and TSP.

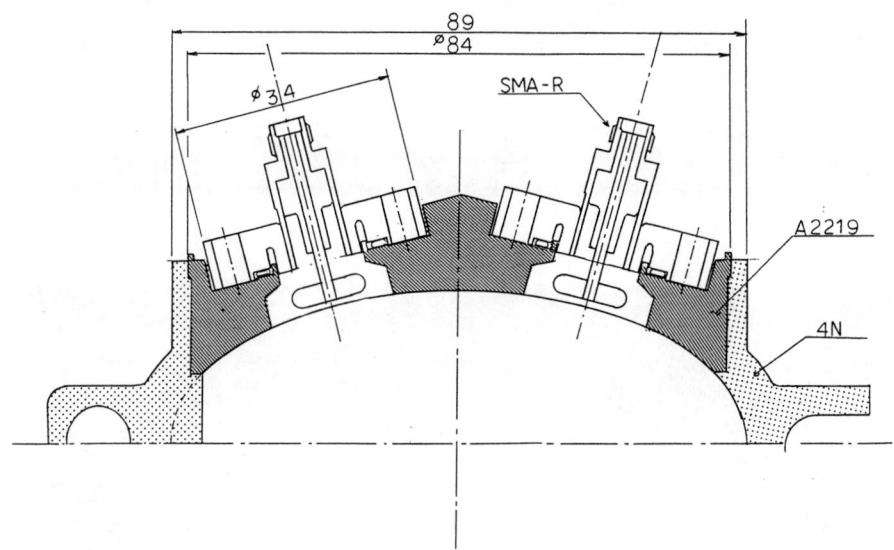

Fig.3 A cross section of a quadrupole magnet chamber with BPM.

12 Upgrade Design for Vacuum System

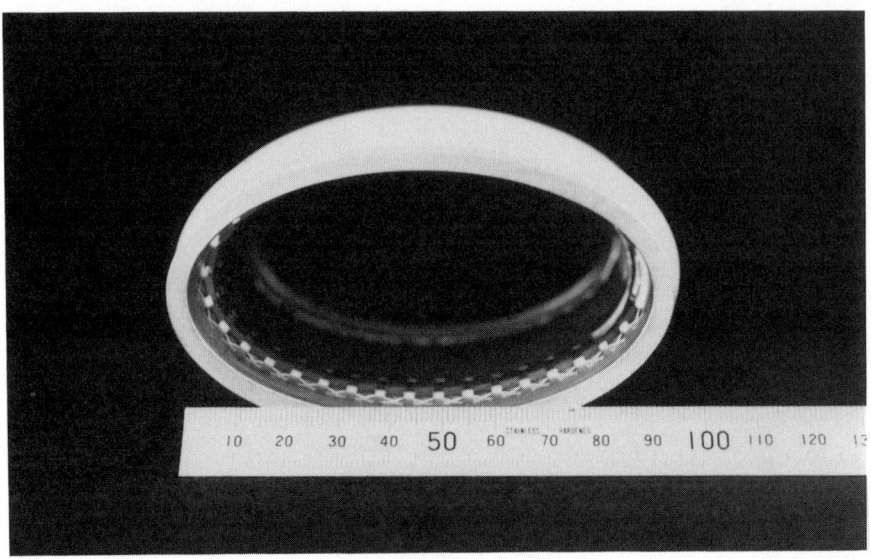

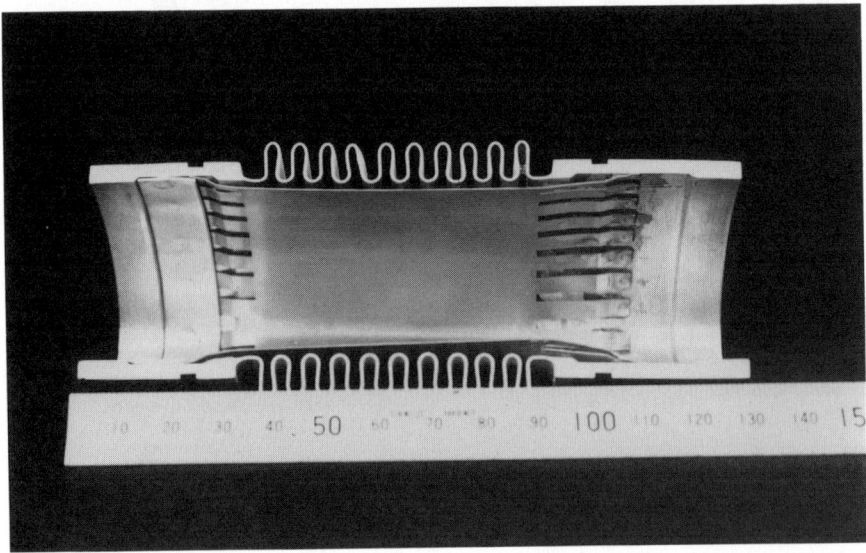

Fig.4 Forming type aluminum alloy bellows with thick ends. An RF shield made of BeCu is installed inside of the bellows. Outer surface is coated with SiO_2.

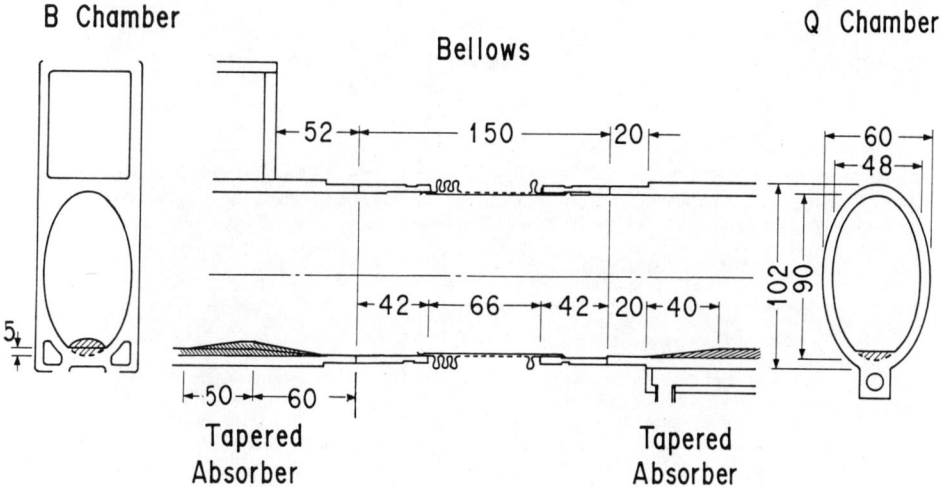

Fig.5 An a absorber for bellows.

14 Upgrade Design for Vacuum System

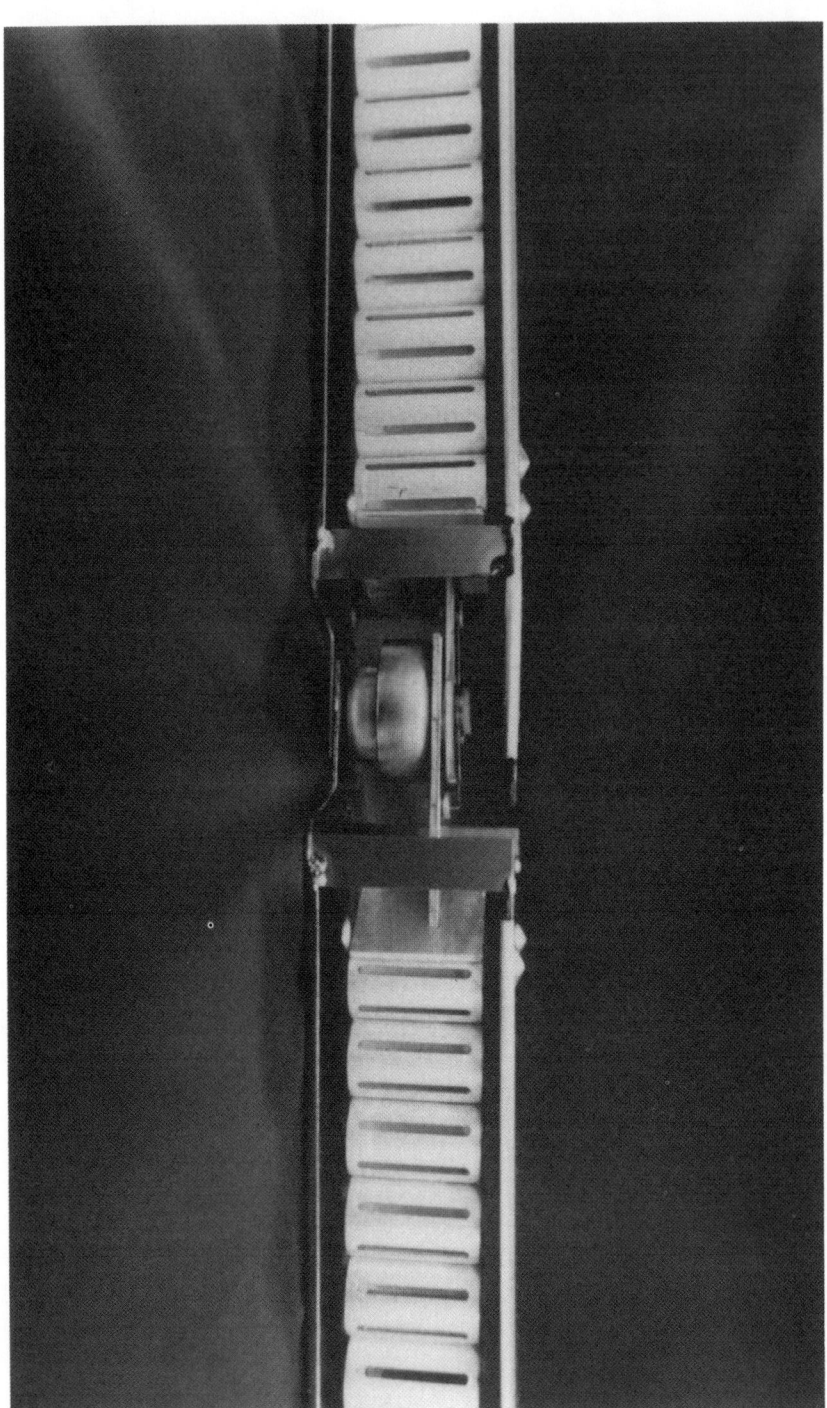

Fig.6 A distributed ion pump unit.

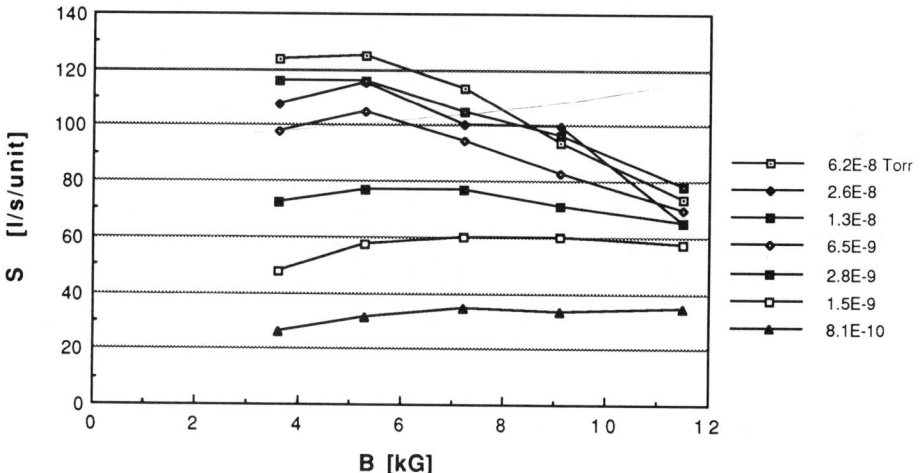

Fig. 7   Magnetic field dependence of pumping speed of the distributed ion pump unit.

16 Upgrade Design for Vacuum System

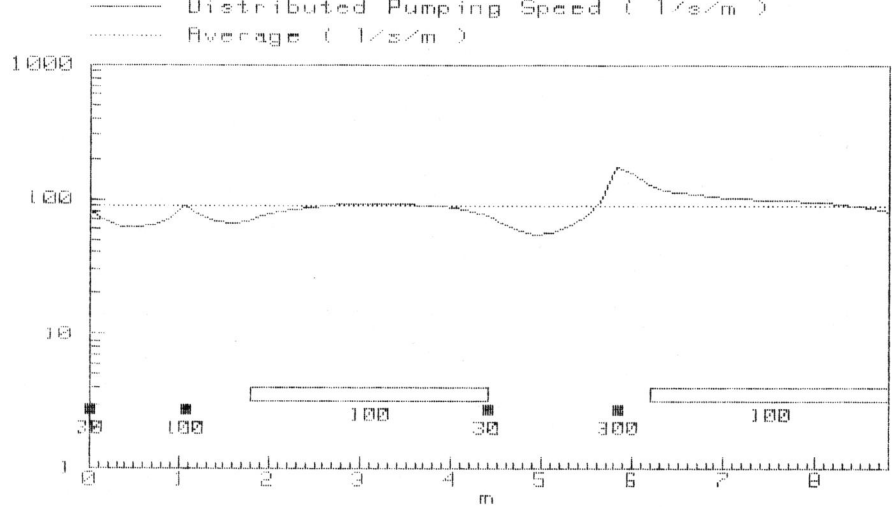

Fig.8 A distribution of pumps, and calculation of pressure and pumping speed distribution.

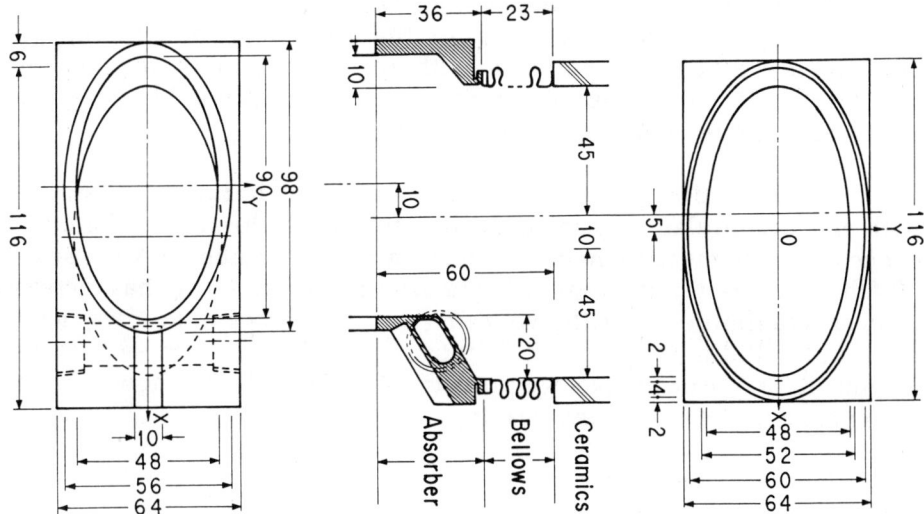

Fig.9 A ceramic chamber with aluminum alloy bellows and absobers.

LEP VACUUM SYSTEM START-UP AND OPERATION EXPERIENCE

O. Gröbner and P. Strubin, CERN, Geneva, Switzerland

ABSTRACT

The 27 km long ultra-high vacuum system of LEP has been successfully commissioned in July 1989. A significant amount of useful experience has been gained during the first year of running. The performance and the experience with the operation of the pumping system of LEP, which uses a combination of non-evaporable getter strip (NEG) and lumped ion pumps are described.

The reduction of the synchrotron radiation induced pressure rise with the accumulated beam dose has followed closely the behaviour expected from measurements on test vacuum chambers. After several months of operation the beam lifetime due to gas scattering has increased from several minutes, for the very first circulating beam, to many tens of hours for colliding beam runs. In most of the 130 vacuum sectors, with lengths up to 474 m, an average pressure of about 10^{-8} Pa is now measured in the presence of circulating beams.

As well as the performance of the vacuum system, a short report on our experience with the vacuum controls is given.

MAIN CHARACTERISTICS OF THE LEP VACUUM SYSTEM

Description of the vacuum system

A detailed description of the vacuum system for LEP may be found in previous notes [1,2], the main characteristics are the following: of a total circumference close to 26.6 km, about 22 km represent a regular, repetitive structure in the bending arcs of the machine and consist of a succession of 3 bending magnet vacuum chambers of 11.7 m length each and of shorter, 3.6 m long chambers in the focusing quadrupoles. The vacuum chambers are connected with stainless steel bellows of 0.168 m length [3].

The standard type vacuum chambers are made from an extruded aluminium profile. The extrusion provides an elliptic beam channel and a parallel pumping duct for the linear NEG ribbon. The NEG pumps through a single row of pumping slots of 20 mm by 8 mm. A single row of 40 slots per m provides a conductance for nitrogen of about 600 l/s/m. The synchrotron radiation power and the heat dissipated by the NEG [4] during activation and reconditioning are removed by 3 cooling water ducts. A lead radiation shield has been applied to the outside of the chamber to absorb the intense synchrotron radiation power when LEP operates at high energy.

In the 8 straight sections, each about 500 m long, accelerating cavities, vacuum tanks for electrostatic separators and special equipment for beam collimation are installed. Wherever possible, the regular structure of the arcs has been maintained in the straight sections permitting the use of standard type of vacuum chambers with integrated pumps.

Near the LEP experiments, in the low-beta quadrupoles and in the region of the RF accelerating cavities, the vacuum chambers are made of stainless steel with a circular aperture of 156 mm and 100 mm diameter respectively. These chambers have no provision for installing the NEG pumps and, because of the lower level of synchrotron radiation power, no water cooling is necessary. Specially shaped cruciform vacuum chambers have been installed in the low-beta quadrupoles to provide the largest possible aperture required for the background in the experiments. The stainless steel sections are pumped by a combination of lumped ion pumps and of titanium sublimation pumps mounted at an average distance of about 10 m.

A detailed description of the vacuum system for the 4 large LEP experiments can be found in [5].

To reduce the cost, a minimum number of ion pumps has been installed in LEP, under the assumption that - apart from the initial phase of LEP operation - the contribution of inert gases, argon and methane, would represent a small fraction of the gas load only. Test measurements in a photon beam line at the DCI storage ring have indeed shown, that methane cleans up more rapidly than the other gases contributing to the dynamic pressure [6] and that the vacuum system, once cleaned, maintains its low outgassing rate after venting to N_2.

Description of the controls

The controls for the LEP vacuum system have been based on distributed intelligence [7]. This approach allows for maximum flexibility and for the easy integration of the mobile pumping and diagnostic equipment used during the pump-down and the commissioning phase.

One particularity of the LEP vacuum system is the extensive use of a large number of mobile equipment during commissioning. It is therefore required that the control system keeps track of pumping groups, bakeout heating machines, thyristors for NEG activation and control consoles which are moved from one sector to another as the installation team progresses [8].

To minimise the cabling costs, up to 8 ion pumps have been connected in parallel onto one power supply. However, as the ion pumps are the primary means of pressure measurements, a device allowing to measure individual pump currents has been developed [9].

Every power supply or equipment controller is able to communicate with the main control room over a local area network. The values returned by the vacuum equipment are processed before they are sent to the control room. For instance, an ion gauge controller returns the pressure already corrected for emission current and X ray contribution.

Powerful work-stations running UNIX are used in the main control room. Because of the pre-processing of the data at the equipment level, it is particularly easy to acquire data on files which can then be transferred to desktop computers (mainly Apple MacIntosh) for further analysis.

Preparation of the components of the vacuum system

A pre-requisite to meet the very tight installation schedule (18 months for the complete ring) was to obtain a very high reliability for all components.

The 2800 bellows were shipped in batches of 100. One additional bellows per batch was tested under simulated LEP conditions, namely 150 °C, with 37 mm stroke and 3 mm radial offset between two flanges, resulting in an average lifetime of 25000 cycles. Some 5% of all bellows have undergone a bakeout and ultimate pressure test.

All electrical feedthroughs have undergone a leak test before and after a bakeout at 300 °C, before being dispatched for installation on the vacuum chambers.

All aluminium chambers were installed on test benches in a dedicated mounting hall. After a bakeout at 150 °C, the NEG pumps were activated and the ultimate pressure reached after 100 hours was recorded. Figure 1 shows the results which were obtained on the 1812 bending magnet vacuum chambers.

The radiofrequency cavity assemblies have been baked for 24 hours before being conditioned to full RF power. Subsequently the units have been stored under N_2 before installation in the LEP machine. The vacuum conditioning and the vacuum performance of the accelerating cavities is outlined in [10].

Due to their complexity, vacuum sectors containing radiofrequency cavities were not baked after installation in LEP.

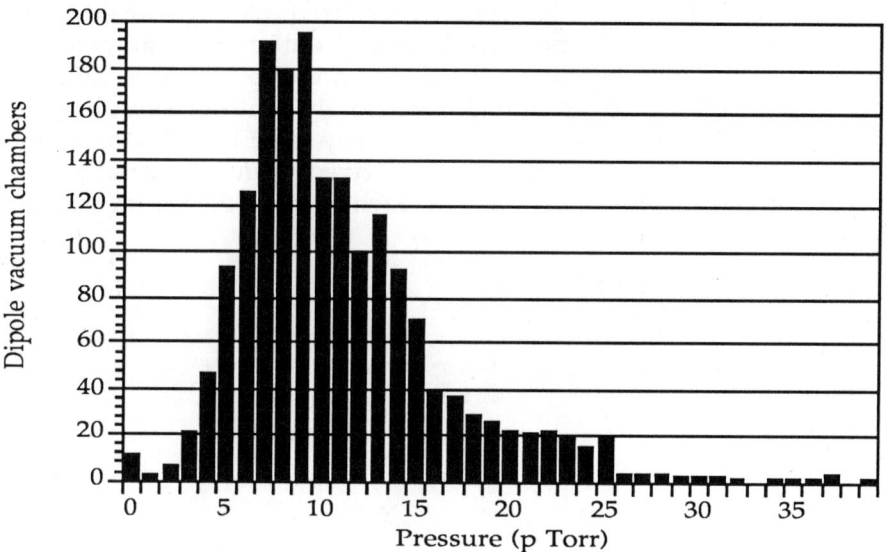

Fig. 1: Vacuum performance of 1812 bending magnet vacuum chambers during acceptance tests. Almost all chambers give a pressure below $2.5 \; 10^{-11}$ Torr.

Commissioning of the vacuum system

The normal commissioning procedure for a standard LEP sector starts with a 24 hour bakeout at 150 °C. This temperature is achieved by circulating hot water at 8 bar in the cooling channels of the aluminium chambers [11]. Ion pumps and gauge heads are heated to 300 °C using electrical heating elements. At the end of the bakeout period the NEG pumps are activated by heating them progressively to 700 °C, with all vacuum chambers still at 150 °C and the mobile pumping stations connected to the beam pipe. After having degassed the ionisation gauges, the ion pumps are switched on, the mobile pumping stations are disconnected from the beam pipe and a final NEG conditioning at 450 °C is done.

A 'static' average pressure well below $2 \cdot 10^{-11}$ Torr has been achieved in all NEG pumped sectors. The static pressure in sectors with conventional lumped ion pumps and titanium sublimation pumps has been below 10^{-10} Torr.

DYNAMIC PRESSURE RISE WITH CIRCULATING BEAM

With the first stored beam circulating in LEP a specific pressure rise of about $1.7 \cdot 10^{-7}$ Torr/mA has been measured. A significant first observation has been that the pressure increased uniformly in all bending sections of the machine, indicating that the cleaning of all 3000 vacuum chambers of the series production had been closely controlled resulting in a uniform surface cleanliness. A pressure survey in LEP with circulating beam is shown in Figure 2. The data are derived from individual pump currents and give the average pressure for each of the 130 vacuum sectors.

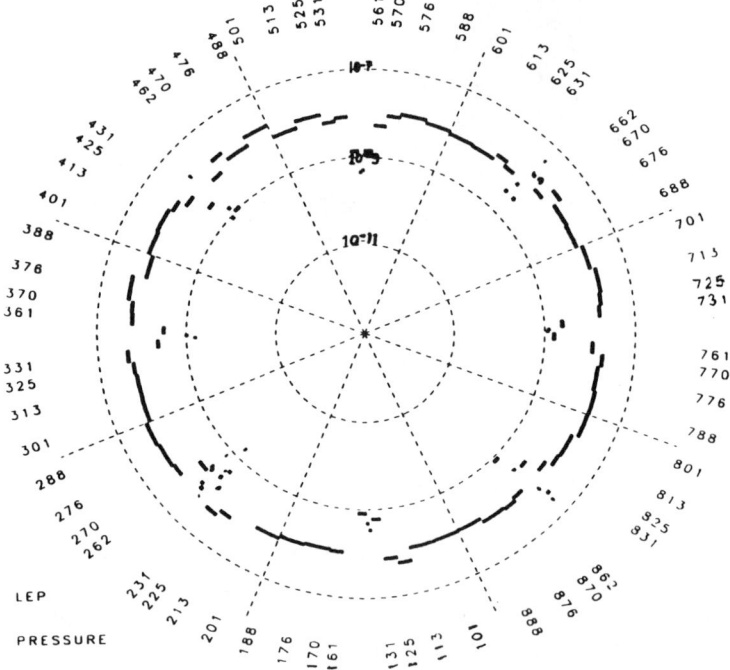

Fig. 2: Average pressure in the 130 vacuum sectors as derived from the ion pump currents. The plot has been taken with 0.4 mA total current shortly after obtaining the first circulating beams.

The plot has been taken with 0.4 mA total current shortly after obtaining the first circulating beams. The sectors representing the long straight sections, far from the bending arcs, show a lower pressure rise due to the reduced synchrotron radiation level. Exceptions from the uniform pressure rise may be found in special locations with higher synchrotron radiation load, e.g. downstream of the wiggler magnets and in the injection regions.

Figure 3 shows a detailed plot of the dynamic pressure in a single vacuum sector. The lower part of the graph represents the initial static pressure before beam is stored. The upper part shows the pressure increase at two current levels, 0.042 and 0.172 mA respectively.

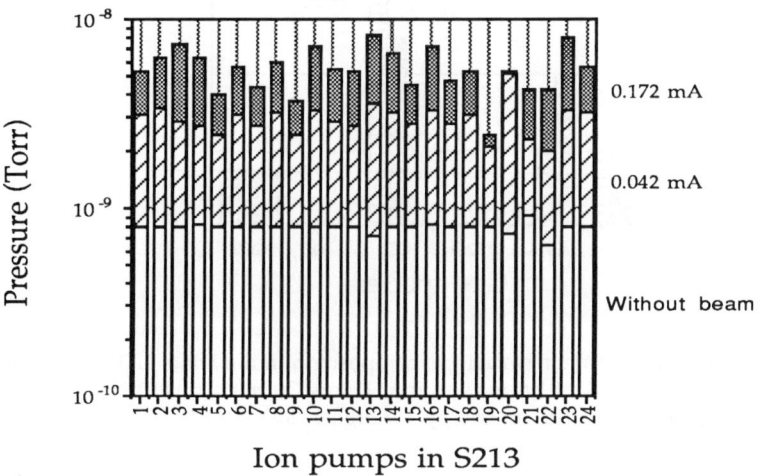

Fig. 3: Pressure distribution in pilot vacuum sector S213 at two current levels during the first runs with circulating beams. The initial static pressure is below the threshold of the measuring system of $6 \cdot 10^{-10}$ Torr.

During the initial beam cleaning phase the residual gas composition has been dominated by CH_4 with relative fractions of typically 5% H_2, 40% CH_4, 40% CO and 15% CO_2. For a baked ultrahigh vacuum this composition is rather uncommon but it can be attributed to the very selective pumping speed from the combined NEG and ion pump system. However, after enough beam cleaning, the gas composition gradually changes, as can be seen on Figure 4.

The gradual reduction of the photon induced desorption during exposure to beam, with a relatively steep slope of about -1, is shown in Figure 5 However, it must be remembered that the beam cleaning curve reflects the global effects of the cleaning of the vacuum chamber surface and of the gradual saturation of the NEG pumps. In order to guide the eye, the solid line joins those points on the graph which correspond to the maximum pumping speed obtained immediately after reconditioning the NEG. The lowest ΔP/I reached after more than 5 Ahour is $5 \cdot 10^{-11}$ Torr/mA.

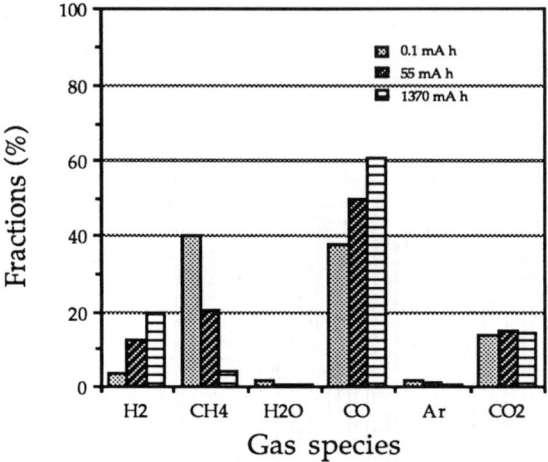

Fig. 4: Evolution of the residual gas composition in the arc of LEP during the initial period of beam cleaning. Below 30 mAh dose the data were taken at 20 GeV beam energy, for higher dose at 46 GeV.

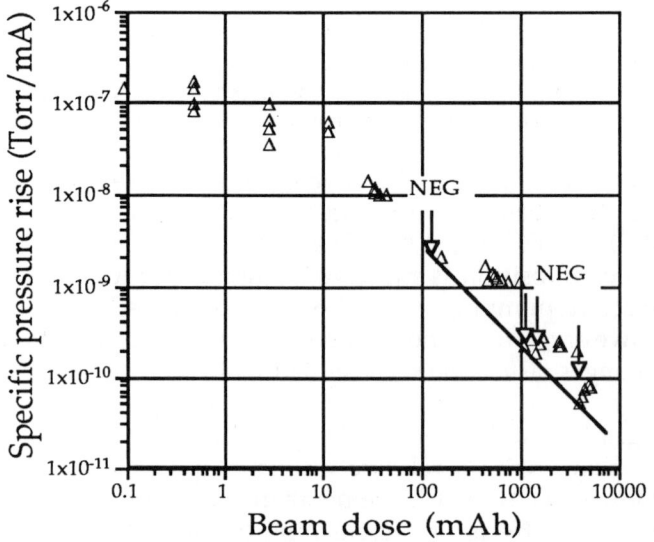

Fig. 5: Dynamic pressure and beam cleaning in a LEP vacuum sector. The arrows point to NEG reconditioning. The solid line joins points with maximum pumping speed.

The dependence of the specific pressure rise on beam energy is shown in Figure 6. The data have been taken after a beam dose of about 4 Ahours. Between 20 and 45 GeV beam energy the specific pressure rise has been found to increase by a factor between 1.4 and 1.6 which is less than proportional to the total number of photons (giving a factor of 2.25). This observed energy dependence is in line with previous estimates and with laboratory measurements [12] which indicate that the desorption efficiency decreases with increasing photon energy.

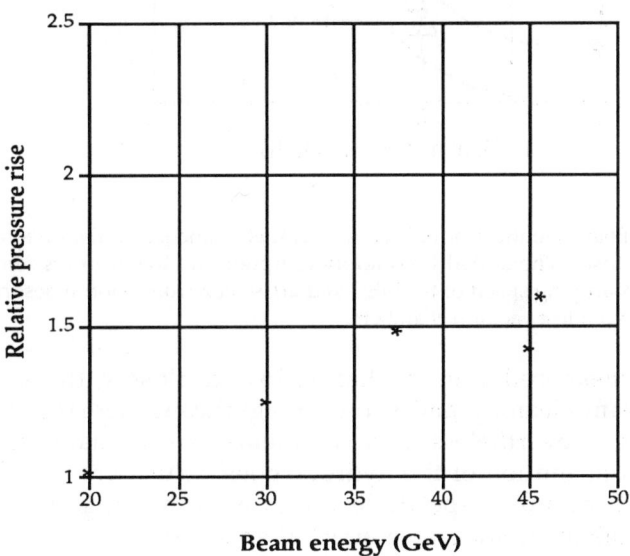

Fig. 6: Dynamic pressure rise in LEP as a function of the beam energy measured after an accumulated beam dose of 4 Ahours.

BEAM - GAS LIFETIME

During the first beam runs the measured beam lifetime has been in the range of 0.2 to 0.3 hours for 1 mA beam current, in line with expectations from laboratory measurements. The gradual improvement of the beam lifetime is shown in Figure 7. Here the dotted curves represent the expected performance scaled from test chamber measurements. The upper curve applies to a hypothetical constant, maximum NEG pumping speed, while the lower curve is for 30 l/s/m, corresponding to a NEG pump after adsorbing approximately 0.2 Torr l/m of CO and CO_2.

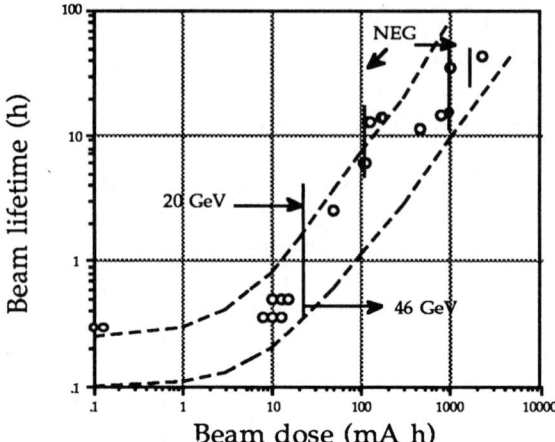

Fig 7: Beam lifetime normalised to 1 mA as a function of the integrated beam dose. The dotted lines apply to upper and lower limits for the linear pumping speed of the NEG and are scaled from photon desorption data from test vacuum chambers.

The measured lifetime has followed closely the expected trend from the beam cleaning and shows a significant increase after each NEG conditioning. Nevertheless, two comments can be made: firstly, that the important contribution of the hydrocarbons, which are not affected by the varying NEG pumping speed, decreases with the on-going beam cleaning and shows the most noticeable effect at small beam dose, secondly, that at high beam dose the measured beam lifetimes tend to be lower than expected from the pressure, which suggests that other effects beside the vacuum contribute to the beam lifetime.

EXPERIENCE GAINED FROM LEP

NEG pumps

The operational experience has confirmed the laboratory performance of the NEG pumps. Due to the high effective pumping speed available, the pressure rise with beam has been low and the beam-gas lifetime has been long and has never limited the machine performance. Since the start-up of LEP in July 1989, the NEG pumps of the complete ring have been reconditioned 3 times. The first time after a beam dose of about 120 mAh, at the beginning of September, the second time after 950 mAh towards the middle of November and the third time after 1.4 Ahour before restarting operation in March 1990. An additional conditioning has been done in sectors open for repairs and in the pilot sectors during the last shut-down.

During the first two pumping cycles the effective linear pumping speed for CO and CO_2 has decreased from the maximum value of about 500 l/s/m to approximately 70 l/s/m. The reconditioning of the NEG pumps has been carried out remotely from the LEP control room. This operation requires closing of the sector valves and switching off the ion pumps to avoid deterioration of the leakage current performance due to the increased hydrogen pressure when the NEG temperature rises above 450 °C. The whole system has been treated in less than 8 hours and this time may be further reduced if necessary. For practical reasons, the reconditioning is now performed towards the end of a long machine stop at close to yearly intervals.

The total quantity of gas pumped by the NEG until May 1990 is estimated to be less than 0.4 Torr l/m for CO and CO_2 and to be about 1 Torr l/m for H_2 (equivalent to about 3 monolayers). However, these quantities are small compared to the total capacity which exceeds 30 Torr l/m for $CO+CO_2$ and 300 Torr l/m for H_2 [4].

Experience with the vacuum components

The first months of operation have confirmed the excellent experience with the LEP-type joints with aluminium to stainless steel flanges and aluminium gaskets during the installation and commissioning phase [2]. The available statistics concerning the 6 leaks which were detected during LEP operation is indeed very encouraging [13]. The items which have developed leaks are: three ion pump feedthroughs, one beam position monitor feedthrough, a beryllium window on a beam monitor and a weld on a stainless steel chamber. Mechanical damage by 'traffic accidents' during access periods have caused a further two broken pump feedthroughs. Leaks by corrosion, on two of the ion pump feedthroughs, have occurred exclusively in places where water leaking into the tunnel has caused excessively high humidity. To reduce the risk of electric leakage currents on the ion pump connectors - but also to avoid corrosion -, small heating elements have been installed.

Among the other components, the mechanical reliability of the gate valves are of concern since several valves have leaked and have developed mechanical defects. Very recently, two ceramic RF couplers developed a large leak while the cavities where powered close to their maximum rating and a leak appeared on a weld of a non standard aluminium vacuum chamber.

A serious problem which is intimately linked with the vacuum chambers and their radiation shield has been identified as the remanent magnetisation of the approximately 9µm thick nickel diffusion barrier which forms part of the lead coating. The effect of the residual field on the performance of LEP and ways to eliminate this magnetisation have been studied. A first section of 270 m in the normal arcs was demagnetised in June 1990. Approximately 2 km will be demagnetised by March 1991.

Vacuum chamber impedance

During the design phase all aspects concerning higher mode losses due to vacuum chamber impedance have been studied with great attention. As a consequence, the vacuum bellows in line with the elliptic beam ducts have been fitted with internal, smooth RF-bridges and the step of a cross section change reduced below 1 mm. The gaps between flanges have been bridged by special RF-finger contacts. The relatively small number of circular vacuum bellows mounted between the RF-accelerating cavities and between circular vacuum chambers have no internal RF-bridge but they have been specially designed with small corrugations only.

Unavoidable cross section change between elliptic and circular vacuum chambers employ tapered transition elements. Direct pump ports to the beam channel have been replaced by connections either to the NEG pump channel, or through small perforations in the wall of the beam pipe. So far no problems have been observed which would be related to higher mode losses in vacuum chamber elements.

Experience with the control system

LEP is the first large machine where distributed intelligence has been used extensively for the vacuum controls. The overall experience is very positive since it allows for rapid modifications and, most important, makes the connection of the many types of mobile equipment very easy.

The complete description of the vacuum system is stored in a relational database from which data is extracted to dynamically define the identification of the various equipment. The database is also used to define the messages to which the equipment has to respond [14].

The possibility to access the vacuum system through general networks is an invaluable tool, which has been of particular help in producing the data for this report.

ACKNOWLEDGEMENTS

The performance of the LEP vacuum system presented here is the result of the common effort of all members of the LEP vacuum group and reflects the global result of a large number of individual and specific contributions in the field of ultrahigh vacuum techniques.

REFERENCES

1. LEP Vacuum Group, Proc. IX IVC, V ICSS, Madrid (1983).
2. H-P. Reinhard et al, Proc. XI IVC, VII ICSS, Köln (1989).
3. W. Unterlerchner, CERN/LEP-VA/89-50 (1989) and Proc. XI IVC, VII ICSS, Köln (1989).
4. C. Benvenuti, CERN-ISR-VA/82-12 (1982).
5. C. Hauviller, O. Gröbner, EPAC, Nice 12-16 June 1990, THP24L.
6. M. Andritschky et al, Vacuum 38, 933 (1988).
7. J. Altaber, P.G. Innocenti and R. Rausch, CERN Report No SPS/85-28, 1985.
8. P.M. Strubin and Nicolas Fietier, EPAC, Nice, 12-16 June 1990, WEP09R
9. P.M. Strubin, Controls for the LEP Vacuum System, J.Vac.Sci.Technol., A 5(4) Jul/Aug 1987.
10. A.G. Mathewson EPAC, Nice 12-16 June 1990, THP28L.
11. H. Schuhbäck, The Cooling and Bakeout System of the LEP main Ring Vacuum Chambers, CERN Report No AT-VA/90-17.
12. K. Booth et al, Proc. IX, IVC, V ICSS, Madrid (1983).
13. W. Unterlerchner, private communication, June 1990.
14. P.M. Strubin, Automatic decoding for the control messages for the LEP Vacuum System, proceedings of the 9th International European ORACLE User Group conference, April 1990.

DESIGN OF THE VACUUM SYSTEM TNK DEDICATED SYNCHROTRON SOURCE FOR X-RAY LITHOGRAPHY

V. V. Anashin, A. N. Bulygin, N. G. Gavrilov,
E. P. Kollerov, V. N. Korchuganov, A. I. Nikitin,
V. N. Osipov, E. M. Trakhtenberg

Institute of Nuclear Physics, 630090 Novosibirsk, USSR

ABSTRACT

The main features of the vacuum system TNK storage ring and its parameters are discussed. Operating vacuum levels are equal to $\sim 3 \times 10^{-9}$ Torr, a total pumping speed of about 75×10^3 l/s. The vacuum chamber is made from an aluminum alloy by the hot extrusion method. The SR-absorber design is also discussed, and the calculation of the middle vacuum level and lifetime is given.

INTRODUCTION

Using synchrotron radiation (SR) it is possible to design and produce chips with the submicron structure. Now in Zelenograd, near Moscow, a scientific technological complex based on a dedicated SR source TNK-electron storage ring is being constructed. The TNK was initiated in 1985, by the INP (Novosibirsk) and is scheduled to be operational in 1991.

This SR-source consists of the following units:
1. Linear accelerator - injector (electron energy 80-100 MeV)
2. Small storage ring-booster (E = 450 MeV)
3. SR-storage ring (E = 1.2 -2.5 GeV)
4. SR-beam lines (39)

The main parameters of the TNK-storage ring are:

1. Maximum electron energy — 2.5 GeV
2. Maximum storage current
    - single bunch mode — 100 mA
    - multibunch mode — 500 mA
3. Spectral range of synchrotron radiation — $0.1 - 2 \times 10^3$ Å
4. Photon critical energy — 7 keV
5. Circumference — 124.13 m
6. Number of SR-lines — 39
7. Total pumping speed
    - getter-ion pumps — 15000 l/s

titanium sublimation pumps	60000 1/s
8. Operating pressure	$3 \cdot 10^{-9}$ Torr

1. GENERAL DESIGN OF THE VACUUM SYSTEM

The TNK storage ring comprises six mirror-symmetrical superperiods each having two straight sections of about 3 m long to accommodate insertion devices (undulators, wigglers), injection and RF cavities. The storage ring vacuum system is separated from the injection beam line by a beryllium foil whose is 100 μm thick. The same foils also separate the storage ring and SR lines from superconducting wigglers. All other SR lines have shutters with a very small vacuum conductivity and operating time of 0.02 s to protect the storage ring vacuum system and vacuum-tight valves [1] with an operating time of ~ 1 s. In addition the storage ring vacuum chamber is divided by 24 similar valves. Vacuum equipment at each superperiod are identical and are shown in fig. 1 (1/2 of the superperiod).

1.1. VACUUM CHAMBERS

Vacuum chambers inside the bending magnets and quadrupole lenses are made from the aluminum alloy AMcS [2] by hot extrusion. The width of the bending magnet vacuum chamber window (5 in fig. 3) is about 200 mm, therefore, SR has no contact with the chamber walls. The cross-section of these two types of the vacuum chambers is shown in fig. 2. The conductivity of the bending magnet vacuum chamber is equal to 40 $l \cdot m \cdot s^{-1}$ and 15 $l \cdot ms^{-1}$ for a quadrupole lens.

1.2. PUMPING AND DIAGNOSTIC BLOCKS

The operating pressure of the storage ring is obtained due to special combined pumps, which have getter-ion and titanium sublimation parts with a total pumping speed of 1000 l/s. There are 60 such pumps at the storage ring. There are also two additional titanium sublimation pumps in each superperiod with a pumping speed of 200 l/s, the latter installed in the lens-type vacuum chambers.

The pumping and diagnostic block are shown in fig. 3, and the above mentioned combined pump (1 in fig. 3) is part of this block. Two vacuum gauges (2 in fig. 3) of PMM-46 type are placed in the same block; it comprises one stationary (3 in fig. 3) and two movable SR absorbers. The last two are necessary to eliminate SR from beamlines. A special chamber-imitator (5 in fig. 3) is introduced in the pumping block to insure the smoothness of the whole vacuum chamber. To remove the ions obtained by collisions of electrons and residual gas molecules a special electrode (6 in fig. 3) is employed in order to switch on the argon-discharge chamber cleaning. Some pumping and diagnostic blocks have pick-up beam-position monitors.

1.3. SR - ABSORBERS

The SR power can be calculated by the formula:

$$P (kW) = 88.5 \cdot E^4 (GeV) \cdot I(A) / R(m)$$

Here E - electron energy;
      I - electron current and
      R - bending magnet radius.

At E=2.5 GeV and I=0.5 A, the total SR power is equal to 352 kW or 1 kW·grad^{-1}. All SR absorbers are located inside the pumping and diagnostic block, thus very close to the pumping area. The main part of the SR power is absorbed by an absorber 4 (see fig 1). This one receives SR from the 10° - arc of the bending magnet, with 500 W/cm density.

The design of this absorber is shown in Fig. 4. To increase the irradiating surface it is mounted at an angle of 15° to the median plane and has special three angle grooves. To increase the water-cooled surface, the grooves are also milled inside the watercooled channel. The SR-absorber has an additional plate, which works as a trap for emitted electrons, thus we minimize the electron stimulating desorption. The calculations have shown that with a total power of 10 kW (500 W·sm^{-1}) and for the water flow of 12 l/min on one SR absorber the maximum temperature will be 153 °C at the groove tops on the irradiated surface. Water heating will be 6 °C and the temperature difference of the absorber-water interface will be 38 °C. Both the SR-absorbers and the aluminum vacuum chamber are cooled by distilled water with pH=6.

1.4. BEAMLINES

1.4.1. HIGH VACUUM BEAMLINES

High vacuum beamlines are intended to transport SR from the storage ring to experimental stations. There are 29 such beamlines.

All parts of beamlines are made from stainless steel. Beamlines are heated up to a temperature of 300-350 °C. Special SR-absorbers protect the channel walls against SR, thus allowing it to reach only the experimental station. Automatically operated vacuum-tight valves and high speed shutters are installed in each beamline to protect the vacuum system of the storage ring. Each beamline has also a delay line system. The number of getter-ion pumps and the low conductivity of the beamlines allow the maintenance of a storage ring pressure level of 10^{-9} Torr with a pressure level in the experimental stations of 10^{-6} Torr.

1.4.2. BEAMLINES FOR HARD X-RAY REGION

The TNK storage ring has 10 hard X-ray beamlines. SR passes through watercooled beryllium foils to the experimental stations via these beamlines. These beamlines are bakeable up to 300-350 °C.

2. LIFETIME

In the electron storage ring the main process which decrease the beam lifetime is the interaction between beam and residual gases. The lifetime of i-th component of the residual gas can be calculated by the formula:

$$\tau_i = 2.12 \times 10 \times X_i/M_i \times P_i \quad \text{(hour)}$$

X_i - radiation length unit (g/cm)
P_i - partial pressure (Torr)
M_i - molecular mass of i-th component

An equation for the effective lifetime is:

$$\frac{1}{\tau_{eff}} = \Sigma \frac{1}{\tau_i}$$

(The summing up is made over the whole spectrum of the residual gas.)
At a vacuum level inside the storage ring of 2×10^{-9} Torr and by the typical residual gas spectrum for the storage ring (70% H_2, 20% CO_2, 10% CO) calculating vacuum lifetime will be about 30 hours.

REFERENCES

1. G. N. Kulipanov, Nucl. Instr. and Meth., <u>A261</u>, 1 (1987).

2. V. V. Anashin et al., High speed automatical vacuum valve, Preprint INP, Novosibirsk, 1977.

34 Vacuum System TNK Dedicated Synchrotron Source

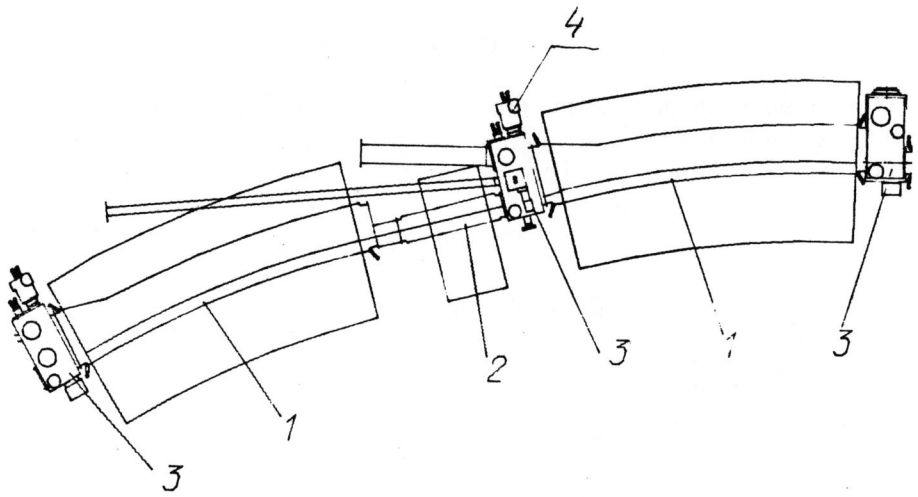

FIG. 1.

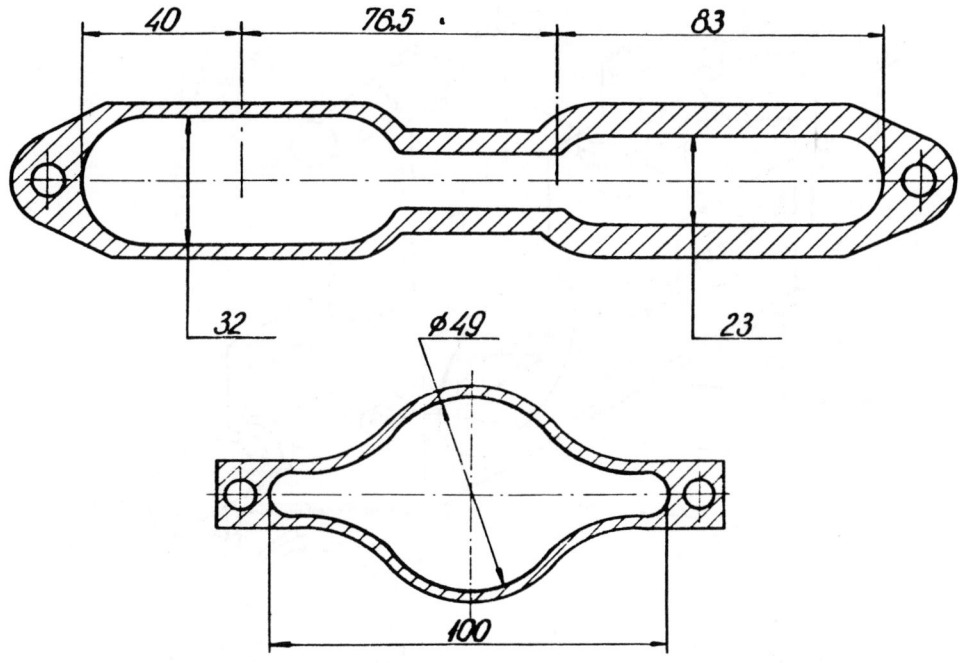

FIG. 2

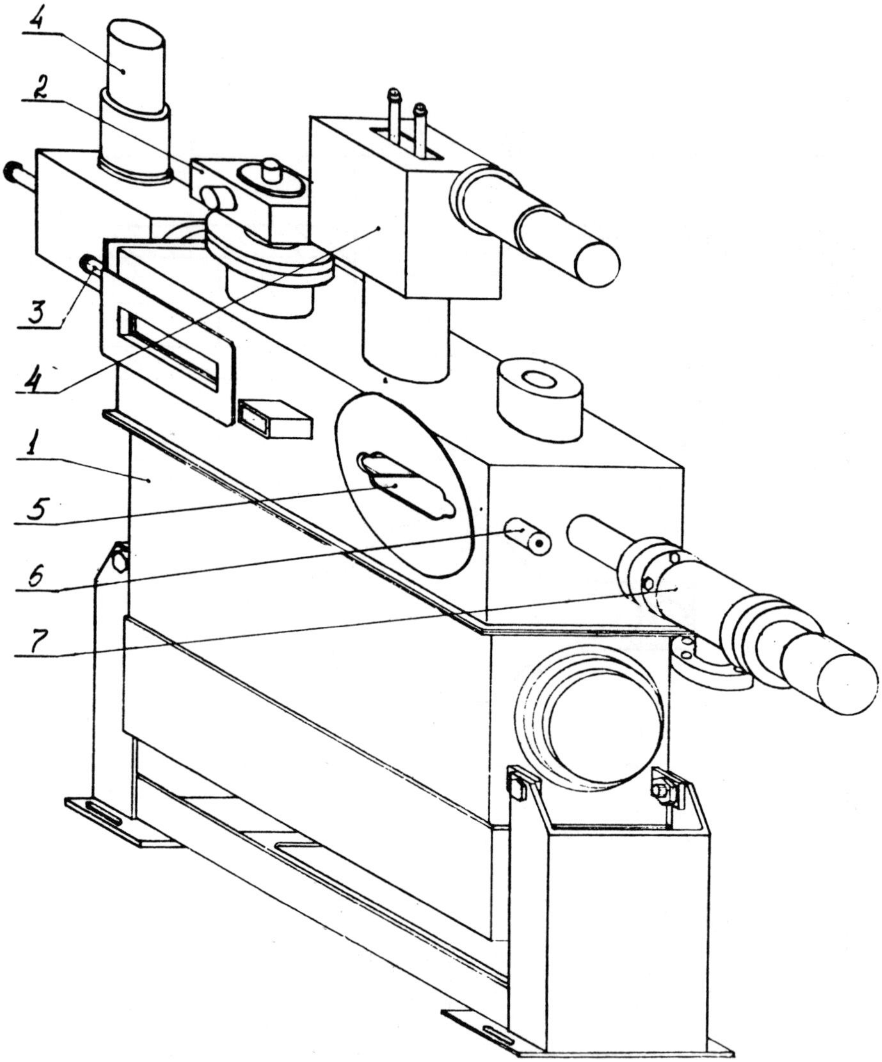

36 Vacuum System TNK Dedicated Synchrotron Source

FIG. 3

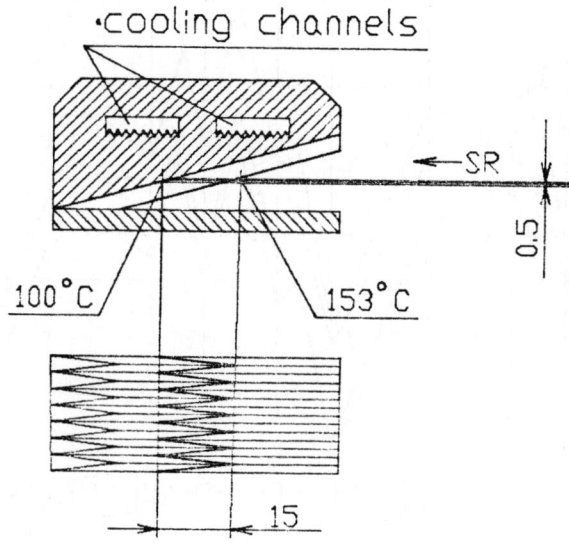

FIG. 4.

38 Vacuum System TNK Dedicated Synchrotron Source

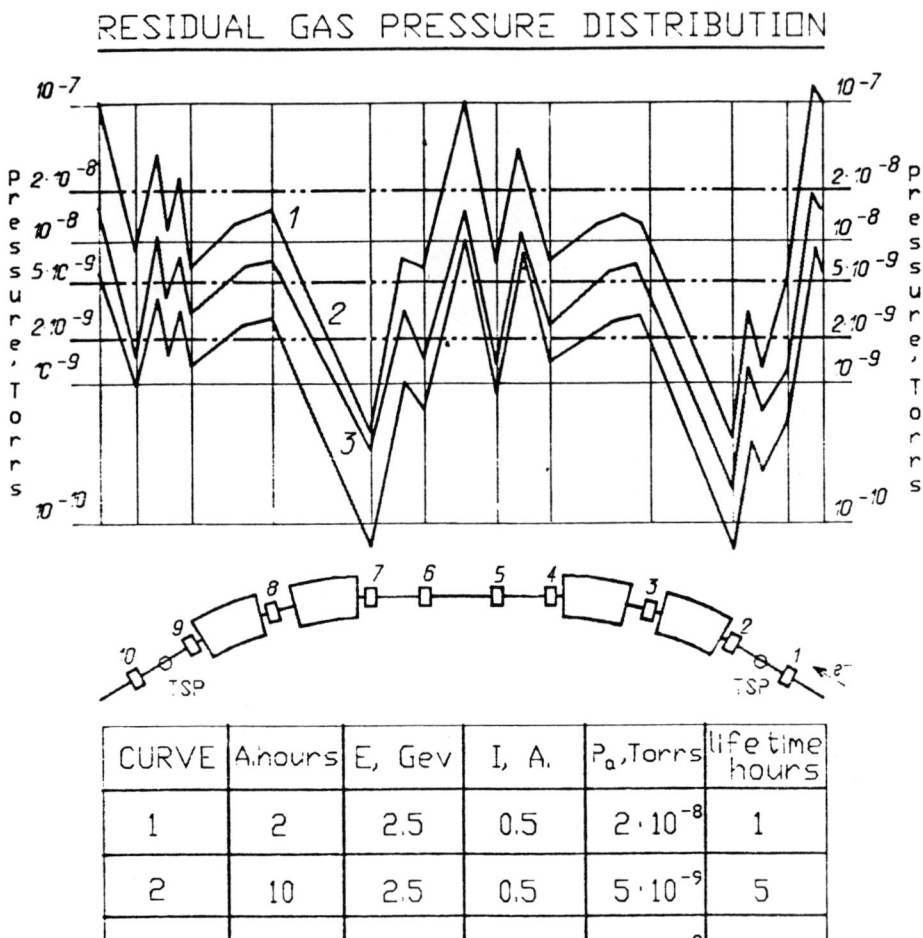

CURVE	A,hours	E, Gev	I, A.	P_a,Torrs	life time hours
1	2	2.5	0.5	$2 \cdot 10^{-8}$	1
2	10	2.5	0.5	$5 \cdot 10^{-9}$	5
3	50	2.5	0.5	$2 \cdot 10^{-9}$	14

FIG. 5.

NSLS VACUUM SYSTEM OPERATING EXPERIENCE CONDITIONING AND DESORPTION YIELDS*

Henry J. Halama
Brookhaven National Laboratory, NSLS - Bldg. 725C
Upton, NY 11973

ABSTRACT

All straight sections in both the VUV and the X-Ray rings have been filled with various insertion devices, most of them fully operational. Beam lifetime in the VUV ring is limited by the Touschek effect to ~ 100 minutes at 800 mA due to the small vertical beam size required by users. With no experiments running, X-Ray beam lifetime is > 35 hours at 220 mA and is limited by beam gas scattering. During the past several years the U10 beam line was used to measure PSD yields from various metals to study their relative merits for light sources. These yields were also compared to those measured in X-Ray ring dipoles during initial commissioning when desorption was high. Despite the large differences in critical photon energies, agreement was quite good. Both rings are now fully conditioned and their pressures and lifetimes have reached equilibrium. Well established conditioning procedures are followed after every intervention into their vacuum systems.

INTRODUCTION

Photon stimulated desorption, PSD, is the most important factor influencing the design of a light source vacuum system. In conventional machines, desorbed gases are removed by an appropriately scaled pumping system close to the electron beam. In newer designs more complex chambers attempt to channel synchrotron radiation to an absorber away from the beam. The most commonly used metals in the construction of beam ducts are aluminum, stainless steel and copper and their PSD yields are discussed in the first part of this paper. The second part deals with conditioning of the X-ray ring and comparison of the data obtained on the U10B beam line with those obtained in the operating X-ray ring. In the third part the present status of the VUV and the X-ray ring is given along with some comments on time-dependent lifetime improvement and conditioning.

*Work performed under the auspices of U.S. DOE under contract DE-AC02-76Ch00016.

I. DESORPTION COEFFICIENTS - U10 BEAM LINE

Three meter long tubes made of aluminum, stainless steel and Cu-plated stainless were exposed to a white photon beam having critical energy of 500 eV at the U10B beam line.[1] Desorption yields, η, were measured in molecules per photon at 10 milliradian incidence and plotted versus accumulated photons. In Fig. 1 tubes having the following surface treatments are compared:

a) 6063 aluminum[1] with standard NSLS cleaning,[2] 2 day in situ 150 °C bake-out, pressure after bake ~1×10^{-9} Torr.

b) 304 stainless steel,[3] pre-baked at 200 °C for 2 days, back filled with dry N_2, installed and pumped for about one week when $P = 1.5 \times 10^{-9}$ Torr was reached (no in-situ bake).

c) Cu-plated stainless steel,[3] pre-baked at 250 °C for 5 days, back filled with dry N_2, installed and pumped for about one week when $P = 1.5 \times 10^{-5}$ Torr was reached (no in-situ bake).

Fig. 2 depicts the lowest η's obtained by glow discharge in nitrogen for Al[3] and in Ar 10% O_2 for stainless steel, respectively. Referring to the above figures and various other experimental data gained on the U10B beam line we can make the following observations:

1) Aluminum exhibits initial yields between 20 and 100 times greater than stainless steel. The steeper slope in aluminum catches up with stainless steel within a dose of ~10^{24} photons m^{-1}.

2) PSD yields of common gases found in vacuum systems, i.e., H_2, CO and CO_2, are similar in Cu and stainless steel for comparable surface treatments.

3) After a week of pumping to a pressure ~1×10^{-9} Torr, the water peak in an unbaked stainless steel tube is desorbed within ~10^{23} photons m^{-1}. In a storage ring this is likely to take longer due to reflected photons, but an in-situ bake out may not be required.

4) A test run using only a pre-baked Al tube was started, but had to be terminated because a very high initial H_2O pressure rise would have required a lengthy low current operation.

In-situ bake is essential in both NSLS storage rings employing aluminum beam tubes after each exposure to the atmosphere. Otherwise, conditioning time would be prohibitively long.

5) The lowest η's are measured on surfaces which are subjected to glow discharge cleaning.

II. X-RAY DESORPTION AND CONDITIONING

The data presented in the previous section allows us to make some predictions as to how much time is likely to be required to reach reasonable beam lifetime which is governed primarily by beam-gas interaction. Since Bremsstrahlung and Coulomb scattering are far more effective in degrading lifetime for CO and CO_2 than for H_2, H_2 is omitted in subsequent figures. To carry out the measurements calibrated BAG and RGA were mounted on the X1BM1 dipole aluminum vacuum chamber containing a Cu crotch through which synchrotron radiation is extracted into beam lines, and a distributed ion pump (DIP). The entire machine including front ends was baked out above 100 °C for one week[4] resulting in pressure of 1×10^{-10} Torr after cool-down. Hydrogen comprised 95% of residual gases. Very large gas bursts (10^{-6} Torr) were observed during the initial stages of commissioning. After one ampere-hour the desorbed gas content depicted in Fig. 3 was 43% H_2, 25% $CO-N_2$, 16% CO and 16% CH_4. $\Delta P/I$ (pressure rise normalized to beam current in Torr/mA) is plotted rather than η, because neither the total pumping speed nor the incident photon flux are known sufficiently well. Assuming $S \geq 100$ ls^{-1} m^{-1} in the dipole, η_{CO} at 1×10^{23} photon/m $\leq 4 \times 10^{-5}$. A rough comparison is given in Table I.

TABLE I

	U10B	X Ray Ring
	3m Al tube	Al + Cu crotch
Bake	150 °C, 24 hrs	<120°C, 5 days
ϵ_c (eV)	500	1700
angle of incidence	10 mr	110 mr
η_{CO} (mol/photon)		
at 10^{23} photons/m*	8×20^{-5}	$< 3 \times 10^{-5}$
at 10^{24} photons/m	2×10^{-5}	$< 1 \times 10^{-5}$

*additional 10^{23} photons/m at 136 eV

Despite many differences, the agreement is quite good.

The fact that the X-ray ring beam energy varies from .75 to 2.5 GeV during each operating cycle permits us to measure $\Delta P/I$s as a function of photon energy from 126 eV to 5 keV. The data was taken several times during machine conditioning and is shown in Figs. 4 and 5 at 10 A-hrs. We also note that $\Delta P/I$ decreases by a factor of 10 and 40 for CO and CO_2, respectively, after 600 A-hrs at 2.5 GeV. Almost no change occurred between 400 and 600 A-hrs. Furthermore, the slope of the curves in Figs. 4 and 5 decreased with conditioning by a factor of approximately 2 between 10 and 600 A-hrs.

III. NSLS STORAGE RINGS

The performance of synchrotron light sources is measured primarily in terms of beam lifetime, beam size, and the recovery of normal operation after a section of the machine has been brought to atmospheric pressure. All the above qualities depend to a large degree on the design of the vacuum system and its flexibility. The vacuum systems of the NSLS storage rings and their performance have been described previously.[4-9] They are routinely operated twenty-four hours a day with scheduled studies and maintenance shifts. Following any intervention into the vacuum chamber, LN_2 boil-off is used for back-fill and standard procedures are followed.[2]

A. VUV RING

The VUV ring[9] operates at 750 MeV (ϵ_c = 500 eV) and to date has accumulated over 8000 A-hours. The highest achieved beam current of 12000 mA was limited by machine components such as heating of ceramics and available rf power. Routinely the ring is filled with 700-900 mA as shown in Fig. 6. The pressure in the machine has reached equilibrium around 1500 A-hours, i.e., subsequent running of 6000 A-hours produced no significant improvement either in pressure or lifetime. This is due primarily to 16 beam lines which are connected directly to the machine and thus dominate its pressure changes. After the first 500 A-hours the lifetime has been limited by:

1) Touschek effect which depends primarily on electron density in the bunch and is significant in lower energy machines only, since $\tau_T \alpha \gamma^3$.

2) ion trapping - where residual gas molecules ionized by circulating electrons tend to neutralize the beam and produce additional beam gas scattering.

As seen in Fig. 7, continuous improvement in pressure during the early stages permits operation with more bunches which further improves both the pressure and lifetime. As the walls of the vacuum chamber are scrubbed by synchrotron radiation, the commissioning time decreases and today represents only a few A-hours after a section has been vented to LN_2 boil-off.

The highest pressure, observed in bending magnets which comprise ~12 m of the circumference, varies linearly with beam current and reads ~11 nTorr at 1000 mA. In the rest of the ring (37 m), including the straight sections, the rf cavity, insertion devices, and injector, the pressure at 1000 mA is ~2.5 nTorr. This results in total average pressure of ~4.5 nTorr with the following partial pressures:

H_2	(2)	= 3	nTorr
H_2O	(18)	< 0.2	nTorr
CO-N	(28)	= 2.4	nTorr
CO_2	(44)	= 0.4	nTorr

which yields pressure dependent lifetime > 5 hours. However, inspection of Fig. 8 reveals a lifetime of 100 minutes only caused by Touschek effect. (Note the good agreement between one and six bunch operation.) The reason that almost no change in lifetime is observed between 600 and 1000 mA is due to bunch shortening which keeps the electron density almost constant. One can obtain longer lifetime at the expense of beam size (emittance) by increasing the horizontal-vertical coupling, as seen in Fig. 7 Curve 5/90. Unfortunately, most experiments require small vertical beam size and therefore small coupling of ~ 1%. Trapped ions also increase the beam size by defocusing the beam. Lifetime could be improved only by lowering electron density, i.e., by filling all 9 available rf bunches. At present the VUV ring can work with up to 7 consecutive bunches due to onset of ion trapping. Reliable operation with 8 or 9 bunches is not possible since the ion clearing system[10] is ineffective, probably because βmin occurs in the center of the dipoles. The only way, then, to increase the lifetime is to increase the bunch length by installing a harmonic cavity which was done last month. Testing is now in progress.

B. X-RAY RING

The X-ray ring operates at 2.5 GeV (ϵ_r = 5 keV) with four insertion devices having ϵ_r up to 20 keV. The first phase of 627 Ampere-hours has been described previously in Ref. 4, 7, 8 and 9. During the ensuing shut-down, major modifications to accommodate insertion devices were carried out.[9] We will cover only the commissioning and the operation following this more than one year long shut-down. Unlike the VUV, which has full energy injection, the X-ray ring is filled at 750 MeV (ϵ_r = 150 eV) and the stacked beam is then accelerated to 2.5 GeV. This fact makes the conditioning significantly more time consuming. The first 28 A-hours are shown in Fig. 9, excluding running at lower energy during stacking and acceleration. After 16 A-hours, lifetime at 100 mA exceeded 5 hours and the beam became more usable for experiments. At 75 A-hours including minor interventions into the vacuum system, I x τ product reached 2000. Since Touschek effect is negligible due to 2.5 GeV energy, lifetime is determined entirely by beam-gas scattering and is inversely proportional to pressure.

The ring (170 m in circumference) is divided into sections according to the gas load. The pressures read on calibrated BA gauges and averaged over respective sections are tabulated in Table II.

TABLE II

Amp-Hours	I mA	Straight Sections $l = 114$ m Torr	Magnets 44 m Torr	Cavities 12 m Torr	Average Pressure Torr
620	0	7×10^{-11}	3×10^{-10}	6×10^{-9}	4×10^{-10}
620	202	3×10^{-10}	2×10^{-9}	8×10^{-9}	1.3×10^{-9}
Shut-down 2/87 - 3/88					
1200	0	5×10^{-11}	$<2 \times 10^{-10}$	4×10^{-10}	1×10^{-10}
1200	230	1.7×10^{-10}	1.1×10^{-9}	1.5×10^{-9}	5×10^{-10}

After 600 A-hours, the residual gas content at high current operation is 61% H_2, 34% CO and 5% CO_2. In Table III we list calculated τ_c and measured τ_m lifetimes along with $\Delta P/I$'s averaged over the entire machine. Desorption in the dipoles is significantly higher.

TABLE III

Amp-Hours	I mA	$\Delta P/I$ Torr/mA	Pressure Torr	τ_m Hours	τ_c Hours
1*	50	5×10^{-10}		~0.5	
10*	50	2.5×10^{-10}		~1	
620*	200	6.4×10^{-12}	1.3×10^{-9}	8	16
+1200	230	2.2×10^{-12}	5×10^{-10}	37	43

*Before 1987 shut-down

The machine commissioning was started in April, 1989 and the A-hour clock was reset to zero. The lifetime at 200 mA in Fig. 10 shows a five fold increase which is due to the following:
1) increase in dynamic aperture.
2) decrease in gas desorption from the walls due to synchrotron radiation scrubbing (Table II).
3) increase in rf power.
4) improvement in rf cavity pressure and performance due to new Helicoflex seals and titanium nitriding.

5) installation of new plate DIPs[11] and NEG modules.
6) improvement in DIP pumping speed after long operation in good vacuum[8].

The shaded strip (1990 in Fig. 10) gives the lifetimes with all insertion devices operating and 4 T field on the wiggler magnet. When the LEGS (Laser-Electron-Gamma-ray Spectroscopy) beam line is open the lifetime is halved. Further degradation in lifetime is observed when the wiggler field is raised to 4.7 T. The best lifetime (B in Fig. 10) is obtained with LEGS and insertion devices not operational and the wiggler field set to zero T. Fig. 11 shows typical "good" operation when one fill (250-110 mA) lasts more than twenty-four hours.

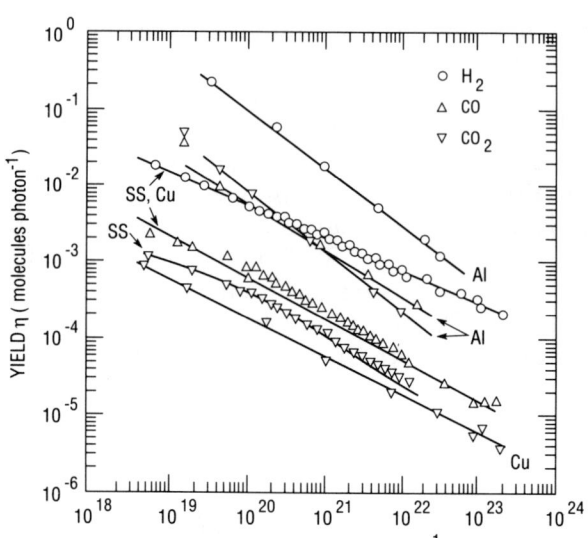

Fig. 1 Comparison of desorption yields from baked aluminum, pre-baked stainless and pre-baked copper-plated stainless

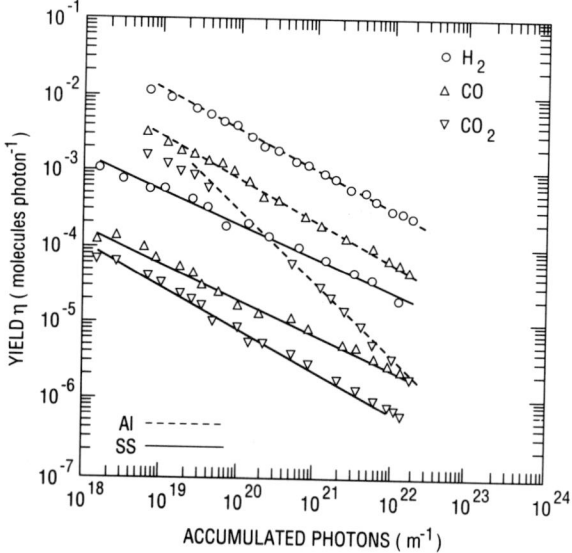

Fig. 2 Comparison of desorption yields from glow discharged aluminum and stainless steel

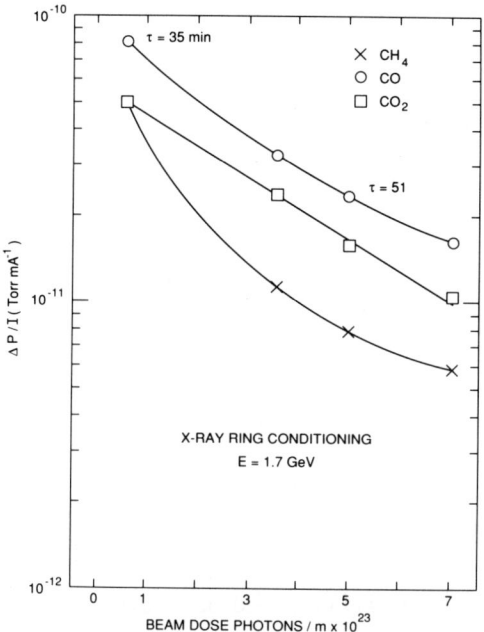

Fig. 3 Normalized partial pressure rise, ΔP/I, in X-Ray dipoles during first 7 A-hours of conditioning.

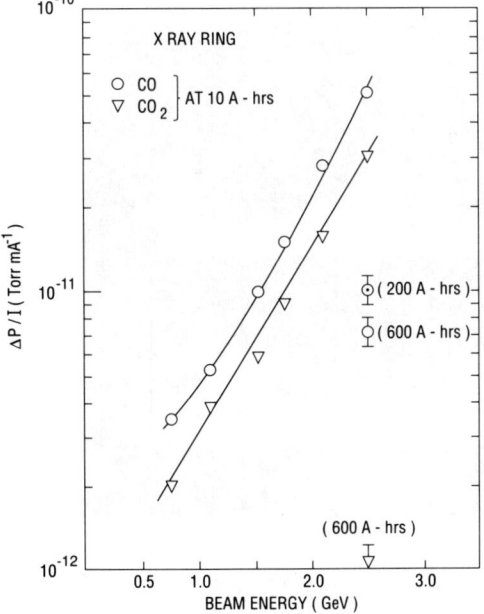

Fig. 4 Normalized partial pressure rise, Δ/PI, in X-Ray dipoles vs beam energy (1 A-hour ≃ 1×10^{23} photons/m)

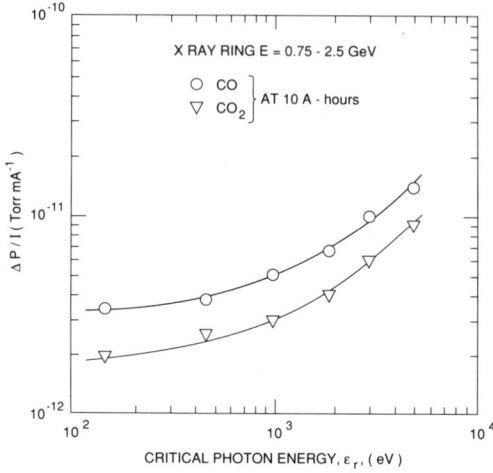

Fig. 5   ΔP/I in X-Ray ring dipoles vs critical photon energy

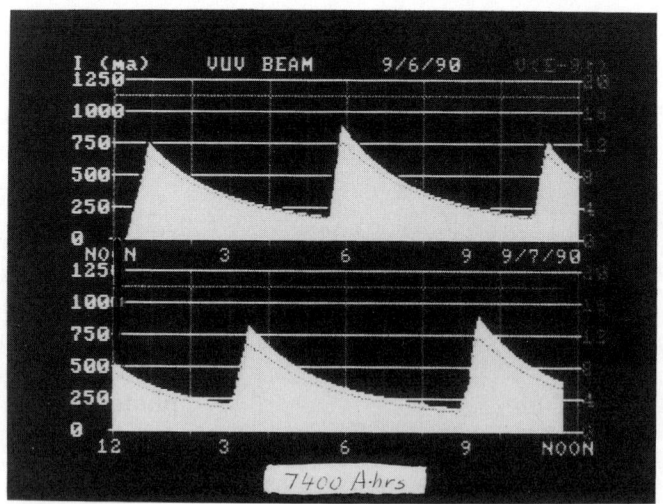

Fig. 6   VUV ring beam current and pressure variations during 25 hours of typical operation

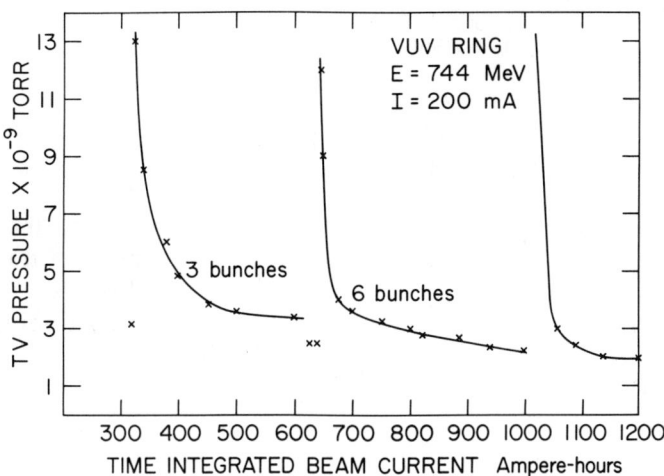

Fig. 7 VUV pressure vs time due to venting of the vacuum chamber during the first 1200 A-hours

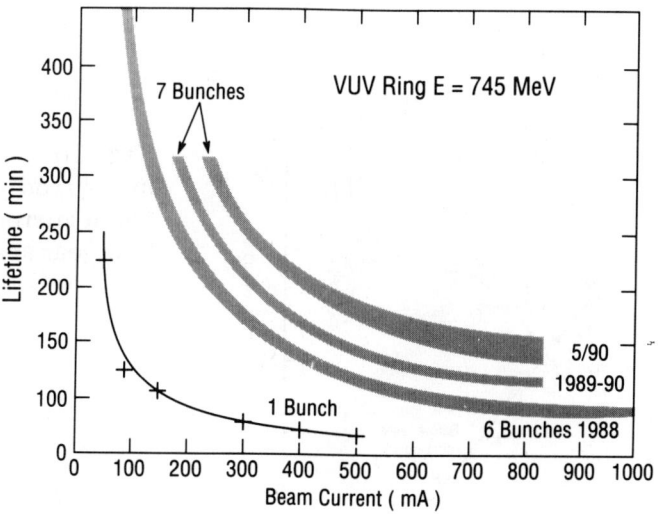

Fig. 8 VUV ring measured beam lifetime vs beam current for several bunch modes

50 NSLS Vacuum System Operating Experience

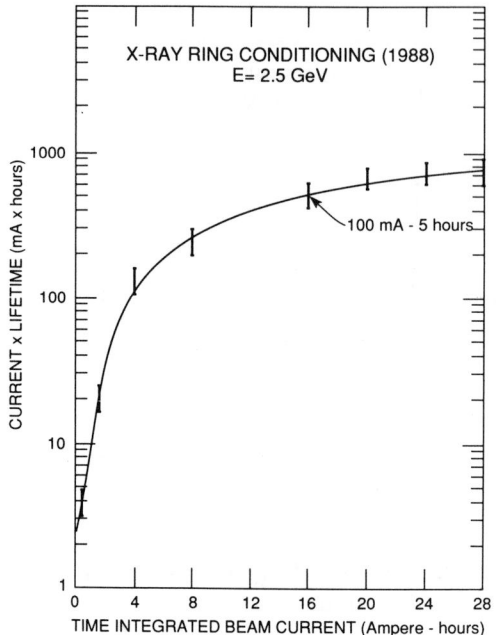

Fig. 9 X-Ray ring current-lifetime product during the first 28 A-hours of conditioning

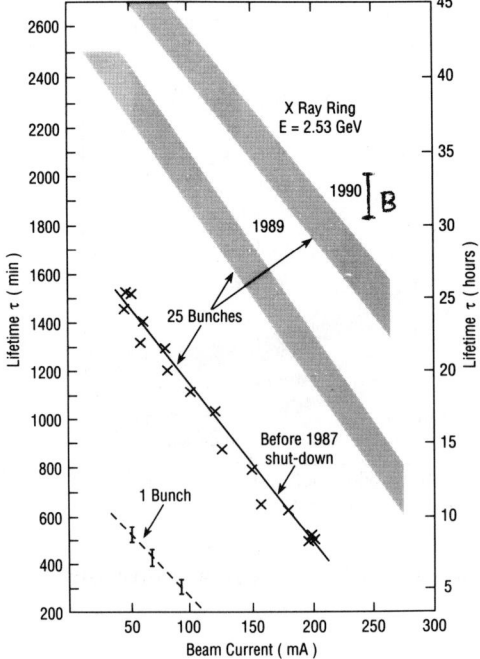

Fig. 10 X-Ray ring measured lifetime vs beam current during a four year span

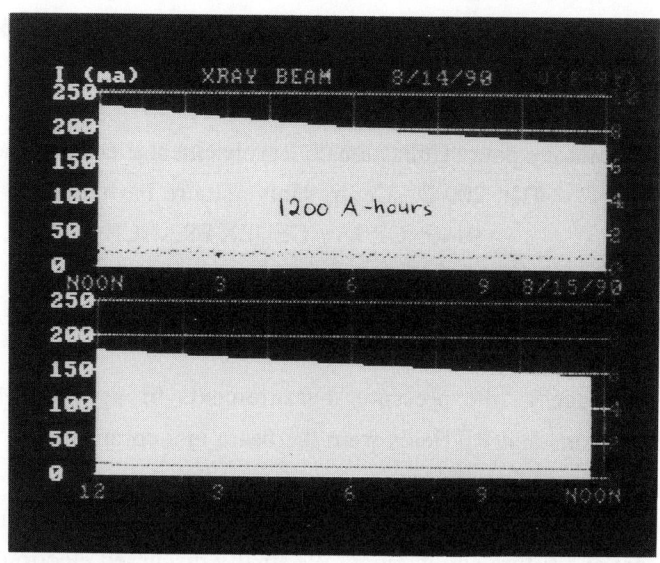

Fig. 11 X-Ray ring beam and pressure behavior during 24 hours of typical operation

REFERENCES

1. T. Kobari and H. J. Halama, J. Vac. Sci. Technol. A5 (4), 2355 (1987).
2. H. J. Halama, AIP Conference Proceedings No 199, 93 (1989).
3. C.L. Foerster, H. J. Halama and C. Lanni, J. Vac. Sci. Technol. A8 (3), 2856 (1990).
4. H. J. Halama, J. Vac. Sci. Technol. A3, 1699 (1985).
5. J. C. Schuchman et al., J. Vac. Sci. Technol. 16, 720 (1979).
6. J. C. Schuchman, J. Vac. Sci. Technol. A1, 196 (1983).
7. H. J. Halama, C. L. Foerster and T. Kobari, J. Vac. Sci. Technol., A5 (4), 2342 (1987).
8. H. J. Halama, BNL Report No. BNL 39768 (1987).
9. H. J. Halama, AIP Conference Proceedings No. 171, 227 (1988).
10. T. S. Chou and H. J. Halama, Proc. 1987 IEEE Particle Accelerator Conference, 1773 (1987).
11. T. S. Chou, J. Vac. Sci. Technol. 5 (6), 3446 (1987).

VACUUM EXPERIENCE AND EXPERIMENTS AT SUPER-ACO*

P. Marin

Laboratoire pour l'Utilisation du Rayonnement Electromagnétique
Bât. 209 D - Centre Universitaire Paris-Sud
91405 ORSAY CEDEX FRANCE

ABSTRACT

Disturbance in the pressure measurements brought by electrons, photo electrons, photons and RF fields from the beam in a compact ring situation is first discussed. Linear and non linear pressure rise are then analysed. The so-called Argon Method for comparing experimental and computed beam-gas lifetime is presented with present and future applications. We finally discussed electron versus positron behavior on Super-ACO as well as the consequences of the Dust Particle Mechanism on micro losses of beam and emittance growth.

I - INTRODUCTION

The ultra high vacuum system of Super-ACO was built in a very compact way. By compactness, on infers here that the sensors and the pumping elements are close to the beam. The problems in discussion arise mainly from the very large number of photons and photo electrons in the vacuum chamber, respectively in the range of 10^{18-19} and 10^{15-16} per meter of orbit length. This average number can still be increased locally by one or two orders of magnitude depending on the angle of incidence on the photon absorbers.

The problem is further complicated, due to the longitudinal compactness of the ring (see **Fig. 1**) which has the effect of producing rather large stray magnetic fields on all the sensors. These bend the trajectories of low energy electrons bringing them in unexpected areas. Scattering of electrons and photons, secondary photon emission are the source of other difficulties in the assessement of the disturbances. The evaluation of the flux of unwanted particles and of their efficiency on the sensors can therefore be wrong by orders of magnitude. However if measured figures are already

---

* Work supported by CNRS-CEA-MEN.

available it is easy to devise steps to reduce these effects by 1 to 3 orders of magnitude.

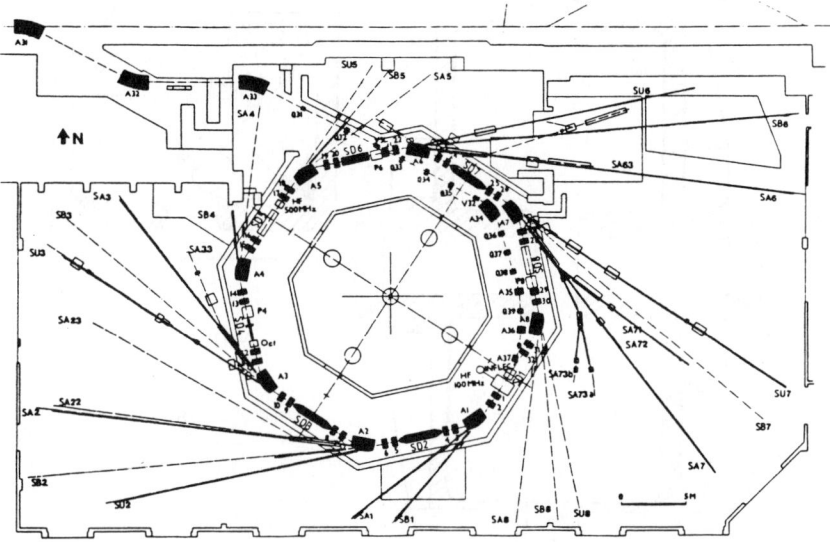

Fig. 1 : Beam lay-out of Super-ACO.

It took two years and a large amount of machine time before all the vacuum problems of Super-ACO could be properly understood. In chapters II and III, we present the various problems encountered, the relevant experimental evidence and the cures which were applied. Several of these problems were found in other S.R. facilities. What is impressing here, is the long list of them. We hope that the new installations in construction can make use of our experience. Chapters IV and VI deal with new ideas developped at LURE.

II - DISTURBANCES FROM ELECTRONS, PHOTO ELECTRONS, PHOTONS AND E.M. FIELDS

1) Electrons

Troubles were experienced with electrons in the RF cavity, even with no beam stored in the ring. A normal B.A. gauge housed in a tube and a diode pump directly connected to the Aluminium cavity, as shown by **Fig. 2**, were used as sensors for pressure measurements. At the operating voltage, 160 kV, the B.A. reading was found to be useless owing to a variable negative contribution of the order of

5×10^{-10} Torr. Moving the gauge further away to Position 2, lowered this figure to a negligible value.

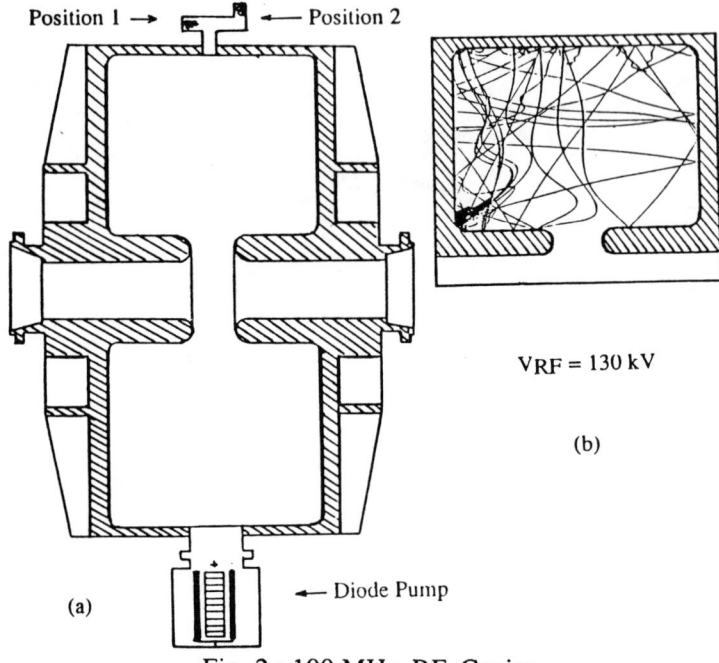

Fig. 2 : 100 MHz RF Cavity
(a) Cross Section and Vacuum Elements, (b) Multipactor Electrons.

In a situation with constant multipactor in the cavity the <u>diode pump</u> measured a large current even when its power supply was switched off ! A cure was obtained by replacing the diode pump by a <u>triode pump</u>, thus repeling electrons up to 5 keV energy. However at the beginning of the vacuum conditionning, it was found necessary to turn off the ion pump (either diode or triode) in order to have a chance to power the cavity. A similar behavior was reported at Chalk River[1]. It has not be possible to decide wether this was due to particles or to photons escaping from the pump discharge. **Fig. 2 (b)** presents electron trajectories in the cavity computed by a multipactor code[2], which illustrate possible deviations of the BA gauge and of the diode pump.

2) <u>Photo electrons</u>

In the commissioning period of the machine pressure measurements were made upstream and downstream each dipole magnet using normal B.A. gauges. It was soon found, that despite the geometry of the system shown by **Fig. 3**, the downstream gauge was recording an electron current an order of magnitude larger than the local pressure, see **Fig. 4**.

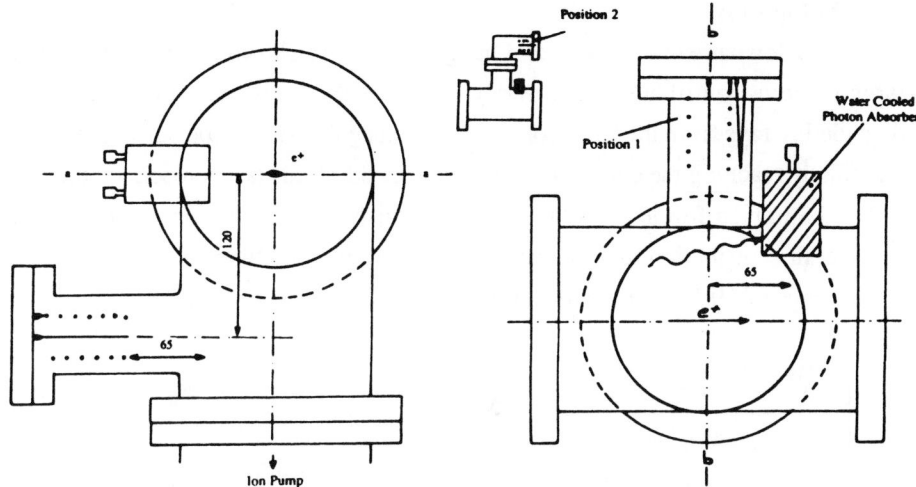

Fig. 3 : Tee Section, Absorber and B.A. Gauge.

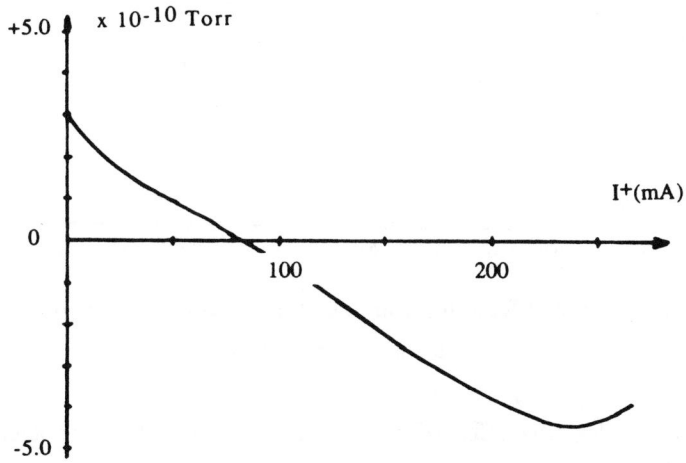

Fig. 4 : Downstream B.A. gauge Reading versus Beam Current.

The replacement of the normal BA gauge by another gauge with "closed" grid suppressed the electron contribution. On the upstream gauge, the effect was in the range of 10^{-11} Torr, owing to the considerably lower X-ray flux in this region, as compared to the incident flux on the copper absorber in the downstream part. As expected, the negative contribution on the downstream gauge is very sensitive to any stray magnetic field.

3) Photons

The downstream "closed" grid gauge located, as shown in **Fig. 3**, presented a large positive contribution from X-rays, see **Fig. 5**. These arise from scattered or degraded X-rays from the Cu absorber, scattered again on the tube housing the gauge and finally reaching the collector. In order to get rid of this last component, the gauge had to be moved further away to position 2 on **Fig. 3**. In this final configuration, the overall contribution from the beam, measured with the filament switched off was in the range of 10^{-12} Torr at large currents.

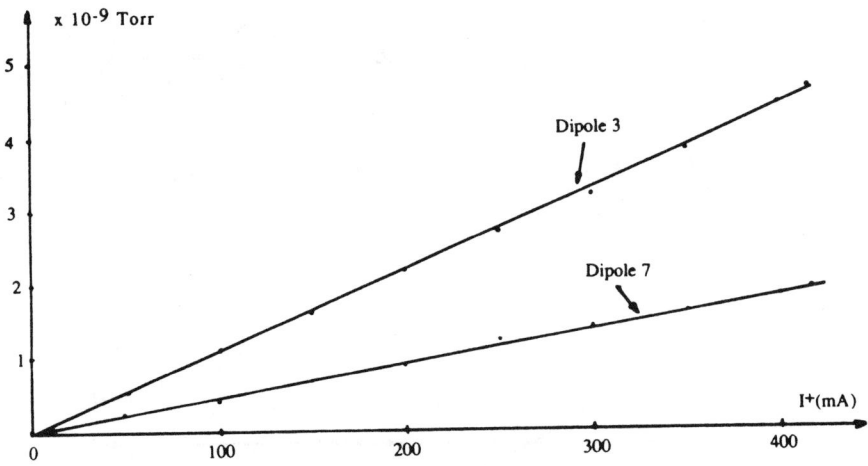

Fig. 5 : Pressure Reading from a Closed Grid Downstream Gauge versus Beam Current.

Another undesirable effect of X-rays was found with the Distributed Ion Pump current (DIP). **Fig. 6** shows the total pump current from the units in 4 dipoles in parallel as a function of the beam current in the 24 bunch mode. One aspect of this curve will be treated in 4). Here we care with the linear rise of the DIP current with the beam intensity. This can be expected both from the contribution of PSD and from X-ray photo current. The latter was measured by lowering the voltage from its nominal setting, 4 kV, down to 300 volts where the discharge no longer occurs. The remaining contribution which comes from X-rays amounts to 90 % of the total in the region of high beam currents.

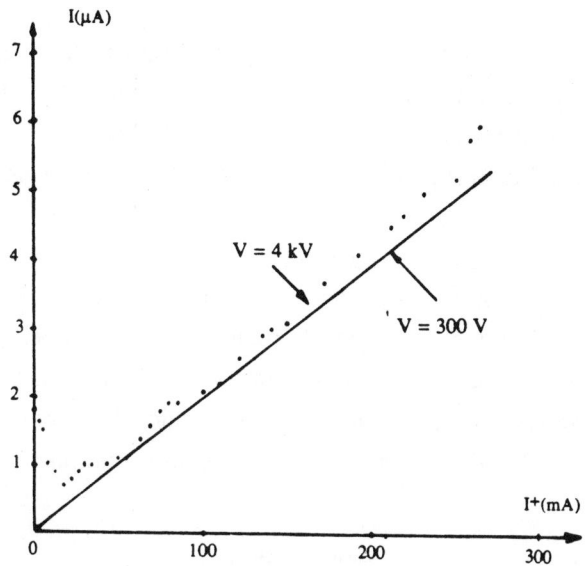

Fig. 6 : Distributed Ion Pump Current from Dipole 4 versus Beam Current.

Fig. 7 shows a cross section of the DIP's in Super-ACO. Photo electrons such as (a) will not contribute to the pump current since they reach the opposite Ti plate and are lost. Photo electrons such as (b) reach the anode and are part of the pump current. The fact that the anodes of the DIP's in Super-ACO are built from

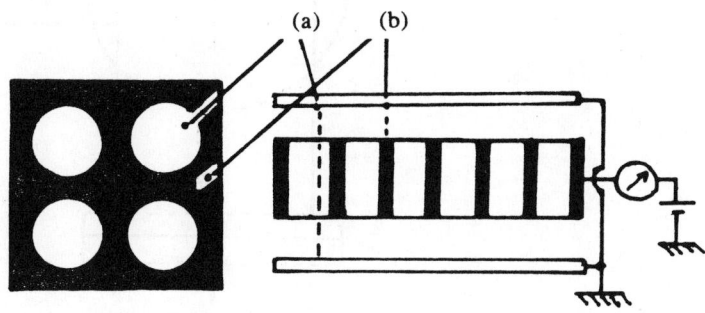

Fig. 7 : Cross section of a DIP. (a) Photo electron lost,
(b) photo electron recorded.

thick plates with bored holes for the cells is unfavorable. The area on which photo electrons can be collected is large as compared to the one from cells normally built out of thin cylinders.

From the above two examples, we see that no presently available pressure sensor is in principle free from X-ray photo current contamination. This is also true for the inverted magnetron gauge, which on the other hand has a small opening, thus reducing the probability for a photon to reach the inside part of the gauge.

4) <u>Electromagnetic fields from the beam</u>

The first problem of great significance encountered was a pick up on the MAS, increasing with the beam intensity, which gradually erased the peaks, starting from mass 2. **Fig. 8** presents a cross section of the MAS and its connexion to the Tee section. The MAS is an RF quadrupole system in which the most critical part is the ionisation chamber. The RF quadrupole and the electron multiplier on the other hand are well shielded.

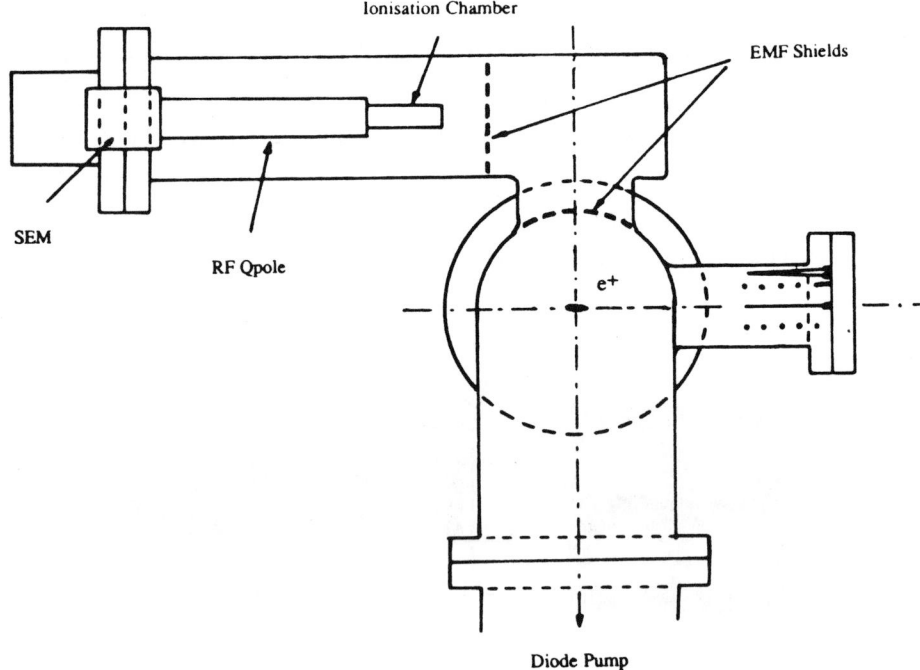

Fig. 8 : Cross Section of a MAS in a Tee Section.

In the initial design of the machine, no precaution was taken in order to faradize the MAS housing. At a beam current of 50 mA distributed in 24 bunches, the mass 2 peak disappeared completely. A limited improvement was obtained using a copper tube with pumping holes around the MAS. A satisfactory situation was reached when two grids, with holes 10 mm in diameter, were provided with good RF contacts in the locations shown in **Fig. 8**.

The criteria was a plot of partial pressures, normalized to the B.A. gauge reading independant of the beam current, see **Fig. 9**. Presently 2 out of the 8 MAS have been faradized according to the above scheme, the others waiting for a vacuum chamber opening.

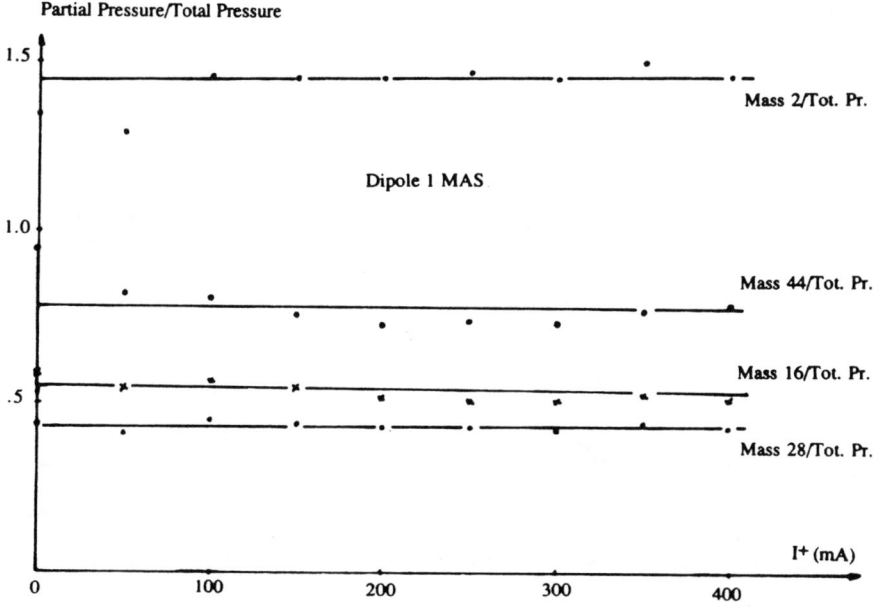

Fig. 9 : Partial Pressure Normalized to Total Pressure versus Beam Current (Faradized Tee Section).

The second problem is seen when looking again at Fig. 6, which shows the DIP current versus the beam intensity. We believe that the initial drop of the pump current and the subsequent wiggles superposed on a linear increase are genuine and reflect E.M. field effect on the pump discharge. One can compute that an isolated electron in the pump oscillates with a frequency in the range of 500 MHz. E.M. fields with frequency components close to it, modulate the electron motion and may kick out particles from the discharge. It has not been possible to establish whether such a pick up was due to the slots or to an uncomplete faradization at both ends of the DIP structure.

III - LINEAR AND NON LINEAR PRESSURE RISE WITH BEAM CURRENTS

Very soon, at the beginning of the commissioning the upstream B.A. gauges displayed a strange behavior shown by **Fig. 10**. Starting from the static pressure, a linear rise due to PSD is observed. Its slope was found to decrease with the beam dose $Q = \int Idt$

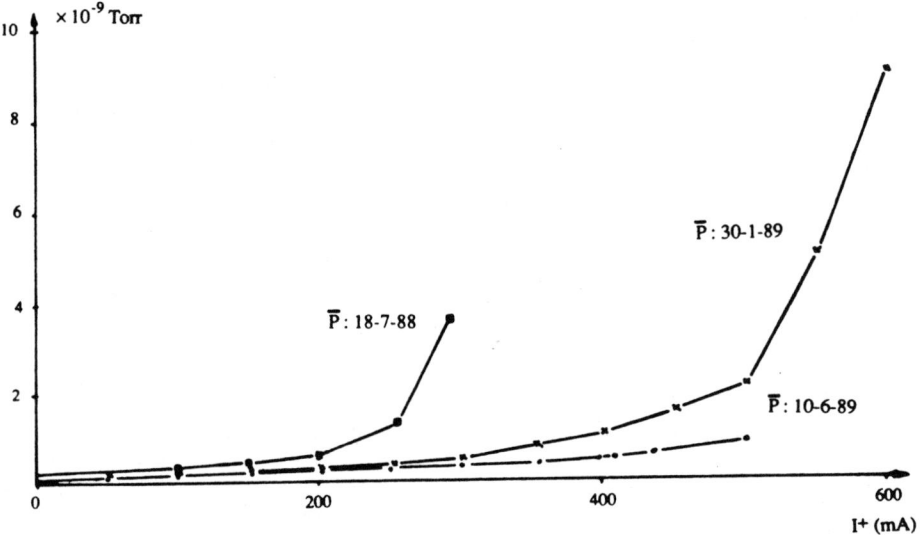

Fig. 10 : Linear and Non-linear Pressure Rise Versus Beam Current.

At a certain current the pressure rise starts to be strongly non linear, the reading being unstable. Fortunately the threshold of the non linear region receded to larger currents, conditioning being gradually observed.

Several possible mechanisms were envisaged.
- Discharges in the RF contacts of the bellow shieldings.

This effect which depends strongly on the bunch current, rather than on the total current was infirmed.

- Multipactor by the beam.

An electron is accelerated by the bunch passing, see **Fig. 11** and hits the opposite wall of the chamber. If the secondary emission coefficient is larger than 1, a resonant discharge may occur. The threshold for this effect in the conditions of Super-ACO is in the range of 600 mA, with 24 bunches and the kinetic energy of the electrons after crossing the full aperture of the chamber around 40 ev. However, killing one out of the 24 bunches should unvalidate the resonant conditions. This was found not to be the case, the non linear contribution to the pressure being more or less proportional to the intensity.

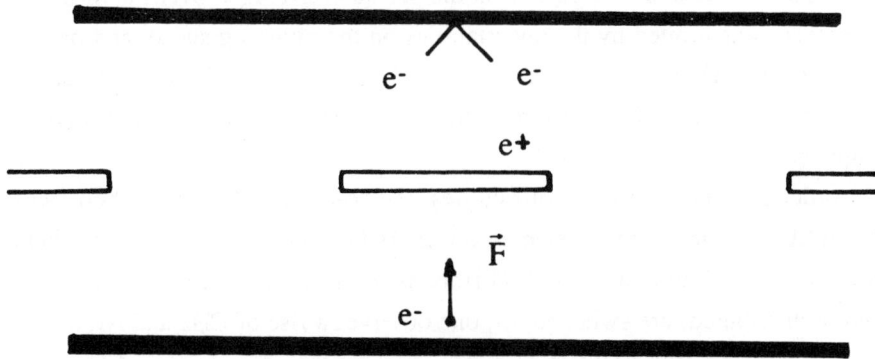

Fig. 11 : Multipactor from a Bunched Positron Beam.

- RF beam excitation of the gate valve body.

At the time Super-ACO was designed, compact, all metal, shielded gate valves could not be purchased and we resorted to unshielded units. RF measurements displayed a broad resonance around 1.5 GHz. The non linear pressure increase is rather well explained by the beam excitation of the cavity of the gate valve body. The RF field accelerates electrons leading eventually to multipactor, with a strong gas release. The threshold character of the effect and the observed conditioning are natural consequences of this phenomenon. Furthermore, when the obturator was pushed in by about 2 cm, the pressure showed a very large increase (1 to 2 orders of magnitude). Finally the presence of a stream of electrons in the vicinity was observed on the near by B.A. gauge when stray magnetic fields were suppressed by a proper shielding.

As mentioned above, conditioning at large beam intensity and with time is observed and the situation is tolerable. Still, whenever an octant of the machine is let to atmosphere (dry N_2), this conditioning has to be done again.

IV - THE ARGON METHOD FOR COMPARING MEASURED AND EXPECTED BEAM-GAS LIFETIMES

At the end of the commissioning of Super-ACO, one of the conclusions was that expected and measured beam-gas lifetimes were differing by a factor 2 to 3[3]. As stated above the pressure measurements were limited to one type of B.A. gauge, the overs being unvalidated. The accuracy on the pressure, averaged over the machine circumference, was limited by the uncertainties on the pumping speed, and on the conductance of various vacuum elements and by the lack of partial pressure measurements with beam. In these circumstances the observed discrepancy is not at all a surprise.

In order to overcome these difficulties, the so-called "Argon Method" was initiated at LURE. The vacuum system of Super-ACO is pumped down with the help of ion pumps, Ti sublimators and NEG ribbons in the ID's. When the ion pumps (lumped or distributed) are switched off, one observes a rise of CH_4 and Ar, all the others species being still maintain at a low level by the other pumps. CH_4 soon saturates, whereas Ar keeps increasing for hours and even days. Within several hours Ar is dominant see **Fig. 12** and the pressure is uniform all over the circumference. This is demonstrated by **Fig. 13** which shows that the ratio of the readings of 7 gauges normalized to gauge 1 is independant of time.

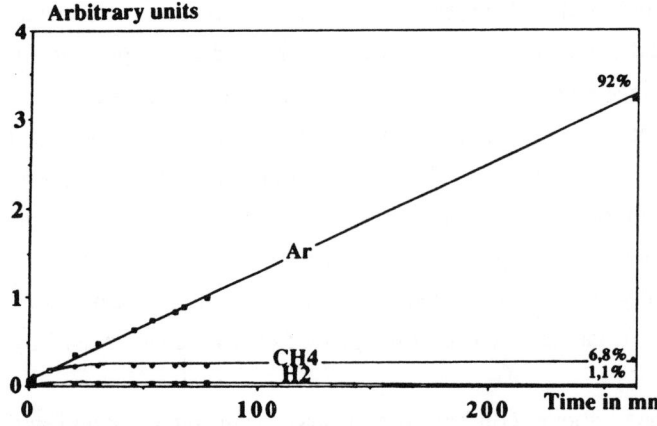

Fig. 12 : Argon Method. Partial Pressure Rise with time (ion pumps off).

Finally the beam-gas lifetime was computed using the average pressure (2 10⁻⁸ Torr), the partial pressure content, the known physical aperture of the chamber and the machine energy acceptance[4]. The lifetime was measured using a weak beam of positrons (15 mA distributed in 24 bunches) so that Touschek contribution could be neglected. The agreement between measured and expected beam-gas lifetime 20 % was indeed much better.

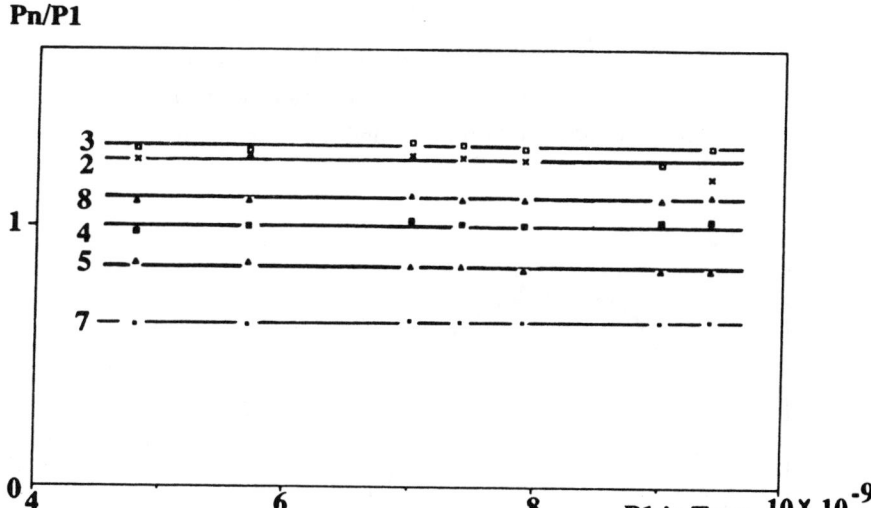

Fig. 13 : Argon Method. Normalised Pressure Rise with time (ion pumps off).

The "Argon Method" was found to be increasingly popular at LURE. It is used whenever a series of many lifetime measurements have to be performed in a short time. **Fig. 14** shows a typical example, where beam-gas lifetime was measured versus the vertical time v_z, the horizontal tune v_x being kept constant. The pressure was in the range of 1 to 2 10⁻⁸ Torr and the beam lifetime around 10 to 5 minutes. We see that the product $\bar{P}\tau$, which is independant of the pressure, decreases when v_z approaches 1.66 (third resonance) and 1.75 (fourth resonance). A dip is also observed close to the coupling resonance $v_x - v_z = 3.0$. This is due to the following effect. Particles from the beam which are scattered vertically are lost in a few turns whatever the tune is. The horizontal aperture being much wider than the vertical one, the particles scattered horizontally are normally damped when the tune is away from the coupling resonance. When the tune is on the coupling resonance, the transverse horizontal momentum is gradually rotated to the vertical direction and the horizontally scattered particles are lost again on the vertical boundary. The cross section for Rutherford scattering which is usualy[3] :

$$\sigma_{sc} = \frac{2\pi\, r_0^2\, z^2}{\gamma^2\, h^2}\, \overline{\beta_z^{cr.} \cdot \beta_z}$$

becomes now :

$$\sigma'_{sc} = \frac{2\pi\, r_0^2\, z^2}{\gamma^2\, h^2}\, \overline{\beta_z^{cr.} \cdot \left(\beta_z \cdot \beta_x\right)}$$

On the other hand, the cross section for Bremsstrahlung on nuclei and atomic electrons and from elastic scattering on atomic electrons is not modified. Altogether the agreement between the observed dip, 0.65 and the expected value, 0.68 is very good.

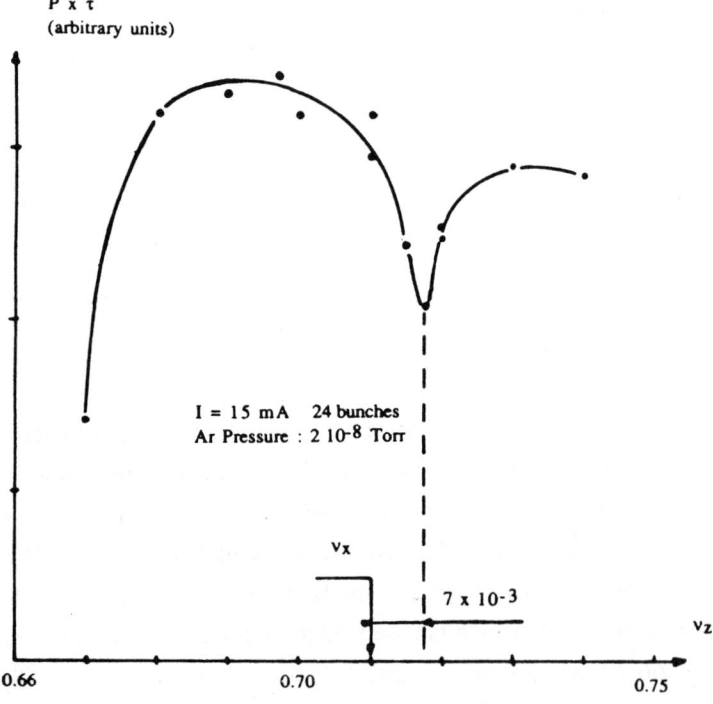

Fig. 14 : Argon Method. Beam-gas Lifetime Versus v_z.

Looking more closely to Fig. 14, one finds that the minimum of the dip occurs above the resonant tune for the damped beam, the difference being $7\, 10^{-3}$. This difference measured for several values of v_x turns out to be constant, see **Fig. 15**. It can be understood by taking into account the tune shift at large amplitudes of oscillation. The particles which are lost from Rutherford scattering have a slight difference of tune as compared to the damped beam, due to octupolar fields in the machine. They sit on the coupling resonance when the damped beam is a little away from it.

The method was also applied to measuring the displacement of the dip as a function of the sextupole setting. It could also be used to <u>test positive ion trapping in an electron beam in ideal conditions</u> : beam intensity, bunch structure, emittance ratio $\varepsilon_z/\varepsilon_x$ and single species for the residual gas.

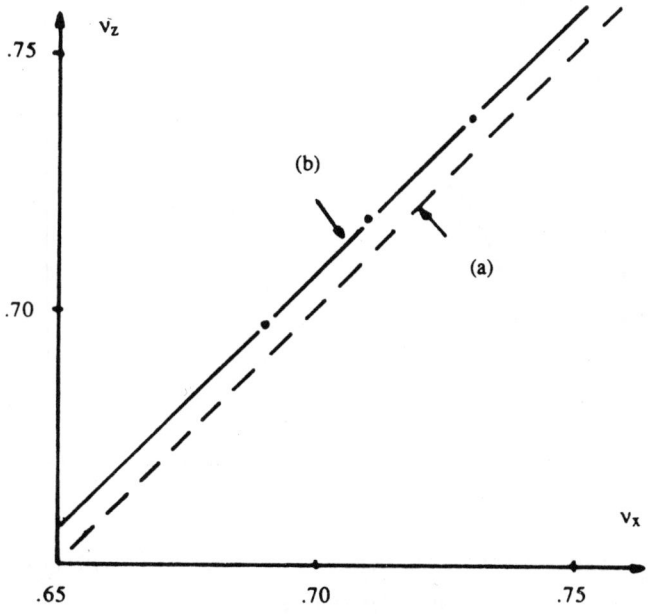

Fig. 15 : Argon Method. Position of the Dip of τ gas Versus v_x.

V - COMPARISON OF ELECTRON VERSUS POSITRON BEHAVIOR

For almost 3 years and anyhow since the very beginning of its life, Super-ACO was operated with positron beams. These are available at Orsay from the time of the e+e- colliders. For this reason, no provision was made for incorporating clearing electrodes in the design of the vacuum chamber, thus reducing its impedance.

In February, this year most part of the vacuum system had been irradiated with a dose corresponding to 400 A.h. Super-ACO was then reversed in polarity and an electron beam was circulated in the same direction as the former positron beam.

The machine was first run with 24 bunches in the usual conditions with a beam emittance $\varepsilon_x = 3.7 \; 10^{-8}$ mrad, $\varepsilon_z = 1.0 \; 10^{-8}$ mrad corresponding to $\sigma_z = 300$ μm in the dipole source. For sake of comparison **Fig. 16** shows the products $I^+\tau^+$ and $I^-\tau^-$ as a function of the total beam current. It is clear that the decay

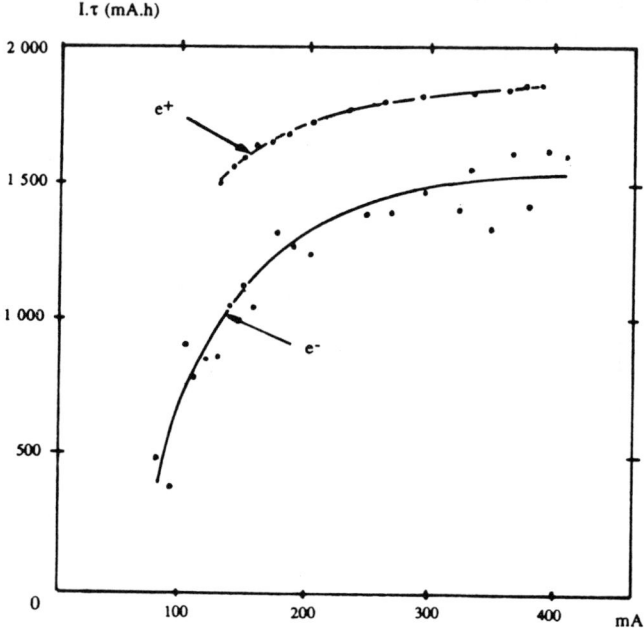

Fig. 16 : Comparison of the Behavior of Super-ACO with Positron and Electron Beams.

of the positron beam is much smoother than the one for electrons. At large intensities, 400 mA, the electron beam lifetime is shorter by 16 %. In the lower range of intensities the situation is worse. Furthermore, abrupt small losses of beam are reflected on the curve although somewhat integrated by the measuring system. Fluctuations in the transverse beam size are also more pronounced with electrons than with positrons, although no definite measurements were recorded.

In the 2 bunch regime which is used for 50 % of the beam time, the situation is much worse. The following table is a summary of the results which were recorded for conditions very close to normal operation.

	σ_μ (μm)	I_{tot} (mA)	$I\tau$ (mA.h)
e+	360	200	550
	"	100	440
e-	330	220	450
	"	105	140
e-	370	200	160
	"	100	64

It is clear from these results that ion trapping is more pronounced with increased vertical emittance and for lower beam current. These observations are in agreement with the predictions of ion trapping calculations and with the experience from other machines.

Super-ACO has to face many situations in order to meet the user's requirements : 2-bunch or multibunch operation, semi flat or fully coupled beam for optimizing Touscheck lifetime and FEL operation. The flexibility offered by operating the ring with positrons was found invaluable in this respect.

VI - MICROLOSSES OF BEAM CURRENT AND THE "DUST MECHANISM"

For many years both DCI and Super-ACO were operated with positron beams and displayed a very smooth decay of the beam current. As soon as Super-ACO was turned to electron filling, in average 6 to 10 microlosses of beam current were observed per fill over several hours. Their amplitude is spreading over a factor 10 and the average value is 0.5 %. The decay time of these losses is shorter than 10 s, which is the resolution of the beam intensity chart recorder.

In a particular case, a 2 bunch beam of intensity 75 mA started trapping something and the lifetime was as short as a few minutes. This situation lasted until the beam intensity reached 300 μA (5 10^8 electrons). In more frequent cases, sequences of poor lifetime separated by temporary recoveries were observed.

These observations do not fit with the usual picture of ion trapping phenomena. They have lead to consider what we call the "Dust Mechanism initiated by photo ionisation".[5] The main aspects of the processes involved in this mechanism are discribed below :

- Stray photons ionize a dust particle supposed at first to be spherical with radius R. The potential reached by the dust particle is limited by the "maximum" energy of the photo electrons eV_0 (in practice 100 volts). We therefore have :

$$V_0 = \frac{N^+ e}{4\pi \varepsilon_0 R}$$

where N^+ is the number of positive charges.

- The electric field from a bunch of electrons passing in front of a dust particle at a distance of 2 cm is, for $I^- = 400$ mA and 24 bunches, $E = 5 \; 10^4$ vm^{-1}.

- The acceleration of the dust particle during the bunch passing time is :

$$\frac{F}{m} = \frac{1}{10^3} \frac{N^+ e E}{\frac{4}{3} \pi R^3 d}$$

where d is the particle density and R is expressed in meters.

- The average acceleration is :

$$\gamma = \frac{F}{m} \frac{t_1}{t_2} - g$$

where t_1 and t_2 are respectively the bunch passing time and the bunch spacing.

- The threshold for raising the dust particle is obtained when $\frac{F}{m} = 42.85$, $\gamma = 0$, but the motion has a very slow initial velocity. In practice $\frac{F}{m} = 50$ is a more suitable value.

On then gets $R_{th.} = 11.7$ µm, $N^+_{th.} = 8.1 \; 10^5$.

- The time duration for a first crossing of a fully coupled beam (round in shape if $\beta_z = \beta_x$) is evaluated to be 600 µs.

- During this time, almost every electron of the beam will hit the dust particle.

- From a gaussian distribution of scattering angles with r.m.s. θ, 0.5 % of the particles have angles larger than 3 θ. In Super-ACO, particles are lost vertically if 3 θ is larger than 3 mrad, therefore $\theta = 1$ mrad.

- The r.m.s. angle for multiple scattering is :

$$\theta = \frac{21}{\xi} \times \sqrt{\frac{X}{X_0}}$$

where ξ is the beam energy in MeV, X_0 is the radiation length of the material and X is the distance crossed by an electron in this particular material. For carbon of density 2, $X_0 = 43$ gcm^{-2} and $X = 310$ µm. Rather than being spherical the required shape for the dust particle will be fiber like with a radius of 12 µm and a length of 300 µm. This new shape has however little effect on the calculations performed above.

- The dust particle will experience a temperature rise following the first beam crossing of 2×10^4 °C which will immediately destroyed it.

VII - BEAM EMITTANCE GROWTH FROM MULTIPLE SCATTERING ON DUST PARTICLES

A similar calculation can be performed for machines with higher energies but more moderate current (100 mA) such as ESRF, APS and SPRING 8. One finds that the temperature rise after the first crossing is independant of the dust particle size and somewhat lower than the above value for Super-ACO, 4 900 °C. Furthermore Möller scattering of high energy particles from the beam on atomic electrons can increase the charge of the dust particle to a value much higher than the one calculated before. The period of oscillation an the time for crossing the beam go down and with them, the average temperature of the dust particle, 1 800 °C. At such a temperature a dust particle of carbon can stay for a very long time oscillating around the beam and one should observe an emittance growth.

To be more specific, using the nominal parameters of ESRF for instance, one finds that one carbon-like dust particle of radius 6 μm, every meter of ring circumference, would double a vertical emittance of 6×10^{-10} m.rad.

CONCLUSIONS

In the design of radially compact S.R. facilities, close attention should be paid to avoid devastating effects of electrons, photo electrons, photons and RF fields from the beam on the pressure sensors or active pumping components. Presently, calculations are much too complicated to be performed, and only rule of thumb starting from the experience of existing rings can be used.

The operation of Super-ACO with positrons has proven to be very flexible in meeting the requirements of the users. The avoidance of dust particle trapping in the beam is favourable both against microlosses of current and beam emittance growth.

REFERENCES

[1] Vacuum and Multipactor Performance of the HERA 52 MHz Cavities, R.J. Burton et al., EPAC 90, Vol. 2, p 1017, Nice June 12-16, 1990.

[2] Private communication, R. Parodi, Istituto Nazionale di Fisica Nucleare, Genova.

[3] Super-ACO, Vacuum Chamber and Related Problems, P. Marin and R. Souchet, Anneaux, RT/88-03, LURE Département Anneaux, Orsay.

[4] Expériences sur l'anneau de collisions AdA, J. Haissinski, Thèse de Doctorat es-Sciences, 1965, n° 81.

[5] Microlosses of Beam Current in Super-ACO Operated with Electrons, P. Marin, RT/90-01, LURE Département Anneaux, Orsay.

DEVELOPMENT OF THE ESRF VACUUM SYSTEM

M. RENIER, D. SCHMIED and B.A. TRICKETT
European Synchrotron Radiation Facility,
BP 220, 38043 Grenoble Cedex, FRANCE.

Abstract

The European Synchrotron Radiation Facility (ESRF) is a high brilliance 6GeV Synchrotron Radiation source which is now being built in Grenoble, France. It will provide intense X-ray beams for the European scientific community for well into the next century, with first experiments scheduled from July 1994. The vacuum system has evolved with the experience gained in the production and testing of prototypes, and some experimental work. The main developments in designs of vacuum vessels and crotches will be described, together with current status including that of the vacuum control system.

Introduction

The ESRF vacuum system has been reported previously /1/, /2/. In this paper we will focus on the evolution of the following:
Crotch absorbers. For the ESRF these have been designed to absorb 8kW of synchrotron radiation with a power density equivalent to 400W/sq.mm at normal incidence. At these powers it is important to reduce "reflected" power, to limit heating of the stainless steel vacuum vessels and gas desorption.
Straight vessels. The distributed absorbers on these vessels present problems similar to those for the crotch regarding the quality of the vacuum brazes and reflected power. Additionally, the reflectivity of the absorber surface is an important parameter since the probability of "specular" reflection for X-rays is increased at the low grazing angles on these absorbers.
Pumping. High gas loads are generated by the intense synchrotron radiation. To cope with this gas load lumped ion

pumps augmented with lumped Non Evaporable Getter (NEG) pumps will be used. The latter are being developed in-house to match different physical constraints.

Bakeout. The bakeout system is designed to heat the storage ring vacuum system up to 200°C in-situ without over-heating the surrounding magnets, cables, etc. In some areas, such as between the poles on dipole magnets, lack of space severely limits the amount of insulation.

Controls. The vacuum system consists of a large number of devices which compounds the remote control and monitoring of them in a fast (for efficient surveillance), user friendly and reliable way. How to develop such a system within the timescale is of prime importance and a major problem.

Crotch absorbers

At 6GeV and 200mA the incident power on some crotch absorbers reaches 7.55kW. In these conditions a flat crotch would re-emit around 320W, mainly by fluorescent radiation /3/, which would necessitate cooling of the vacuum vessel. Another undesirable effect would be photo-desorption increasing the local pressure. For these reasons, a partially enclosed structure with a flattened "C" type section /2/ has been designed, Fig.1.

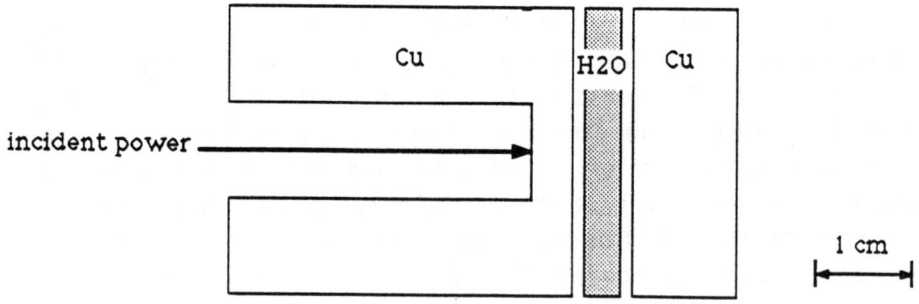

Fig.1. Section of the crotch absorber.

The absorber is a sandwich with three distinct sections. The front part, with a flattened "C" profile absorbs most of the incident power from the intense synchrotron radiation X-ray beam which strikes the centre of the "C", whilst the wings of the "C" profile absorb the "reflected" power. Here, "reflected" power includes the total flux from fluorescent, coherent, and incoherent radiation generated at the absorber. The central part is for water cooling and consists of an array of square copper tubes (6 x 6 sq.mm). The back part absorbs the high energy photons so that the power transmitted is negligible. It also mechanically stabilizes the structure particularly during vacuum brazing.

Calculations using the program "PHOTON" /4/ applied to this structure have shown that 0.24% of the 0.78W/mA/mrad of incident power are transmitted through the front section, whilst only 0.001% remain after transmission through the back plate. The efficiency of the two wings of the "C" profile is such that of the 320W emitted by fluorescence and back scattered radiation, only about 20W escapes the absorber, assuming isotropic radiation from the effective line source.

Two very important aspects have still to be resolved, namely, the choice of material and the procedure for brazing the different parts together.

For the material, the choice is restricted by the maximum stress it has to sustain at high temperature, its thermal conductivity and outgassing characteristics. Two materials offer the mechanical characteristics we need: GlidCop (Cu + Alumina) /5/ and DSCop (Cu + Zirconia) /6/. Both of these are dispersion strengthened copper and sustain a maximum stress of 250MPa at 300°C, compared with 50MPa for OFHC copper. The main problems are: the outgassing rate of GlidCop which was high in Argon on some samples, and the homogeneity of DSCop was poor in the samples tested.

After brazing our specification calls for a minimum of 90% contact area between the absorber and the cooling tubes, to ensure good heat transfer and reduce the probability of creating virtual leaks. Tests are currently being conducted in order to specify the necessary brazing procedure.

At this time an intensive study of both materials is in hand to enable a decision to be made. The default is OFHC copper which would solve both outgassing and brazing difficulties, but reduce the maximum current capability in the storage ring to around 50mA. Such a solution would be a temporary one for the first phase of Machine operation, allowing time for the development of a reliable crotch for operating at 200mA.

Straight vessels

The design of vacuum vessels for straight sections in the storage ring allows for "reflected" power from the distributed flat absorbers, Fig. 2.

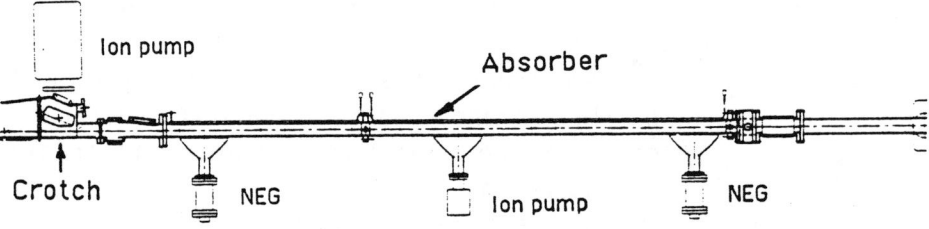

Fig.2. Vacuum vessel design: distributed absorber downstream from a dipole vessel.

Calculations /3/ show the sum of fluorescent, coherent and incoherent radiation may reach 10% of the incident power. This power is re-emitted to first order isotropically and therefore has a local heating effect on the vacuum vessel. In the first "upstream" part of the vessel, it may create a temperature gradient of the order of 90°C across the vessel section and generate high stresses in the vessel walls /7/. For example a stress of 480MPa if fully restrained against bending, or a bend radius of curvature equal to 32m if left unrestrained. To suppress this effect it was decided to add two cooling channels on the opposite face to the distributed absorber on the outside of the vessel. In these conditions the maximum temperature gradient predicted across the vessel section is reduced to about

28°C and the maximum stress by a factor of 5. Also the radius of curvature of the vessel, if free to bend, increases to around 2250m.

The problem created by "specular" reflection on the distributed absorbers has to be solved, since synchrotron radiation strikes the absorbers at glancing angles in the range from 0.4° to 1.5°.
At these angles and considering the very high photon energies, from 22% to 4% of incident power could be specularly reflected from the distributed absorbers. Therefore from each achromat vessel some 200W could be reflected into the following dipole vessel where no cooling is foreseen. The situation is similar in the Insertion Device straights which receive reflected power from the vessels "upstream". It is therefore of prime importance to reduce the reflectivity of the distributed absorbers by about an order of magnitude, so that reflected power will not be a problem.

To reduce the reflectivity of OFHC copper four methods have been tried. These all effectively roughen the surface locally to increase the angle of incidence of the X-rays on the surface. The reflectivity of the samples was measured using Cu Kα radiation from a Cu anticathode operated at 25 kV and 10 mA /8/. The detector was a Si(Li) diode coupled to a multichannel analyser. The full spectrum of the source was measured successively without and then with the sample mounted on a rotatable sample holder. Dividing the second set of intensities by the first then gives the reflectivity.

The first method was to prepare several well defined machined surfaces. This turned out not to be a practical solution, because defining the status of the cutting tool exactly, ie. its rotational speed, advance and cutting angle, are very difficult and the reflectivity may reach 10% or more if the machining conditions are not exactly respected.

The second method was to machine a toothed profile. Measurements then confirmed the predicted low reflectivity.
However, because of high thermal stresses on the tips of the teeth, dispersion strengthened copper would have to be used. The machining of this hard copper would be very difficult and hence time consuming making such absorbers prohibitively expensive,

with the possibility also of introducing brazing difficulties.
The third method was to bead blast the surface, which in principle is fast and simple to do. However an exact definition of the process to ensure repeatability in the results is not easy to establish. It can also produce a large amount of surface damage and introduce impurities. For example, six samples were prepared by bead blasting with glass and alumina beads, and then analysed using a Scanning Electron Microscope together with an Energy Dispersive X-ray spectrometer. This revealed the presence of aluminium in the surfaces after alumina bead blasting, and with both alumina and glass the surfaces presented a chaotic aspect with implanted bead particles. Such surfaces would be unsatisfactory for the ultra-high vacuum required in the storage ring.
The last method was a simple chemical etch of the copper surface, using Nitric acid (65%) diluted in 4 times its volume of water at room temperature for 20min.
The results for the acid etched surface are shown in Fig.3.

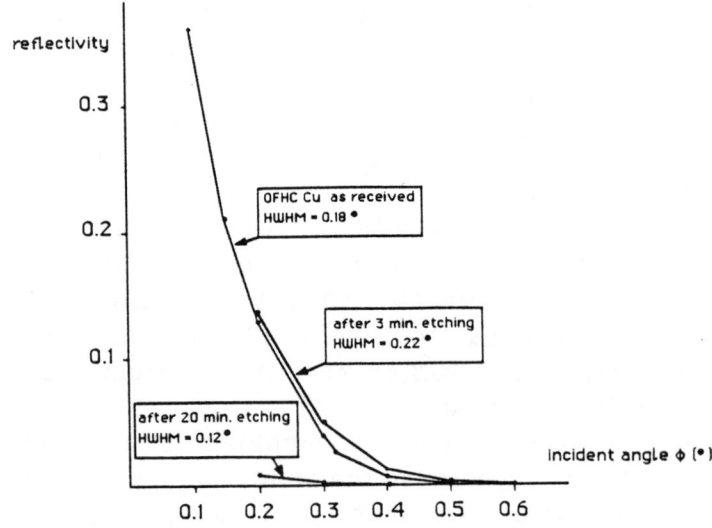

Fig.3. OFHC copper reflectivity after chemical etching.

The figure shows reflectivity versus angle of incidence Ø with respect to the surface in the range from 0.1° to 0.5°. It is of interest to note that for cold rolled OFHC copper, the reflectivity increases if the etching time is too short, ie. 3min instead of 20 min., which may be due to an initial surface cleaning effect.

Our conclusion is that an appropriate chemical etch on the copper absorbers will reduce their reflectivity by a factor of 10, so that reflected power will not be a problem on the ESRF. A big advantage of this method is that little extra cost is involved, since a similar process would be necessary anyway in preparing for brazing the absorber to its stainless steel support.

Pumping

The storage ring will be pumped by a combination of some 400 lumped triode ion pumps together with a similar number of lumped NEG pumps. The expected pressure profile has already been reported /2/. At each crotch, an St101 /9/ commercially available module is installed in the ion pump. This pump is supplied on a NW150 Conflat flange with a feedthrough for electric heating for activation and regeneration. Smaller lumped NEG pumps are installed on the dipole and straight vessels. In order to meet our specific requirements for pump speed, capacity and physical constraints, we decided to develop our own design based on St707 strip /9/. The prototype, Fig.4, uses a 600W internal electric heater for activating and regenerating the 10m of NEG strip, which is necessary for the gas capacity required. For first tests the pump was mounted on a prototype straight vacuum vessel, to simulate as

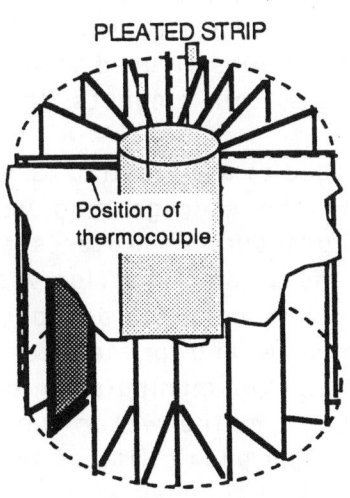

Fig.4. NEG pump design.

closely as possible the operating conditions in the Machine. After activation at 550°C, during which time the vacuum was maintained with an auxiliary turbo pump, the pressure fell from 2E-9mbar to 4E-10mbar. During this time the 16amu peak was increasing relatively on the Residual Gas Analyser (RGA) scan. This peak corresponds to methane which must be produced by a chemical reaction of hydrogen in the NEG element. First results demonstrate a pumping speed of around 100l/s. The pump is now mounted on a speed test system for more accurate measurements with different gases. The pumps capacity will also be measured to confirm the design life of around 10 years in the operating conditions predicted for the storage ring.

Bakeout

The bakeout system comprises 5 differents items:

i) <u>Thermocouples.</u> Each cell (1/32 of the Storage Ring) will be equipped with a total of 64 thermocouples evenly distributed along the vacuum vessels. Two types of thermocouple are required, the K-type (Chromel-alumel) and the T-type (Copper-constantan). The K-type is necessary where the temperature exceeds 400°C and or where spot welding the thermocouple is useful, eg. in the NEG pumps. Unfortunately, this type of thermocouple could disturb magnetic fields if used to monitor the temperature of vacuum vessels within magnets. In these cases the T-type must be used.

ii) <u>Heating system.</u> It has been decided to use electric tapes and collars throughout for bakeout of the storage ring vacuum system, in place of super-heated pressurized water systems originally intended for bakeout of the straight section vessels. The idea to use the three cooling channels for bakeout was abandoned because it was expensive and presented many difficulties. A major problem would be the manipulation needed to disconnect the cooling channels from the water supply to connect the boilers, and vice versa. The possibility of leaks of hot or cold water was considerable with the risk of water then entering electric cable trays. With the electric system most connections are permanent and therefore manipulations required for bakeout will be a minimum.

iii) Thermal insulation. Fig.5 shows the arrangement for dipole and straight section vessels. In the dipole the gap between the magnet poles and vacuum chamber is only 3mm. It is only sufficient for one layer of insulating hard board /10/ with a thermal conductivity of 0.024W/m°K.

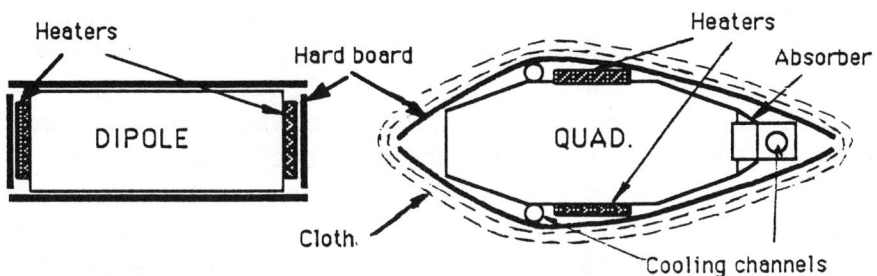

Fig.5. Disposition of the insulating material wrapping the dipole and the straight vacuum vessels.

For straight sections it is foreseen to complete the thermal insulation with three layers of insulating clothe /11/. This is possible because of the larger gaps between the vessels and poles in the quadrupole and sextupole magnets. The thermal conductivity of each layer is 0.054W/m°K.

iv) Temperature controllers. Two sets of mobile controllers will permit bakeout of two cells which equates to one sector of the storage ring as defined by the remotely operated gate valves. Each cell will be divided into six sub-sections, the temperature of each being controlled independently. The controllers will be connected locally to the corresponding sub-sections while the temperature of the whole storage ring will be monitored continuously in the Control Room.

v) Thermocouple interfacing. All the thermocouples will be permanently monitored through the vacuum control system, thus permitting a temperature map of the storage ring. This gives the possibility to localize hot parts of the storage ring vacuum chambers during Machine operation, as well as during bakeout. Therefore we believe temperature maps will provide very useful diagnostics in times of vacuum problems in the storage ring.

Control system

The successful operation of a Machine like the ESRF implies a reliable communication infrastructure between host computers, workstations (in the control room) and, in racks close to the Machine, complex VME based structures comprising of minicomputers and microprocessors in Machine instrumentation.

The storage ring is divided into 32 nearly identical cells which includes vacuum equipment eg. ion and NEG pumps, Inverted Magnetron Gauges (IMG's) and Pirani gauges, RGA's, valves, etc. The instrumentation for this equipment is linked to VME controllers with the operating system OS9 via RS422 serial lines or through digital logic. The VME controllers are connected over the ETHERNET network to the Host computer with the UNIX operating system.

	synchrotron	transfer line 2	storage ring	beam port
Ion pumps	87	5	391	58
Controllers	24	2	130	58
NEG pumps	0	0	442	0
Controllers	0	0	442	0
Pirani gauges	9	2	64	58
IMG's	12	4	224	58
Controllers	9	2	65	58
RGA's	9	1	64	0
Thermocouples	0	0	1920	0
Controllers	0	0	32	0
Pneumatic valves	11	2	18	0
Man. valves (limit switches)	0	0	46	58

Table I. Number of instruments to be remotely controlled through the computer system on each part of the Machine.

Table.I, Fig.6 and Fig.7 give an idea of the volume of data to be treated.

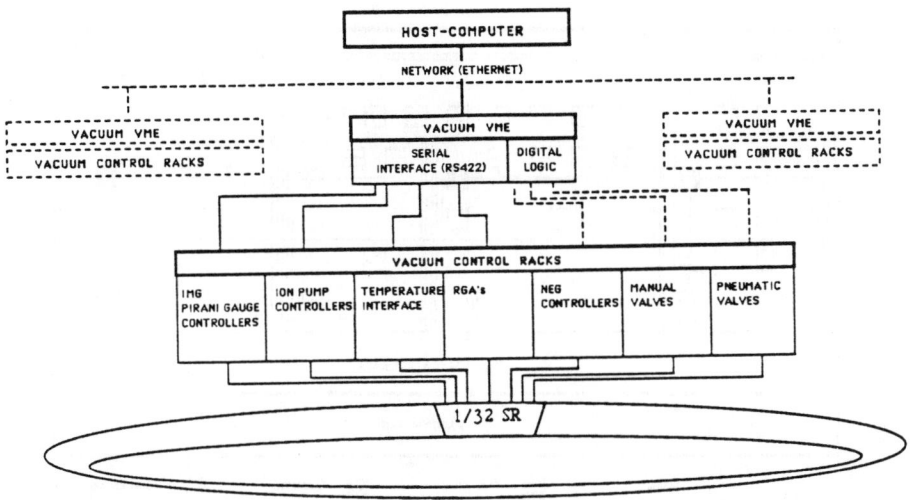

Fig.6 Structure of vacuum control for storage ring.

With the control structure proposed, it has been estimated that the response time to collect the pressure signals from all 200 or so IMG's in the storage ring, for a snapshot of the pressure profile, will not exceed 2 seconds. This information will be complemented by current readings from each ion pump, which will also be used to give vacuum levels around the Machine and therefore provide pressure information in regions where there are no gauges.

In the choice of controllers, we have tried where possible to have modular systems to allow for future developments. In the case of the ion pump supplies different sizes of pump have to be catered for. The system chosen supplies and monitors up to four pumps from one main high voltage unit, with a plug-in unit for each pump. For the gauges each instrument controls up to four IMG's and two Piranis. The controllers are designed to accept global commands for control and data acquisition. This allows simple and fast programming, which enables quick snapshots of the pressure profile in the storage ring.

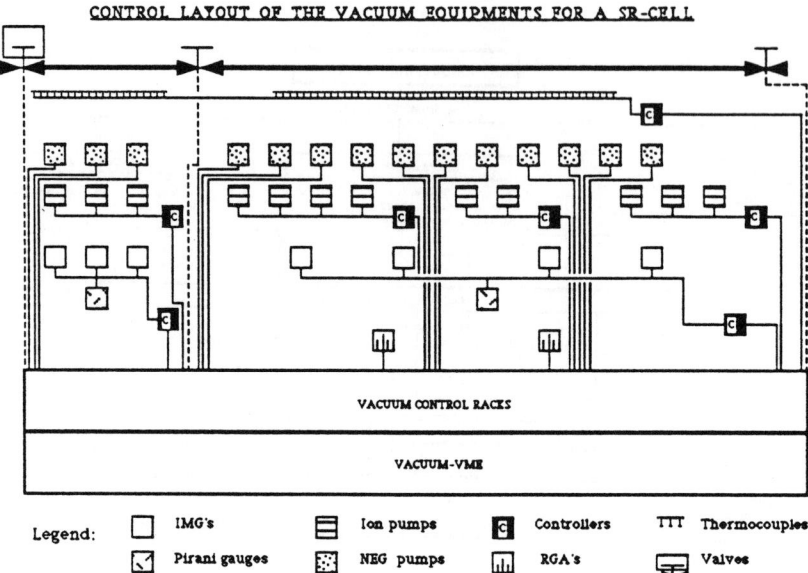

Fig.7. Detail of the vacuum control for 1 cell (1/32 of Storage Ring).

Conclusion

The ESRF is at an intensely active period marked by ongoing design and development, series of tender exercises, quality assurance inspections and tests. Contracts for most of the components for the vacuum system including the vessels will be placed by the end of 1990.

The LINAC installation will start in January 1991 and the synchrotron between February and June 1991. The storage ring installation is planned to start in February in parallel with the synchrotron and should be completed in late 1991. In order to follow this very ambitious planning, intensive effort and rapid progress are required in all areas.

Acknowledgements

The authors are indebted to C. Burnside and G. Le Flem for their skilful contributions in developing and testing prototypes.

References

/1/ B A Trickett, Vacuum, 38,(1988), 607.
/2/ B A Trickett and M Renier, Proceedings of EVC2-AIV X1, Trieste, May 1990.
/3/ P Elleaume, F de Bergevin and D Schmied, Internal ESRF note (1990).
/4/ PHOTON, BNL report 34934 (1984).
/5/ GlidCop, SCM Metal Products Inc., USA.
/6/ DSCop, OUTOKOMPU OY, Finland.
/7/ R J Bennett, Internal ESRF note (1990).
/8/ M Brunel and F de Bergevin, Internal note, CNRS Grenoble, (1990).
/9/ SAES Getters S.p.A., Milan, Italy.
/10/ MICROTHERM Europa NV., Sint-Niklaas, Belgium.
/11/ HEXCEL-GENIN, Lyon, France.

APS STORAGE RING VACUUM SYSTEM

R. C. Niemann, R. Benaroya, M. Choi, R. J. Dortwegt,
G. A. Goeppner, J. Gonczy, C. Krieger, J. Howell,
R. W. Nielsen, B. Roop, R. B. Wehrle
Argonne National Laboratory, Argonne, IL 60439

ABSTRACT

The Advanced Photon Source synchrotron radiation facility, under construction at the Argonne National Laboratory, incorporates a large ring for the storage of 7 GeV positrons for the generation of photon beams for the facililty's experimental program. The Storage Ring's 1104 m circumference is divided into 40 functional sectors. The sectors include vacuum, beam transport, control, acceleration and insertion device components. The vacuum system, which is designed to operate at a pressure of 1 nTorr, consists of 240 connected sections, the majority of which are fabricated from an aluminum alloy extrusion. The sections are equipped with distributed NeG pumping, photon absorbers with lumped pumping, beam position monitors, vacuum diagnostics and valving. The details of the vacuum system design, selected results of the development program and general construction plans are presented.

INTRODUCTION

The Advanced Photon Source (APS) incorporates a 7-GeV positron storage ring approximately 1104 m in circumference. The storage ring vacuum system is designed to maintain a pressure of 1 nTorr or less with a circulating current of 300 mA to enable beam lifetimes of greater than 10 hours.[1,2,3] The vacuum system employs Non-evaporable Getter (NeG) strips as the primary source of distributed pumping throughout the ring. Lumped NeG modules and/or ion pumps are used at photon absorber locations. Significant parameters of the Storage Ring Vacuum System are listed in Table 1.

Table I Storage Ring Vacuum System Parameters

CIRCUMFERENCE	1104	m
FORMED SHAPE	39.896	m INSIDE BEND RADIUS
VACUUM CHAMBER MATERIAL	6063 ALUMINUM EXTRUSION 2219 ALUMINUM FLANGES AND PORTS	
HORIZONTAL APERTURE (BEAM CHAMBER)	85	mm
VERTICAL APERTURE (BEAM CHAMBER)	42	mm
CHAMBER WALL THICKNESS	12	mm
OUTGASSING LOAD AFTER 150 A·h	3.77×10^{-5}	Torr l/s
POSITRON BEAM LIFETIME	> 10	h
BEAM-ON OPERATING PRESSURE	1	nTorr or less (average) after 100 A·h

STORAGE RING SECTORS

The storage ring is divided into forty 27.6 m long sectors. Each sector consists of 6 sections, 5 of which contain magnetic elements for beam transport with the other section containing either rf (3), diagnostics (1), injection (1), abort (1) or insertion devices (34). A total of 69 experimental facility photon beams are possible, 34 of which are generated by insertion devices and 35 by bending magnets.

The storage ring sector layout is as shown in Figure 1. Sector components include the crotch photon absorbers and pumps at the end of the two dipole curved chambers. The straight chambers contain downstream end absorbers and pumps. The largest gas desorption loads are located at the photon absorbers and are pumped locally by high capacity pumps. Large lumped ion and lumped NeG pumps are mounted at these locations in order to capture the bulk of the desorbed gases. Significant amounts of these and other photon and thermal desorbed gases permeate the chamber and are subsequently pumped by the distributed NeG pumps located in the pump antechamber. The distributed NeG pumping strips are essentially continuous in the chambers and are shown in Figure 2.

86 APS Storage Ring Vacuum System

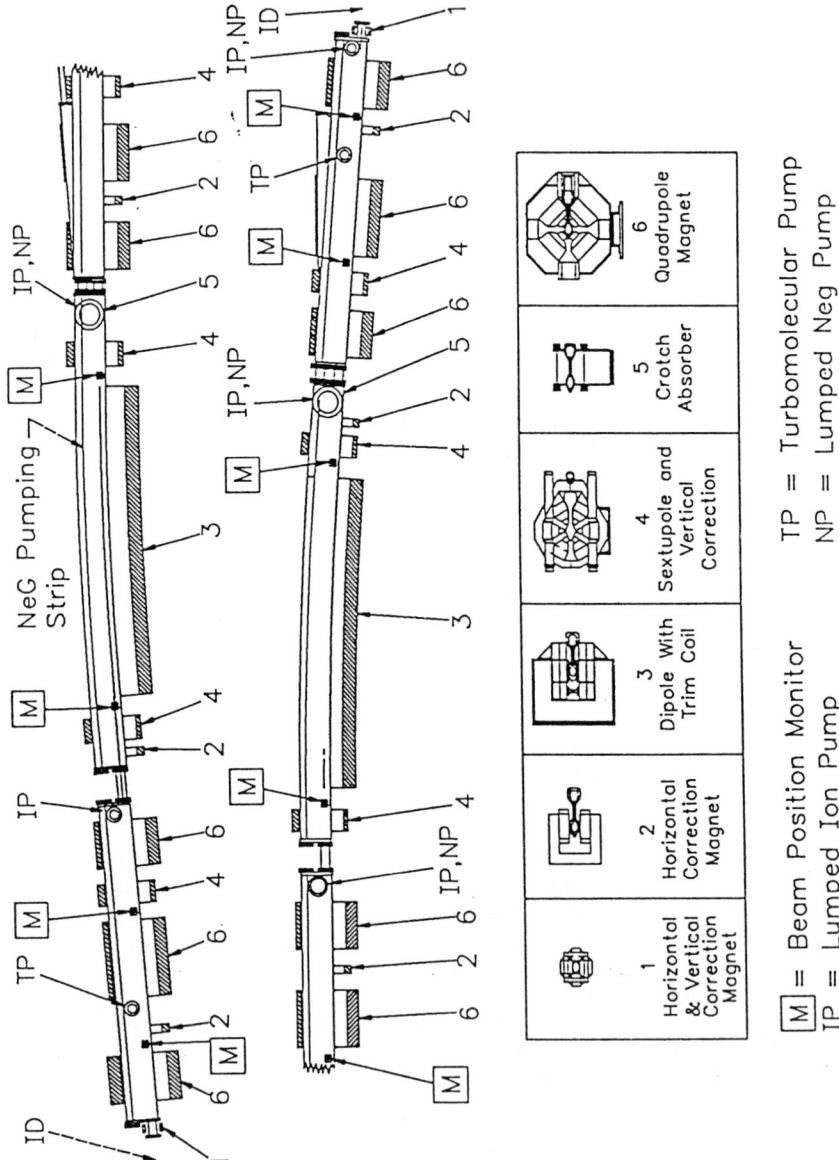

Figure 1
APS STORAGE RING SECTOR LAYOUT

VACUUM CHAMBER

The vacuum chamber consists of three main regions, the positron beam chamber, the pump antechamber, connected by the photon beam channel. The 10 mm photon beam channel gap is high enough for the photon beams to pass through to the extraction channel, but narrow enough so that rf leakage is negligible. The antechamber entraps the outgassing that permeates from the absorber locations. The high-speed pumping in the antechamber, i.e., initially, for both upper and lower pumping strips is 270 l/sm for CO and 650 l/sm for H_2, assures efficient removal of both photon and thermal desorbed gases. The enlarged cross-section of the pump antechamber improves the conductance of desorbed gases to the pumping surface of the NeG strips.

The chamber is a 6063 T5 aluminum extrusion. It contains water passages for cooling and bakeout. A 150°C bakeout is achieved with portable water-heating units. The chamber is covered with thermal insulation to reduce heat losses. The lumped pumps, gauges, valves and bellows are baked with heating electrical tapes.

The extrusions will be chemically processed after machining and prior to welding to provide a clean vacuum surface. Alkaline degreasers and etchants have been tested, and Auger analysis of the cleaned surfaces has indicated comparable results to those of Mathewson at LEP.[4,5] Since the 6063 aluminum alloy is magnesium based, the extruded surface consists of a magnesium oxide layer, which contains most of the surface contamination. The first cleaning step employs a degreasing agent as well as an agent to remove the magnesium oxide layer. Once the magnesium oxide is removed, the second step uses an etchant (potassium hydroxide) to reduce the aluminum oxide layer. A final water rinse and blow drying completes the cleaning procedure. The 2219 aluminum end flanges, utility flanges and photon exit ports are cleaned with solvents or detergents after machining and prior to welding.

The 2219 T851 aluminum end flanges are joined to the chamber extrusions with full penetration weldments. The inside surface of the weld bead is even with the beam chamber inside surface for low rf impedance. The ends of the sector each terminate in an elliptical tube beam chamber, which also has a full penetration weldment with a similar inside surface. The photon exit port block is attached by a full penetration weldment to the side of the chamber. The utility flange weldments are inside vacuum welds, either of a fillet or butt type. The large, 12" and 8" diameter crotch and distributed absorber flange welds follow the contour of the extrusion. These exacting weldments are done by an automated computer-controlled TIG welding machine.

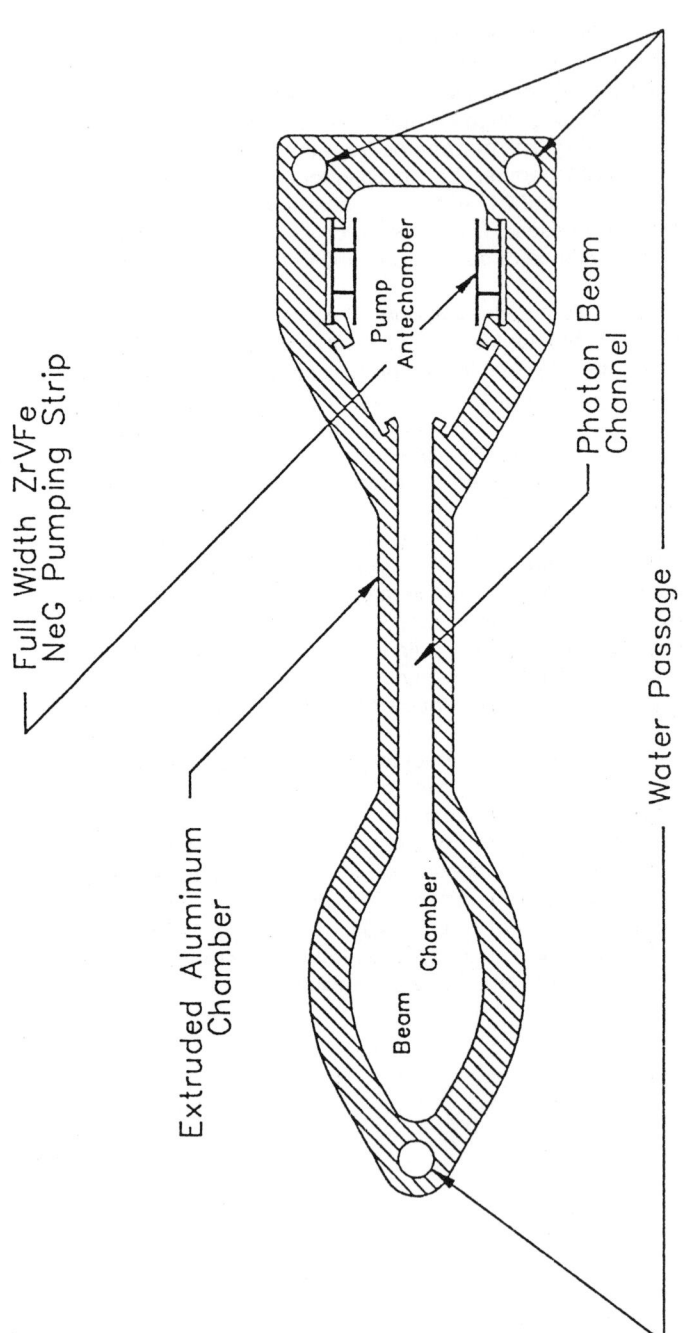

Figure 2

STORAGE RING VACUUM CHAMBER CROSS SECTION

VACUUM CHAMBER SUPPORTS

Three supports are used per straight section. One support is rigid and does not allow chamber motion in any direction. Its position is located approximately at the center of the section. The other two supports are each located near the ends of the chamber. These two supports are leaf springs permitting chamber thermal expansion in the beam line direction during the bake cycle.

The curved sections also use three supports. Two leaf spring supports are located at the ends of the chamber to allow thermal expansion in the approximate beam direction. The center support, unlike the rigid support of a straight section, allows for the motion of a changing bend radius that occurs during the bake cycle. The midpoint of the curved section is maintained in a level position and is constrained in the beam direction. The section midpoint is allowed to move radially in the direction of its natural radial expansion by means of a pin and slot guide.

The salient aspect of both support systems is that the chambers are free to thermally expand with negligible forces placed on them. The support system ensures that the chambers are not stressed significantly and return to their original location, size, shape and geometry after each bake cycle.

PHOTON ABSORBERS

Non-experimental facility synchrotron radiation is absorbed by photon absorbers in order to reduce the number of photons striking the vacuum chamber surfaces.

The storage ring incorporates a series of such absorbers in each sector as shown by Figure 1 and as listed in Table 2.

Table 2 Storage Ring Photon Absorbers

LOCATION	NAME	INCIDENT HEAT LOAD*[KW]
Section 1, Downstream End	End Absorber (EA1)	0.23
Section 2, Downstream End	Crotch Absorber (C1)	13.3
Section 2, Downstream End	Crotch Wedge Absorber (W1)	0.51
Section 3, Downstream End	Distributed Absorber (A1)	4.32
Section 4, Downstream End	Crotch Absorber (C2)	11.5
Section 4, Downstream End	Crotch Wedge Absorber (W2)	3.3
Section 5, Downstream End	Distributed Absorber (A2)	3.67
Section 6, Downstream End	End Absorber (EA2)	1.08

*At positron beam current of 300 mA.

Photon absorber development is continuing with emphasis placed on the design of the crotch absorbers. Due to the high normal incident linear power density, i.e., 149 W/mm, the crotch absorbers encounter the most severe thermal and structural loads and generate large photon induced desorption gas loads.[6] Several crotch absorber geometries have been evaluated, both analytically and experimentally, with electron beam welder (EBW) beam testing. The experiments, using the EBW beams, evaluate thermal and structural performance, including deformation and fatigue due to the cyclic heating and cooling. The most promising crotch absorber geometry, the compound-V as shown in Figure 3, is being developed in detail. Initial EBW beam test measurements agree well with predicted performance. Tests with photon beams will be conducted.

The density of desorbed gas will be high in the crotch areas, and high speed lumped pumping is provided in these regions. Lumped 220-l/s ion and 1000-l/s NeG pumps are employed to remove a large portion of the desorbed gases. The remaining gas permeating into the chamber is pumped by the distributed NeG pumps.

Distributed absorbers are located at the end of the straight sections following the upstream bending magnet sections in order to absorb the radiation that passes between the crotch absorbers and the positron beam. About 25% of the bending magnet radiation is absorbed by distributed absorbers. As a result, a significant amount of gases will be desorbed. As in the case of the crotch absorbers, most of the gases are trapped and pumped locally by lumped 220 l/s ion and 250 l/s NeG pumps.

The absorber vertical access flange penetrations result in discontinuities in the chamber cross section. In order to minimize the rf effects due to such discontinuities, rf screens are installed in these areas. The screens are made of aluminum sheets formed to correspond to the inner contour of the chamber and are provided with access slots. The slots provide pumping access for lumped pumps in the absorber locations while minimizing Wakefield effects on the beam and controlling rf leakage from the positron beam. The screens are welded to the extrusion and have electrical continuity to the absorber structure through metallic spring finger contacts.

PUMPING SYSTEM

The use of integrated ion pumps employing the dipole fields is impractical in the storage ring, because only 24% of the ring is occupied with bending magnets. The vacuum system, therefore, relies on Non-evaporable Getter (NeG) strips as the primary source of distributed pumping in both the bending magnet and straight sections.

The NeG[7] is a constantan strip coated with an alloy of Zr V Fe. This alloy forms thermally stable chemical compounds with most of the active gases (O_2, N_2, CO and CO_2), while the absorption of H_2 is thermally reversible. To become effective as a pump, the strip is activated after pumpdown from atmospheric pressure. This procedure consists of heating the getter, which results in diffusion of the saturated surface layer into the bulk of the material. This heating also reduces the H_2 content in the getter whenever the H_2 dissociation pressure of the getter exceeds the H_2 pressure in the vacuum

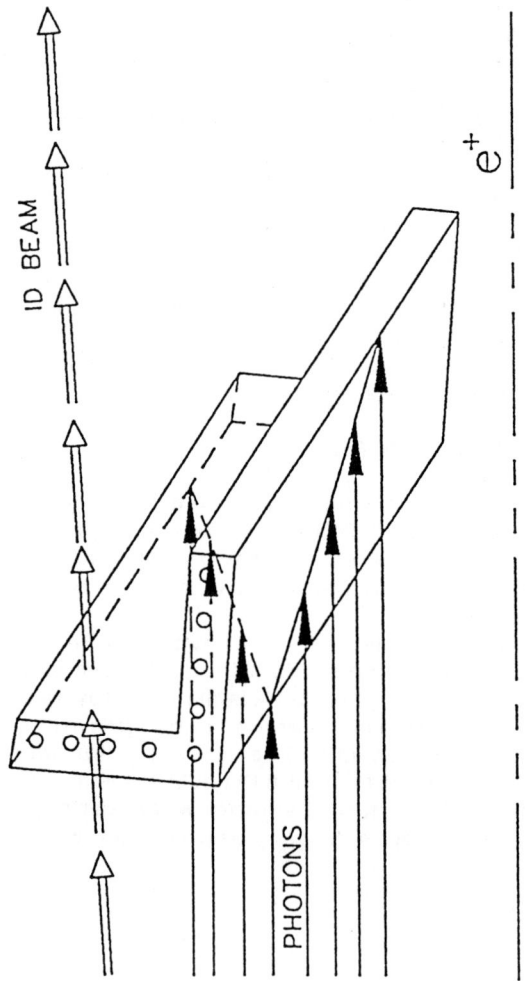

Figure 3
STORAGE RING COMPOUND-V SHAPE ABSORBER CONCEPT

system. After activation, the strength of the gettering action depends on the temperature of the getter and on the amounts and molecular species of the gases that have been pumped. The pumping speeds immediately after activation are high, but they decrease progressively as the getter surfaces saturate. Before saturation, which is dependent on the gas composition, the NeG strip is conditioned by heating it again to restore the pumping speed. Heating of the NeG strips is done electrically and only during activation and conditioning. During normal pumping, the strips are not powered and operate at ambient temperature.

Two NeG strips (SAES ST707) are mounted in the chamber. Details of the NeG strip mounting are shown in Figure 4. The base mounting strip is constrained to the top and bottom antechamber walls within grooved tracks. The NeG strip is supported by stainless steel clips, which are attached to but electrically isolated from the base mounting strip by ceramic insulators. This mode of mounting the NeG strips is similar to that used in LEP.

As noted previously, lumped pumping is installed at photon absorber locations to pump the desorbed gas loads. The NeG strips do not pump CH_4 and noble gases such as Ar and He, therefore, ion pumps are required. The lumped pump distribution is as given by Table 3.

Table 3 Lumped Pumping

LOCATION	PUMPS
Section 1, End Absorber	30 l/s Ion
Section 2 & 4, Crotch Absorber	1000 l/s Lumped NeG 220 l/s Ion
Section 3 & 5, Distributed Absorber	250 l/s Lumped NeG 220 l/s Ion
Section 6, End Absorber	220 l/s Ion

94 APS Storage Ring Vacuum System

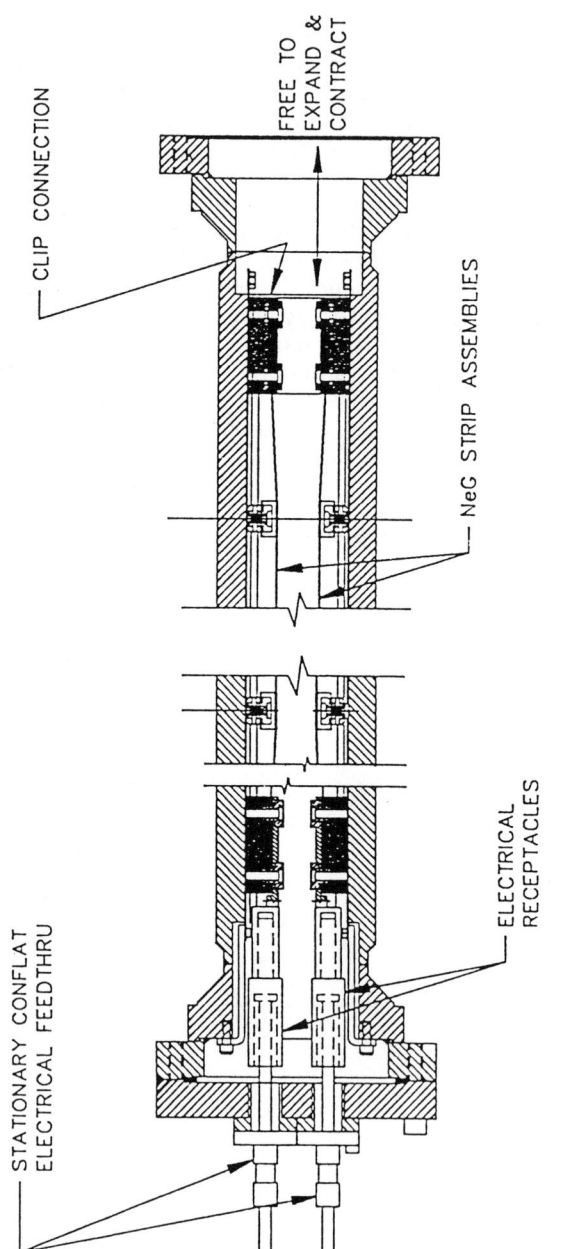

Figure 4
NeG Strip Mounting Details

The ion pumps are started after bake-out and prior to isolation of the turbomolecular pumps as the activated NeG begins to dominate the pumping.

For initial pumpdown from atmospheric pressure, oil-free pumps evacuate the system to turbomolecular pump starting pressure. Turbomolecular pumps further reduce the pressure to ion pump starting pressures.

The roughing and turbomolecular pumps are integrated into a portable pumping station which is employed for sector pumpdown.

VALVING

The storage ring is equipped with valving for both transient and steady-state operations. The valves are of all metal construction, contain no organic materials and are bakeable to 300°C.

Each sector is equipped with two ring isolation valves. These 4 inch rf gate valves have an open valve geometry identical to the elliptical shape of the vacuum chamber and include rf contacts that maintain electrical continuity between the valve and it's mating connections.

Each experimental facility beam line is equipped with a 4 inch gate isolation valve.

Each sector is equipped with two 4 inch right angle pumpdown isolation valves.

Access for vacuum monitoring and control is through 1 1/2 inch right angle diagnostic isolation valves.

BELLOWS

Bellows are installed between sections around the storage ring and between the experimental facility beamlines and the storage ring. The bellows are required for installation operations, alignment flexibility and bake/cool thermal motion tolerance. The bellows are formed from non-magnetic stainless steel and will be furnished with stainless steel flanges welded on each end. The positron beam chamber bellows will be equipped with rf liners to replicate the specific impedance of the elliptical beam chamber.

FLANGES & SEALS

Flanges welded to the storage ring chamber are aluminum alloy 2219 T-851 Conflat flanges. They are coated on the sealing surfaces with a nitride or a carbide coating to increase knife edge hardness and prevent sticking of the seal.

All seals are metal seals. The aluminum Conflat flanges require an aluminum Conflat seal. These seals are made of A1050-H24 aluminum alloy that has been annealed.

MONITORING

Ionization gauges are distributed around the storage ring. Since the ion pumps are situated in areas of highest desorption rates, their currents are monitored continuously and should provide adequate pressure measurements down to 1 nTorr. Gas analyzers, strategically placed around the ring and permanently connected to the control system, are used to monitor the composition of the residual gas. High-pressure gauges are installed in each sector to shut down the NeG power supplies in the event of a vacuum failure during activation and conditioning periods.

SPECIAL FEATURES

Transition connections are required in the positron beam channel between the storage ring and insertion device vacuum chambers. These chambers provide a gradual rf transition between the two chambers that have different cross sections. The transitions are water cooled to absorb photon wall interaction heat loads and may incorporate NeG pumping to absorb a portion of local photon induced desorption gas loads. The preliminary design of a transition chamber is as given by Figure 5.

Special chambers for beam injection, abort, and accelerator diagnostics will be incorporated in the storage ring. The designs of these chambers are being developed.

PERFORMANCE ANALYSIS

The analysis and calculations for gas loading for the storage ring of 2.26×10^{-5} Torr·l/s has been previously reported.[2] As then reported, the desorption coefficient, η_γ is 2×10^{-7} mol/photon after 150 Ah of beam bombardment. This machine desorption efficiency is based on considerations of the major gas desorption occurring at discrete machine locations wherein the synchrotron radiation is absorbed on a relatively very small area of copper absorber accompanied by immediate intense pumping. Gas loading for a similar type geometry machine[8] from experience with an electron storage ring

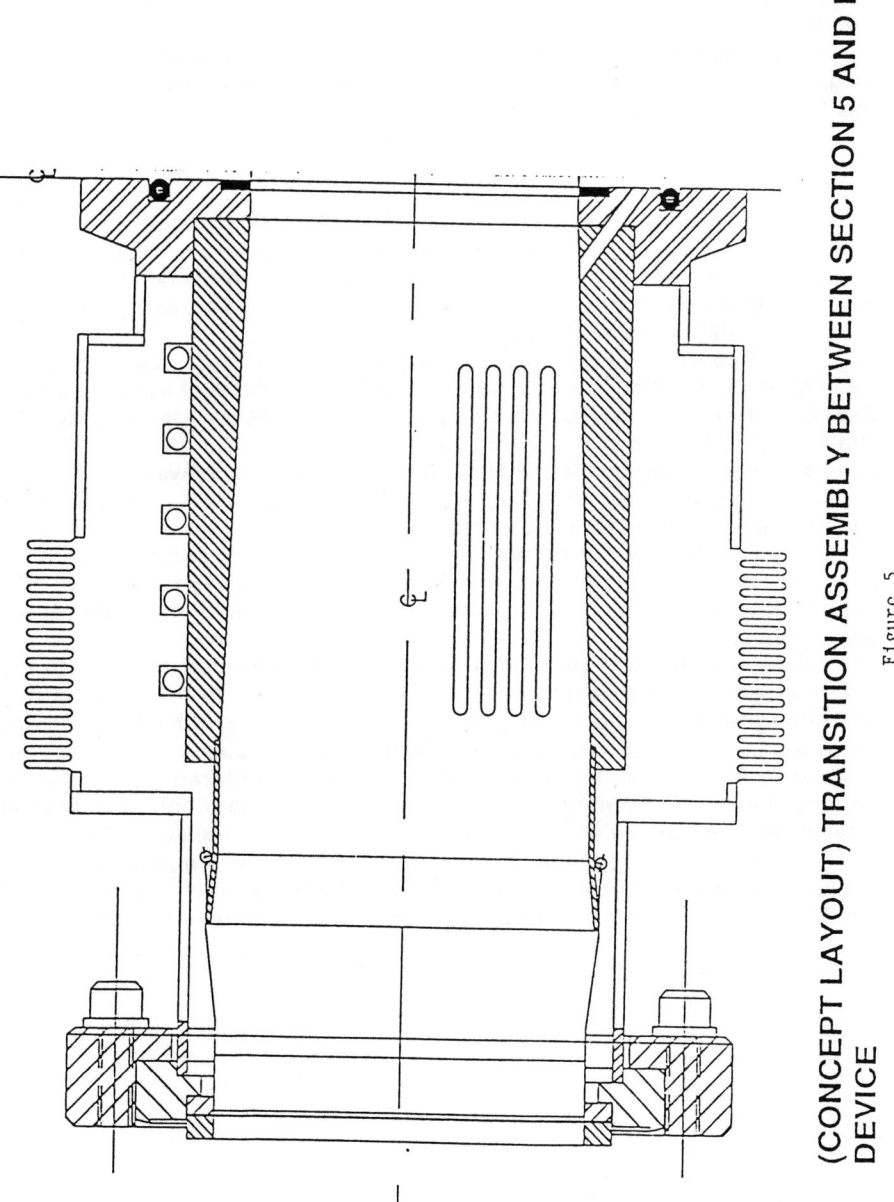

(CONCEPT LAYOUT) TRANSITION ASSEMBLY BETWEEN SECTION 5 AND INSERTION DEVICE

Figure 5

after bakeout utilizes the expression for photon desorption coefficient as $\eta_\gamma = 5 \times 10^6 \, D^{-.67}$ mol/photon (D=Ah) resulting in 1.7×10^{-7} mol/photon after 150 Ah.

Vacuum system performance analysis[2] utilizes a finite-element-analysis computer program developed to calculate the pressure profile around the ring.[9,10] The storage ring sector is divided into 34 elements for which lengths, volume, conductance, pumping speed, and thermal and photon desorptions are the program input parameters.

The predicted pressure gradient through a sector determined by the program is shown in Figure 6. The curves display typical pressures after 1, 10 and 100 A·h of operation.

DEVELOPMENT PROGRAM

Beam Transport Sections 1-5 of a sector of the storage ring vacuum system is under construction and will consist of the working vacuum system with diagnostic equipment. Construction of this facility provides a means for testing fabrication and assembly methods and evaluating vacuum performance at full scale.

The five vacuum chamber sections were manufactured during the machining and welding development program and include all system details. Chambers have been evaluated on an individual basis for vacuum integrity and base pressure. Base pressures of 4×10^{-11} Torr, after baking, with ion pumps operating along with the NeG strips, have been achieved. (Details of these measurements are presented in a companion paper at this conference.[10]) Full sector tests will follow.

The achievement of reliable sealing of the Conflat joints after 150°C bakeout has not been consistent. A program is underway to improve this condition to system design requirements. Alternate sealing systems are being evaluated.

Cleaning studies continue. The addition of ultrasonic agitation to the chamber extrusion cleaning process was included. In order to reduce machining time and thus cost associated with the currently planned dry machining, studies of machining with lubricants and subsequent cleaning have been made. Machining with lubricants is considered to be a viable option. Detergent cleaning of the 2219 aluminum components, rather than solvent cleaning with its associated safety concerns, is being evaluated.

The alignment thermal stability of the chambers and supports subjected to multiple bake/cool cycles has been evaluated with two sections (Sections 1 and 2) connected under storage ring conditions. Initial results indicate that

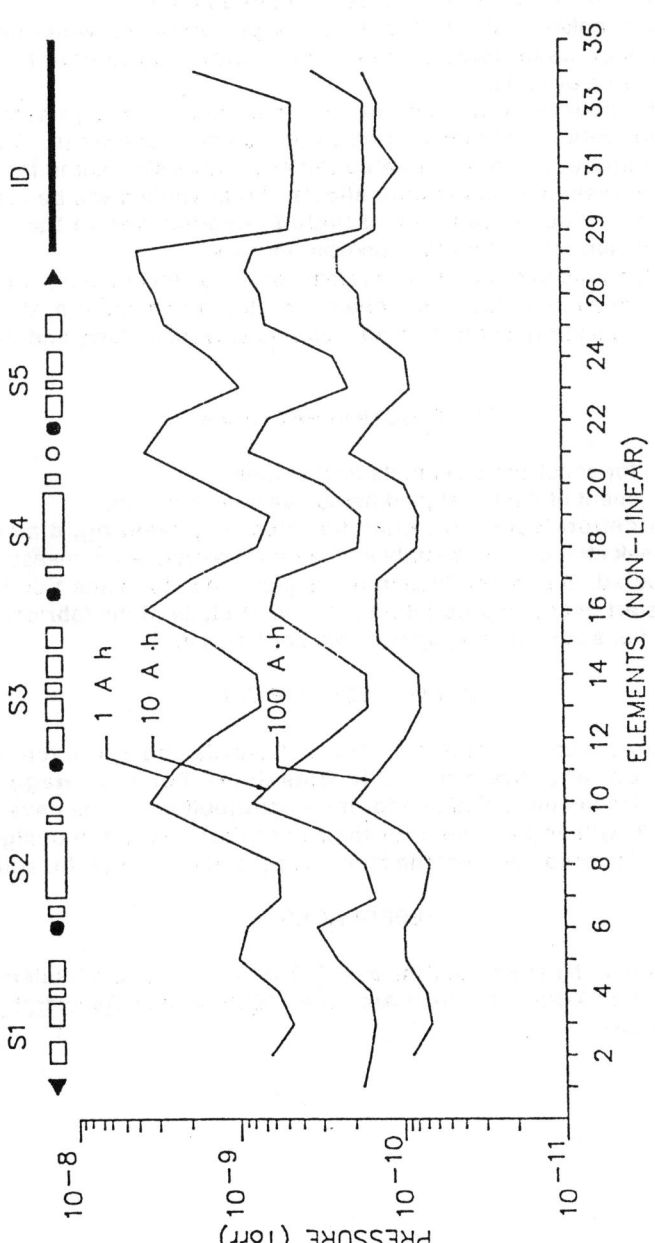

Figure 6
Storage Ring Pressure Gradient Profile for Prototype Sector

position reproducibility is within the acceptable limits of beam position monitor alignment. The two section chamber tests are continuing. The evaluations will be expanded to a full sector configuration.

Chamber bakeout studies continue. A prototype hot water bake heat source has been assembled and is in operation. Candidate insulation systems are being evaluated.

Crotch absorber studies continue. Included are analysis, materials selection, fabrication methods and performance measurements. Candidate absorber configurations are evaluated initially with EBW beams for thermal and structural response and fatigue effects. Final studies will be made with photon beams. These studies will include evaluations of the machine desorption efficiency in the crotch absorber regions.

Rf evaluations of the vacuum system continue. Included are impedance analysis and measurements of the chamber, absorber structure, absorber rf screens, beam position monitors, ring isolation valves, bellows and transition sections.

CONSTRUCTION PROGRAM

Design for construction is in its final stages.

Procurement of final designed items has been initiated.

A fabrication facility for chamber cleaning, welding, dimensional inspection, leak checking, assembly and vacuum performance measurements is being prepared. A pilot production run is planned to fabricate 1 or 2 sectors prior to the start of actual production activity. (Details of the fabrication plan are presented in a companion paper at this conference.[11])

ACKNOWLEDGEMENTS

The authors are grateful to Y. Cho, J. Galayda and R. Kustom for their encouragement and direction; to R. Blaskie, R. Ferry, K. Haggerty, M. McDowell, R. Prien and E. Wallace for their contributions to the development program; to T. Gill and R. Piech for their contributions to the design; to A. Salzbrunn for the preparation of the manuscript; and to Y. Amer for editing.

REFERENCES

1. R. Wehrle, J. Moenich, S. Kim, and R. Nielsen, "Vacuum System for the Synchrotron X-Ray Source at Argonne," **IEEE particle Accelerator Conf.**, Vol. 3 (1987).

2. R. B. Wehrle and R. W. Nielsen, "Design for APS 7 GeV Storage Ring Vacuum System at ANL": American Institute of Physics Conference Proceedings No. 171 American Vacuum Society Series 5, Upton, NY, 1988.
3. R. Wehrle, R. Nielsen and S. Kim, "Vacuum System Development Status for the APS Storage Ring". ANL, Argonne, IL.
4. A. G. Mathewson, "The Temperature and Time Dependence of the Cleaning Efficiency of the Alkaline Detergent Almeco 28," LEP Vacuum Technical Note (January 15, 1986).
5. A G. Mathewson, "The Effect of Amklene on the Surface Composition of Extruded Al Alloy at Different Temperatures," LEP Vacuum Technical Note (1986).
6. M. Choi, "Development of APS Photon Absorber Design," APS Light Source Note. ANL, Argonne, IL, 1990 (In preparation).
7. B. Ferrario, L. Rosai, and P. Della Porta, "Distributed Pumping by Non-Evaporable Getters in Particle Accelerators," IEEE Trans. Nucl. Sci., NS-28, 3333 (June 1981).
8. B. A. Trickett, "The ESRF Vacuum System", American Institute of Physics Conference Proceedings No. 171 American Vacuum Society Series 5, Upton, NY, 1988.
9. J. F. J. Van Den Brand and A. P. Kaan, "Design Study of the Vacuum System for the E.S.R.F.", European Synchrotron Radiation Project Report, ESFP-IRM-61/84 (1984).
10. J. Kneuer, unpublished information (1985).
11. G. Goeppner, "APS Storage Ring Vacuum Chamber Fabrication," presented at the Topical Conference of the Vacuum Design of Synchrotron Light Source Conference, ANL, Argonne, IL, November 1990 to be published.

VACUUM SYSTEM FOR SPring-8 STORAGE RING AND TEST EXPERIENCE

S. H. Be, S. Yokouchi, T. Nishidono, Y. Morimoto*, K. Watanabe,
S. Takahashi, S. R. In**, H. Daibo and Y. Oikawa

RIKEK-JAERI Synchrotron Radiation Design Team,
Wako-shi, Saitama, 351-01 Japan

ABSTRACT

We present the vacuum system design for the SPring-8 storage ring and our experience with a series of its tests and R & D. We manufactured a leaf spring and slide guide as chamber mounts, and verified whether the chambers are distorted in any manner and return to their original locations even after each bake cycle or not. Deformation of a 4m-long aluminum-alloy chamber during evacuation and bakeout was also investigated. A bending magnet chamber was built by means of stretch forming. We evaluated the aluminum-alloy flanges of various sizes. We investigated how many flanges reveal leakage in an experimental setup and what is the origin of the leakage. We manufactured all-metal gate valve with RF contact. Its lifetime was investigated together with the leakage reliability at the seals. Bellows assembly was manufactured, and contact force together with contact resistance of the slide fingers was measured. The results in a variety of topics mentioned above will be described in detail.

INTRODUCTION

The SPring-8 (Super Photon Ring-8 GeV)[1] is a highly brilliant synchrotron radiation source which is presently under construction, and scheduled for completion in 1997. The vacuum system forms approximately 455m-diameter ring, and consists of two differently shaped aluminum-alloy (A6063-T5 whose strength is equivalent to that of T6) chamber extrusions, two types of absorbers and the various chamber components such as bellows, flanges and valves.

To achieve a beam lifetime of approximately 24 hours, the vacuum chamber with its pumping system should be designed so as to maintain the beam-on pressure of 1 nTorr or less. The main pumping system is based on non-evaporable getter (NEG) strips, which are used in the straight and bending chamber. In addition to the NEG strips, a distributed ion pump is installed in the bending magnet chamber. Lumped NEG pumps, sputter ion pumps are used at the crotch and absorber locations. The performance characteristics of the lumped NEG pump will be published in this proceeding by S.R. In et al[2].

The most important task for the vacuum system should be considered as the development of the flanges whose reliability, especially of leakage, is still maintained through the multiple extreme operation like baking. The other important tasks are : 1) Design of crotch. 2) Design of the mount or

* Present address : Kobe Steel, LTD, Machinery Division
** on leave from KAERI, Korea

guide of the chamber for a beam position monitor (BPM) which can ensure the displacement of the chamber within the accuracy of 0.05 mm or less. 3) Design of the chamber components such as bellows, flanges and gate valves including stepchange which introduce a finite RF impedance. We manufactured the flange, the mount of the chamber for BPM and chamber components, and investigated their performances.

In this paper we present a series of investigations on their performances and test experience in the development of the vacuum system.

VACUUM CHAMBER

In our storage ring, we are considering to use an extruded aluminum-alloy as the materials of chambers. The chemical composition is 0.55 w/% of Mg, 0.44 w/% of Si and the balance is Al. The alluminum- alloy chamber is extruded in the atmosphere of $Ar+O_2$.

A cross-sectional view of the vacuum chamber for the straight sections is shown in Fig.1. The chamber consists of an electron beam chamber and a

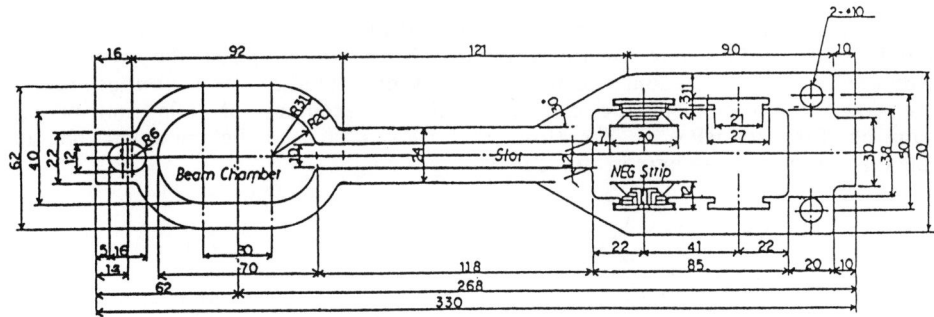

Fig.1 Straight Section Chamber   Unit in mm

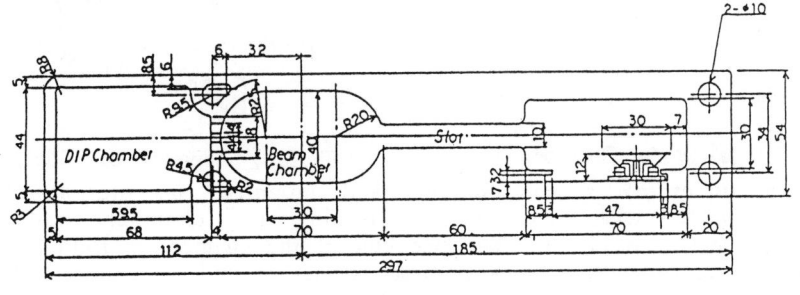

Unit in mm

Fig.2 Bending Magnet Chamber

slot-isolated antechamber in which NEG strips are installed. The cross-sectional view of the vacuum chamber for the bending magnet is shown in Fig.2. This chamber consists of a beam chamber, the slot-isolated antechamber in which NEG strips are installed, and the chamber for installation of a distributed ion pump. This bending magnet chamber is built by means of stretch forming. The vacuum performance characteristics of these chambers with and without NEG strips will be published in this proceeding by S. Yokouchi et al[3].

The initial design[4] of the cooling and bakeout system for the chambers were based on water cooling system and flexible sheathed heaters, respectively. This initial design was abandoned because it is impossible to carry out the bakeout and NEG activation simultaneously. The present design is discussed to be used the super heated water system which has been employed at LEP[5].

SR light is almost all intercepted by crotches and absorbers placed just downstream and upstream of a bending magnet. A photon emission for energy less than 10 eV having an larger angular spread than 1.5 mrad in the vertical plane is intercepted by a slight part of slot wall in the straight vacuum chamber.

MOUNTS OF BEAM POSITION MONITOR CHAMBER

Because of vacuum chamber deformation of about 0.2 mm at the locations of BPM's due to the pressure difference between the atmospheric pressure and the vacuum, using the present chamber it is impossible to keep the deformation of the chamber with the accuracy of 0.05 mm or less required for the BPM's. Therefore the BPM's are to be mounted on any special monitor chamber in a real storage ring.

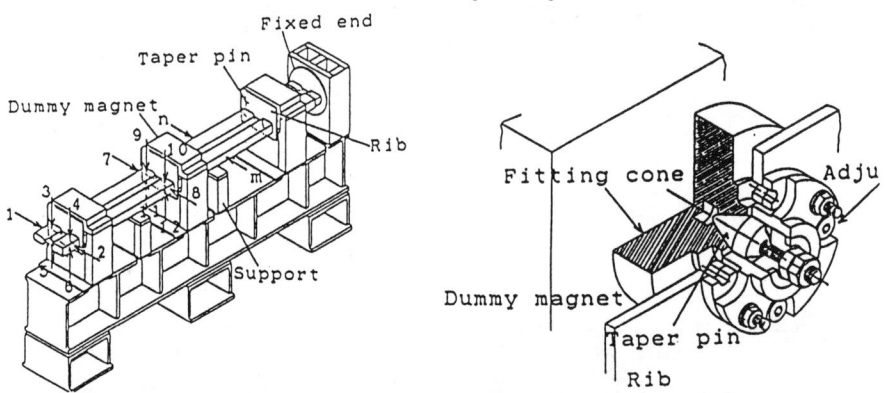

Fig.3(a) Model Chamber   Fig.3(b) Schematic of Slide guid

We manufactured a new mount, and mounted the present straight section chamber with ribs for suppressing the chamber deformation mentioned above (see Fig.3). The one end of the chamber is rigid and does not allow chamber motion in any direction before and after bakeout. Three slide guides allow the chamber motion in any direction during bakeout. Therefore, the chamber is distorted in some manner during the bakeout, but returned to original locations after each bake cycle. The measurement results are shown in Fig.4. From this figure we can find that the locations of the BPM's are ensured within the accuracy of 0.04 mm. For the same chamber with leaf springs[6] instead of slide guides, preliminary test result showed that the locations of BPM's are maintained within the accuracy of 0.03 mm.

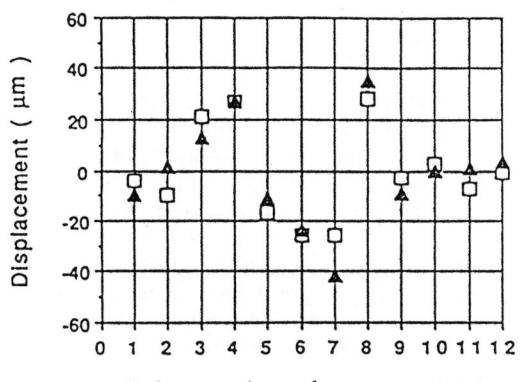

Fig.4 Displacement of Chamber after bakeout

CHAMBER COMPONENTS

A normal section cell SPring-8 is shown in Fig.5. Chamber components such as bellows, flanges and valves are designed so as to minimize their impedances. The bellows are shielded by RF fingers, and a step changes in the cross-section of the chambers are provided by means of a tapered transition. The flanges with RF contact are used . In the gate valve, the RF contacts are directly connected with the valve mechanism. The apertures of these components have larger width than that of a normal chamber to avoid an interception with SR light. The details of the crotch will be published in this proceeding by Y. Morimoto et al[7,8].

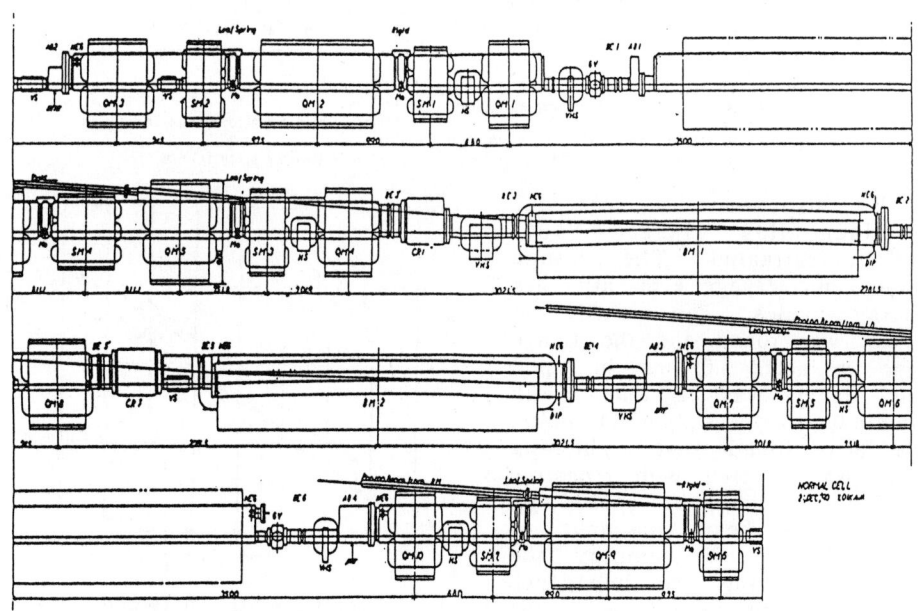

Fig.5 Plan view of a normal section cell   Unit in mm

ALUMINUM-ALLOY FLANGE [9]

Since the chambers should be baked out at about 150 °C to minimize their thermal outgassing rate, aluminum-alloy flanges as well as stainless-steel flanges are required to be leakless even after many cycles of baking. We investigated how many flanges leak in an experimental setup and what is the origin of leakage. We baked out the experimental setup 11 times for a year. Sixteen flanges out of forty in the setup showed leakage. Most of these leaking flanges were heterogeneous pairs of aluminum-alloy and stainless-steel as shown in Table 1. All flanges in the setup were assembled with aluminum-alloy (A1050H18) gaskets. Not applying additional torque(load) to the joint bolts, but new softer gaskets (A1050H24) were found effective to protect the flanges from leakage. It is also worth pointing out that aluminum-alloy flanges and gaskets should be standardized for avoiding problems arising from their scattered dimensions.

Table 1   Leak statistics for the experimental setup.

Flange size	Material of flange									Total		
	Al+Al			Al+S.S.			S.S.+S.S.					
	Quan-tity	Leak	%	Quan-tity	Leak	%	Quan-tity	Leak	%	Quan-tity	Leak	%
034	--	--	--	2	1	2.5	3	0	0	5	1	2.5
070	7	3	7	10	4	10	1	0	0	18	7	7.5
114	--	--	--	2	2	5	--	--	--	2	2	5
152	1	1	2.5	1	1	2.5	1	0	0	3	2	5
203	5	0	0	6	4	10	1	0	0	12	4	10
Total	13	4	10	21	12	30	6	0	0	40	16	40

Note :
1) Leak rate is above $2 \sim 3 \times 10^{-10}$ Torr·l/s .
2) The specifications are shown as follows:
Flange:
  Material is aluminum alloy (A2219-T852).
  Surface treatment is CrN ion plating.
Bolts:
  Material is aluminum alloy (A2024-T4).
Gasket:
  Material is aluminum alloy (A1050H18).

We also made a larger-size aluminum-alloy flange(ICF406) and investigated. The flange showed no leakage even after six baking cycles. The profile of gasket was round and the flange edge was about 0.05 mm in depth. The flange was tightened with aluminum-alloy (A2024T4) bolts. We also found that tightening force was lowered with repeating baking; this may be caused by material or heat treatment of bolts (Fig.6). We employ the method reducing the impedance by bridging the gap between the flanges with a metallic seal.

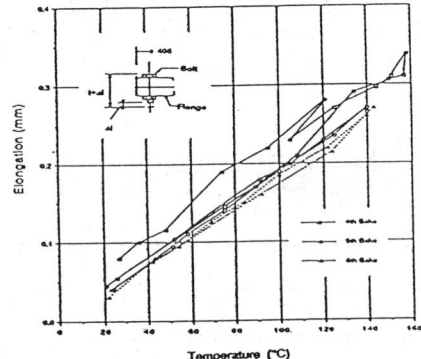

Fig.6   Relationship between baking temperature and bolt elongation. Tightening force is calculated from the elongation.

BELLOWS ASSEMBLY

A bellows with RF sliding contact fingers, which is shown in Fig.7, must permit variations in the length direction of the chamber caused by thermal expansions or contractions as a result of baking and fluctuation in the cooling water and room temperature. In addition, they must compensate for manufacturing tolerance and alignment errors in the longitudinal and transverse direction. These errors are considered to be 2.5 mm in the longitudinal direction and up to 1 mm in the transverse direction. The length of the Cu-Be slide finger and its force must be chosen so as to ensure the contact force of a few ten kg/cm^2, which is equivalent to that of the fingers in the RF cavity of RIKEN Ring Cyclotron. Calculated contact force are 35g at narrow fingers and 354g at wide ones, and preliminary experiment values are in the range of 30g to 50g and 150g respectively. The total contact resistance of 43 fingers is $0.8 m\Omega$. The results do not satisfy our requirements. Therefore, a new design is in progress.

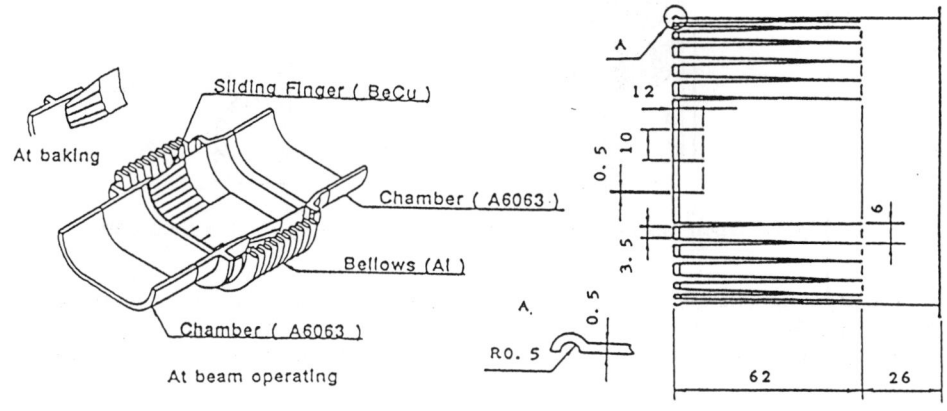

Fig. 7 Bellows assembly   Unit in mm

ALL- METAL GATE VALVE WITH RF CONTACT [10]

An open position mechanism of the all-metal gate valve manufactured is shown in Fig.8. The present valve body was made of stainless-steel, but that to be eventually used is made of aluminum-alloy except a part of valve seats and plates. A technique which provides an enclosed evacuated space between double seals is employed, thereby reducing the pressure difference across the seals, and decreasing the leak rate of the seals. In the open position, the RF contact which has the same geometry as the beam chamber, can bring about a direct electrical connection between two body flanges. A sealing force is obtained by pushing the disks coated with silver to tapered seats polished like mirror (Fig.9). To investigate a leakage reliability at seals , we actuated the valve 1000 times. The results showed that the performance is not yet satisfied completely because there are leak of the order of 10^{-8} Torr l/s over several times during actuations of these first 1000 times. Further plans are to reduce this high leak rate to less than

1×10^{-10} Torr l/s. However, the present leak rate can be reduced to less than 1×10^{-10} Torr l/s by evacuating the enclosed space between double seals.

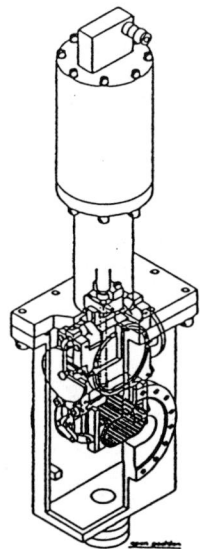 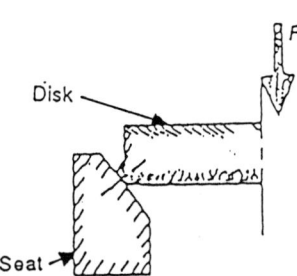

Fig.8 Open position mechanism.   Fig.9 Valve Seat part.

Finally, impedance of chamber components remains to be measured, but the measurement will be started by the end of the year. A modified crotch is under design to reduce production costs, and thermal analysis also in progress.

CONCLUSIONS

Preliminary tests on two types of chamber mounts, i,e, slide guide and leaf spring, were carried out. The results showed that the locations of the BPM's are ensured within the accuracy of 40 µm for the slide guide and 30 µm for the leaf spring. Chamber components such as bellows assembly, flange and gate valve are under tests, and the tests should be completed by the end of March, 1991.

The authors are grateful to Dr. H. Kamitsubo for his support of this work.

REFERENCE

1. M.Hara, S.H.Be, R.Nagaoka, S.Sasaki, T.Wada and H.Kamitsubo, Proc. of the 1989 IEEE Particle Accelerator conf., Chicago, March (1090). p.476.
2. S.R.In, S.Yokouchi and S.H.Be, to be published in this proceeding.
3. S.Yokouchi,S.R.In,T.Nishidono,H.Daibo and S.H.Be, to be published in this proceeding.
4. S.H.Be, S.Yokouchi, Y.Morimoto, H.Sakamoto, Y.P.Lee and Y.Oikawa, Proc.

of the 1989 IEEE Particle Accelerator conf., Chicago, March (1090). p.577.
5. H.Schnhback/SL-MR, CERN-AT-VA/90-17.
6. R.Wehrle, R. Nielsen and S.Kim, Proc. of the 1989 IEEE Particle Accelerator conf., Chicago, March (1090). p.581.
7. Y.Morimoto, H.Sakamoto, S.Yokouchi and S.H.Be, American Vacuum Society Series 5, Vacuum Design of Advanced and compact synchrotron Light Source upon,NY(1988).p.327.
8. Y.Morimoto, T.Shirakura, K., S.Yokouchi and S.H.Be, to be published in this proceeding.
9. T.Nishidono, S.Yokouchi,Y.Morimoto, H.Sakamoto, Y.P.Lee and S.H.Be, J. of The Vacuum Society of Japan, 33, 241 (1990).
10. S.Yokouchi, S.H.Be, K.Yagi, T. Ohbayashi and K.Yshida, to be published in J. of The Vacuum Society of Japan 34, No.3 (1991).

DESIGN OF THE CROTCH FOR SPring-8

Y.Morimoto, T.Shirakura and K.Konishi
Kobe Steel, Ltd., 3-18, Wakinohamacho 1-chome, Chuo-ku
Kobe-shi, Hyogo, 651, Japan

S.Takahashi, S.Yokouchi and S.H.Be
RIKEN (The Institute of Physical and Chemical Research)
2-1, Hirosawa, Wako-shi, Saitama, 351-01, Japan

ABSTRACT

Power density deposited by photons at just downstream of the bending magnet reaches approximately 35 kW/cm^2 in the storage ring of SPring-8. This high power density causes high metal wall temperature and resultant high thermal stress. We made three-dimensinal finite element analysis for several models to minimize the temperature and thermal stress.

In parallel with the analytical work, we reformed the geometry of the crotch to compact the size and reduce the weight. Especially the length along the electron beam orbit should be shortened by reason of the space limitation due to the various magnets and other components around the crotch. Reformed crotch geometry is presented with some experimental results regarding to the vacuum performance characteristic obtained from the prototype crotch before the reformation.

INTRODUCTION

We proposed a new type of the crotch which has a structure in which particles such as photons and their associated photo-electrons and gas molecules by photon-induced outgassing are efficiently trapped.[1] This type of the crotch has a good vacuum performance characteristic due to the trapping structure which limits a chance for gas molecules to bounce into the electron beam chamber. We manufactured a prototype model of the crotch to find problems that may occur in manufacturing and to confirm its vacuum performance. The space around the crotch becomes tight in the newest lattice alignment. We designed newly the crotch for the purpose of the reduction in size and weight to keep clearance between other components and make installation work at site easy.

The crotch receives the high power density of radiation, which is approximately 35 kW/cm^2 at just downstream of the bending magnet. It is required to reduce the wall

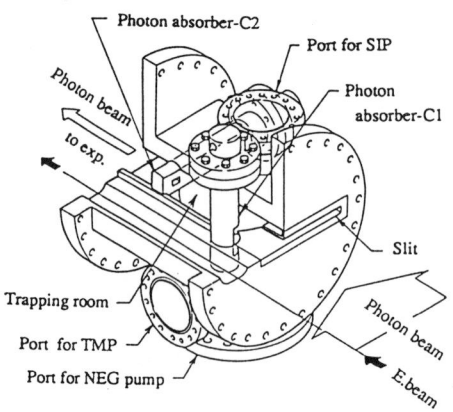

Fig. 1. Isometric view of the reformed crotch.

temperature and resultant thermal stress for repeated operations. By using the finite element analysis, the temperature and thermal stress for several models having different geometry and material are calculated to investigate the heat transfer performance.

Some experimental results regarding to the vacuum performance characteristic for the prototype model are also presented.

CROTCH GEOMETRY

An isometric view of the reformed crotch is shown in Fig.1. Main changed items compared with the previous one are as follows.
- The size of the main body is reduced from $420^W \times 420^H \times 323^L$ mm to $336^W \times 80^H \times 274^L$ mm.
- The distance between two large flanges for connecting with the normal storage ring chamber is reduced by 31 mm.
- The flanges of ports for the pumps and photon absorbers are connected to the main body by short pipes. (The flanges of the previous one are mounted to the main body directly.)
- The port for TSP(Titanium-sublimation pump) is removed.
- The flange size of the port for SIP(Sputter-ion pump) is reduced from 203 mm to 152 mm in diameter.

The design concept is the same as the one up to now, that is, to maintain ultra-high vacuum of the storage ring, particles should be trapped in the space other than electron beam chamber, and to evacuate gas molecules before they have a chance to bounce into the electron beam chamber. The feature of present design is to give the equivalent trapping effect with the more compact and lighter structure than previous one.

The horizontal and vertical cross sections of the crotch are shown in Fig.2. About 20 % of photons deposited on the crotch hit a photon absorber-C1(ABS-C1) and about 80 % of the rest move into the trapping room through the slit and are absorbed by a photon absorber-

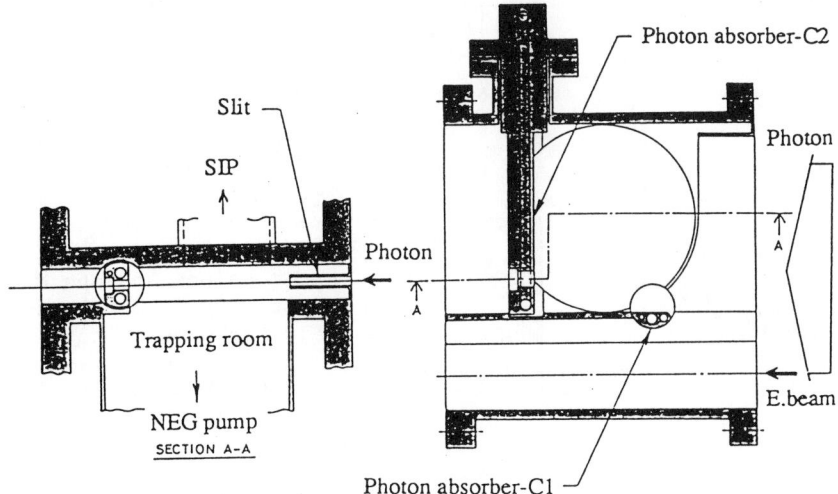

Fig. 2. Horizontal and vertical cross sections of the reformed crotch.

C2(ABS-2) positioned at the back of the trapping room. These absorbers are cooled by flowing water. The photon passageway is separated from another space by the slit and ABS-C2 itself with low conductance (The space between the slit and ABS-C2 is called "trapping room".), and also isolated from the electron beam channel by the wall. Consequently almost particles produced in the trapping room are efficiently kept in it. Photon beam to experimental facility is passed through the aperture of the ABS-C2. The photon absorbers and main part of the body are made of oxygen free copper-class 1. The flanges are conflat type ones made of aluminum alloy which are connected to the copper parts with the transition pieces by explosion bonding. A 1500 l/sec(for CO) non-evaporable getter(NEG) pump and a 60 l/sec sputter-ion pump (SIP) are to be mounted on the trapping room.

THERMAL ANALYSIS

The long distance between the bending magnet and the crotch would be great advantageous to make the power density at the crotch as small as possible. In the original lattice alignment, the crotch is located at just downstream of the bending magnet, but we kept the crotch away from the bending magnet by exchanging the position of the crotch and that of the steering magnet, so that maximum power density perpendicular to photon beam at the ABS-C1 is decreased from 34.3 kW/cm^2 to 27.0 kW/cm^2.

As mentioned above, photon beam is stopped at the two kinds of photon absorbers installed in the crotch. ABS-C1 receives the photons having higher power density than ABS-C2, therefore we pay more attention to the analysis of ABS-C1. Three models for ABS-C1 (model C1-1, C1-2 and C1-3) in three dimensions and one model for ABS-C2 (model C2-1) in two dimensions because of the inherent symmetry of the thermal problem, which are shown in Fig.3, are considered. Models C1-1 and C2-1 are the basic models whose shapes are the same as those of

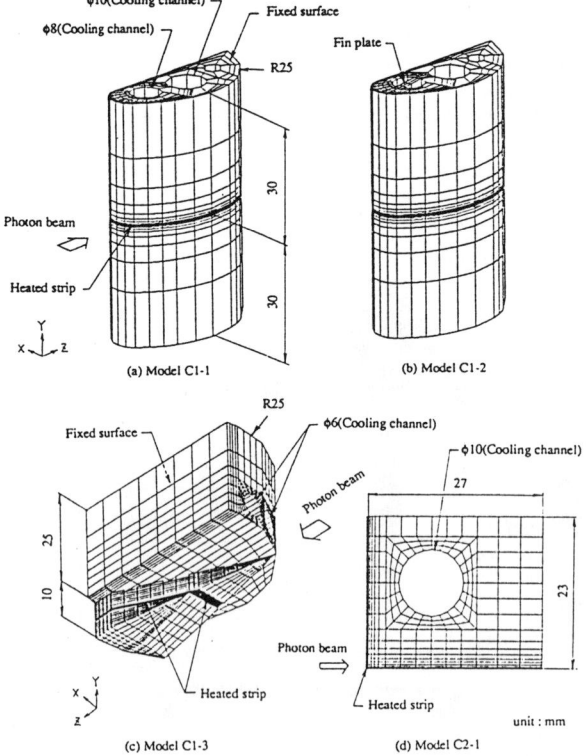

Fig. 3. Simplified heat transfer model to calculate the temperature and stress.

the ABS-C1 and C2 which have been manufactured for the prototype crotch. Models C1-2 and C1-3 are the alternative ones to decrease the wall temperature and thermal stress. Model C1-2 has a fin plate for increasing the heat transfer area at the place where maximum temperature would be observed. Model C1-3 has two surfaces inclined with respect to the photon beam to reduce the power density. One surface is vertically rotated and another one horizontally to the direction of the outside of the ring so that reflected photons and gas molecules would not tend to go into the electron beam channel.

Thermal analysis was made using a finite element program ANSYS developed by Swanson Analysis Systems, Inc,. The heat loading was determined from the photon beam spectrum on each element and was deposited just on the surface. The maximum power density perpendicular to photon beam for ABS-C2 is 17.9 kW/cm^2, and absorbed power for ABS-C1 and ABS-C2 are 1.2 kW and 4.3 kW, respectively. Heat transfer is set so as to occur only at the surfaces along which cooling-water flows, while other surfaces are insulated. In calculation of the deformation and thermal stress, upper surface of the model is considered to be fixed.

For forced convection region, the flux per unit area, Q_{FC} at the water-cooled surface is given by

$$Q_{FC} = h(T_W - T_C) \tag{1}$$

where h is the heat transfer coefficient for the turbulent flow, T_W the wall temperature and T_C the water temperature. The Sieder-Tate correlation[2] was used to calculate h, it is given by

$$Nu = 0.023 Re^{0.8} Pr^{1/3} (\mu/\mu_w)^{0.14} \tag{2}$$

where Nu = Nusselt No. = hd/λ, Re = Reynolds No. = ud/ν, Pr = Prandtl No. = $C_p\mu/\lambda$ with d the diameter of the duct carring the coolant, λ the thermal conductivity, u the coolant's average velocity, ν the kinetic viscosity, C_p the specific heat and μ the viscosity. All physical properties shown above are the values at the bulk temperature of coolant except μ_w at the wall temperature. The values of h obtained from Eq.(2) are depended on the wall temperature and duct diameter which are shown in Fig.4. The water bulk temperature and velocity are assumed to be 30 °C and 3 m/sec, respectively.

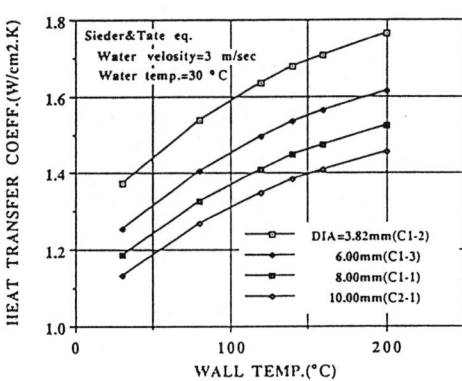

Fig. 4. Heat transfer coefficient calculated by Sieder & Tate eq..

The nucleate boiling at the cooling-surface is permitted in our analysis to design the absorber having higher heat transfer efficiency. In the nucleate boiling region, the heat flux at the cooling-surface increases rapidly by the bubble motion. In this region, the total heat flux per unit area, Q is given by

$$Q = Q_{FC} + Q_B \tag{3}$$

where Q_B is the heat flux due to boiling. Q_B was calculated using the Nishikawa-Fujita correlation[3]

$$\begin{aligned} Q_B &= 243 f_s^2 f_p^2 \lambda_1^3 z (T_W - T_S)^3 \\ &= 7.7 \times 10^{-3} (T_W - T_S)^3 \quad (W/cm^2) \end{aligned} \tag{4}$$

where f_s is the bubble coefficient, f_p the pressure coefficient and T_S the boiling point. Subscript l presents the liquid condition. z is given by

$$z = \frac{1}{M^2 P} \frac{c_{pl} \gamma_1^2}{\lambda_1 \sigma \gamma_v \Delta i} \tag{5}$$

where M and P are the dimension constant, γ the density, σ the surface tension, Δi the latent heat. Subscript v presents the vapor condition. The calculation was made using two kinds of material constants for oxygen free copper(OFHC-Class 1) and strengthened copper containing aluminum(GlidCop AL-35). The water pressure is set to be 4 kg/cm² absolute(boiling point is 143 °C).

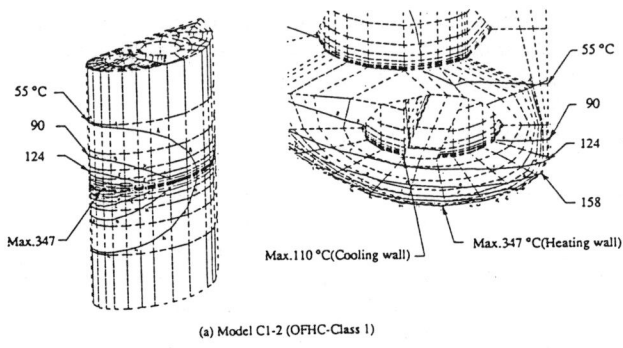

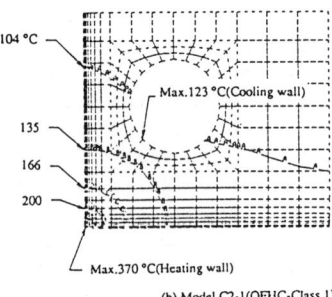

Fig. 5. Temperature distributions computed for model C1-2(ABS-C1) and C2-1(ABS-C2).

Figure 5 and Table I show the calculation results for the temperature distributions(The analysis for model C1-3 is in progress). The calculated temperatures for GlidCop AL-35 are higher than those for OFHC-Class 1 owing to its lower thermal properties. The lowest temperature distributions are observed in model C1-2 for OFHC-Class 1. in which maximum temperatures on the surface irradiated with the photon beam and on the cooling wall are 347 °C and 110 °C, respectively. The effect of the fin plate installed in model C1-2 is big, thereby temperatures on both surfaces are decreased

Table I. Computed maximum temperature.

Model	Material	Max.temperature (°C)	
		Heating wall	Cooling wall
C1-1	OFHC-Class 1	365	141
	GlidCop AL-35	408	151
C1-2	OFHC-Class 1	347	110
	GlidCop AL-35	387	117
C2-1	OFHC-Class 1	370	123

by 20~30 °C. Only maximum temperature of 151 °C on the cooling wall in model C1-1 for GlidCop AL-35 shows that nucleate boiling occurs on a part of the cooling wall. But the heat flux at that place of 182 W/cm^2 is small enough, compared with the critical heat flux of 2677 W/cm^2 calculated by Katto correlation[4,5] for the sub-cooled forced convection nucleate boiling. If the heat flux exceeds the critical value, the wall temperature increase rapidly because of the reduction in heat transfer. It is caused by film boiling and at worst the wall may be melted down. This is called "burn out". To avoid "burn out", we must pay much attention not to give excessive heat flux through the wall-water interface.

Table II shows the calculation results for the thermal stress distributions. These resulting stress values should be judged by the following two kinds of the criteria. One is the static yield strength whose allowable stress is less than twice the copper yield strength at the operating temperature. The other is the fatigue strength due to the thermal cycle, which should be judged by the fatigue curve (S-N diagram). The equivalent stress, σ_{eq} is evaluated by Tresca's theory ($\sigma_{eq} = \sigma_1 - \sigma_3$, $\sigma_1 > \sigma_2 > \sigma_3$), which are also listed in Table II. The values of the 0.2 percent offset at 400 °C for OFHC-Class 1 and GlidCop AL-35 are approximately 5 kg/mm^2 and 30 kg/mm^2, which correspond to 10 kg/mm^2 and 60 kg/mm^2 as the allowable stress, respectively. The calculated equivalent stresses for each model are 25~35 kg/mm^2, which exceed the allowable stress for OFHC-Class 1.

Table II. Computed maximum thermal stress and Tresca's equivalent stress σ_{eq}. Minus values mean compressive stress.

Model	Material	Max.thermal stress (kg/mm^2)			σ_{eq} (kg/mm^2)
		X direct.	Y direct.	Z direct.	
C1-1	OFHC-Class 1	-37.4	-12.4	-23.7	-25.0
	GlidCop AL-35	-49.6	-14.5	-28.1	-35.1
C1-2	OFHC-Class 1	-30.4	-7.2	-17.1	-23.2
	GlidCop AL-35	-39.9	-9.3	-23.1	-30.6
C2-1	OFHC-Class 1	-	-	-	-

Figure 6 shows the best fit fatigue curve and design fatigue curve for normal oxygen free copper[6] and OFHC-Class 1. The best fit curve is obtained by various experimental data for normal oxygen free

Design of the Crotch for SPring-8

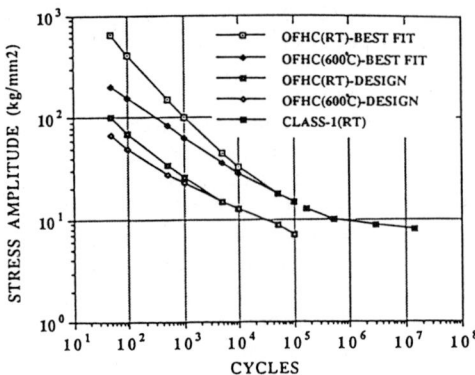

Fig. 6. Fatigue failure strength of OFHC subjected to cycles of alternating tension and compression.

copper. The design curve is decided using the safety factor 2 for the stress and 20 for the cycles in accordance with ASME Boiler and Pressure Vessel Code Sect. III and Sect. VIII Div.2. If the equivalent stress of 30.6 kg/mm^2 whose stress amplitude is 15.3 kg/mm^2 obtained in model C1-2 for GlidCop AL-35 is estimated by the design curve, the design life would be 5000 cycles. It corresponds a ten year life assuming that the thermal cycle in operation is 500 cycles/year (2 times/day x 250 days/year).

VACUUM PERFORMANCE OF PROTOTYPE CROTCH

Some experiments to confirm the vacuum characteristics were made using prototype crotch manufactured for the previous design. This prototype crotch is basically the same as a reformed one, which has a trapping structure made of OFHC-Class 1 and OFHC-aluminum alloy joint by explosion bonding. The schematic diagram of a test device is shown in Fig.7. The test stand made of stainless steel is connected to a crotch with a flexible tube. A 400 l/sec SIP, a TSP and two extractor-type BA nude gauges (BAG 1, BAG 2) are mounted on the crotch. BAG 1 and BAG 2 are for pressure measurements in the trapping room and the electron beam channel, respectively.

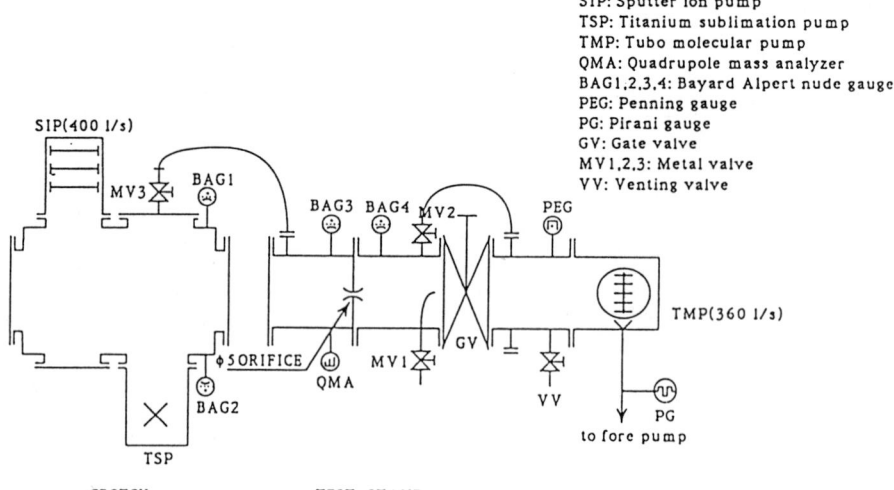

Fig. 7. Schematic diagram of the test device.

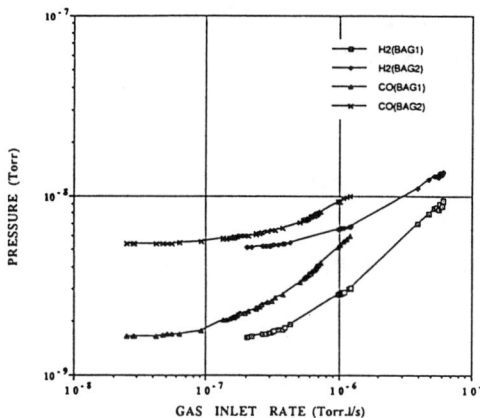

Fig. 8. Measured pressures in the trapping room(BAG 1) and electron beam channel(BAG 2) when the gas is introduced.

After two cycles of baking(150 °C x 24 hrs), the pressure is decreased down to 2.8×10^{-11} Torr in the electron beam channel(both SIP and TSP are in operation). Under the same baking conditions, the outgassing rate of the crotch including SIP and TSP which are made of stainless steel was found to be 2.1×10^{-13} Torr.l/sec.cm^2. The measured pressures when the gas is introduced to the trapping room through a metal valve MV3 are shown in Fig.8(Only SIP is in operation). The gas inlet rates were simulated to the gas load produced by the photon-stimulated desorption in the operating ring whose beam dose is 1~100 A.hr. Large pressure difference between BAG 1 and BAG 2 indicates that the trapping structure is effective on the extension of a beam life time.

REFERENCES

1. Y.Morimoto, H.Sakamoto, S.Yokouchi and S.H.Be, American Institute of Physics Conference Proceedings No.171, (1988), p.327.
2. E.Sieder and G.Tate, Ind.Eng.Chem. 28, 1429(1936).
3. K.Nishikawa and Y.Fujita, Int.J.Heat Mass Transfer Vol.20, No.3, 233(1977).
4. Y.Katto, Trans.Japan Soc.of Mech.Engrs.Vol.44, No.387, 3865(1978).
5. Y.Katto, Science of Machine Vol.34, No.5, 552(1982) and Vol.34, No.6, 671(1982).
6. JSME Data Book, Fatigue of Metals IV-Low Cycle Fatigue Strength, (March 1983), p.41.

VACUUM PERFORMANCE OF THE LNLS INJECTOR LINAC - PRESENT STATUS

Paulo Alberto Paes Gomes

Laboratório Nacional de Luz Síncrotron/CNPq

Cx. Postal 6192

13081, Campinas, SP, Brasil

ABSTRACT

The Brazilian Synchrotron Light Laboratory (LNLS) completed the construction of the first stage of the Injector LINAC in December 1989. The Vacuum System of this 50 MeV LINAC is described. Brazilian stainless steel was used in all chambers, pumped by small (20 l/s) ion pumps, achieving the required performance. We also describe the present status of the vacuum system of this injector and the energy spectrometer installed at the electron beam exit port.

INTRODUCTION

The LNLS is a national laboratory whose aim is to provide the Brazilian researchers with a VUV and soft X-ray photon source.
This will be based on a 1.15 GeV electron storage ring under construction and scheduled for operation in 1994. The injector will be a 100 MeV LINAC.

The first stage of the Injector LINAC is ready. Table I presents the main measured parameters.

The aim of this work is to present the vacuum system design and the present status of this LINAC.

Table I Measured parameters of the 50 MeV LINAC

Maxium energy	54 MeV
ΔE/E	2%
Pulse length	100 ns
Current	220 mA
Repetition rate	0,5 to 33 Hz
Emittance	1.4 π.mm.mrad

VACUUM DESIGN

In order to avoid scattering of the electrons by residual gas molecules and also RF breakdown, a working pressure smaller than 1×10^{-6} mbar is needed [1,2]. Next to the RF windows it is desirable to have a pressure of 1×10^{-7} mbar so that no material deposition will occur on them. In order to increase the cathode lifetime we would like to have 5×10^{-8} mbar in the electron gun. Another requirement is the absence of hidrocarbons in the residual gas. The presence of these contaminants can attenuate the Q of the accelerator structures. We have chosen cryosorption pumps for roughing and ion pumps for high vacuum to reach these requirements.

Because of the low vacuum conductance of the accelerator structures and the higher outgassing rate of copper, compared to stainless steel, the maximum pressure point of our LINAC is at the center of the structures. Using a simple model of a tube pumped at both sides [3], a structure conductance of 0,5 l/s and a copper outgassing rate of 1×10^{-11} mbar l/s, we have the following expressions for the pressures near the pump, in the center of the structure and the average pressure as functions of the pumping speed (S).

$$P_{pump} = 8/S \times 10^{-8} \text{ mbar} \qquad (1)$$

$$P_{center} = \frac{8}{S}\left(1 + \frac{S}{2}\right) \times 10^{-8} \text{ mbar} \qquad (2)$$

$$P_{average} = \frac{8}{S}\left(1 + \frac{S}{3}\right) \times 10^{-8} \text{ mbar} \qquad (3)$$

Where S is the pumping speed and where the second term in the parenthesis is the conductance term.

We can see that the system is conductance limited and then a large pumping speed is useless in this case so we have chosen 20 l/s ion pumps for the structures. For the electron gun we use a 20 l/s plus a 60 l/s ion pump.

The vacuum components of the LINAC are shown in fig. 1. All the tubes are made of 304 L Brazilian stainless steel and are connected by ConFlat flanges with copper gaskets. Pressurized aluminium RF wave guides connected to the structure by dielectric windows were chosen, so they do not belong to the vacuum system.

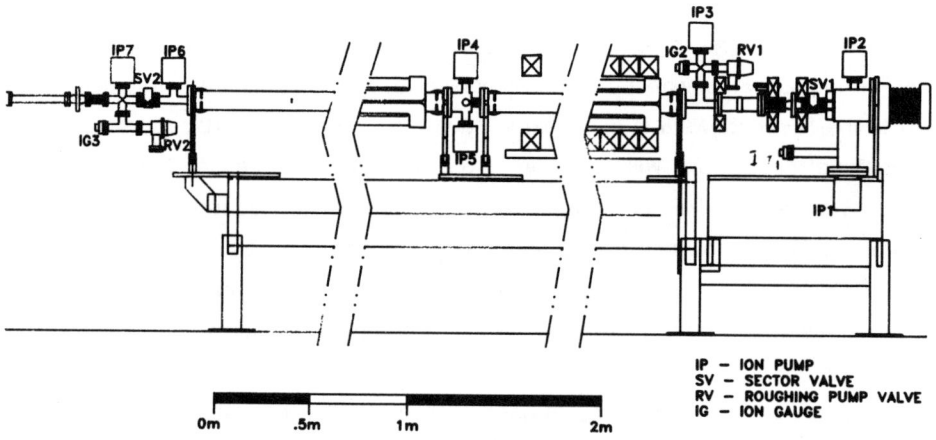

Fig. 1 - Vacuum system components of the 50 MeV LINAC.

ASSEMBLING AND COMISSIONING THE LINAC

All the components were machined, cleaned and welded following LNLS standard procedures. Cleaning procedures for Stainless Steel consists of:
- Acetone cleaning with paper;
- Degreasing in a tricloroethane vapor bath (15 min.);
- Tap water rinse;
- Ultra sound with alkaline detergent pH=11(15 min.);
- Tap water rinse;
- Deionized water rinse;
- Dry in a 150°C air furnace (at least 2 h).

The components were then all leak tested separately before assembly and a pressure of the order of 10^{-8} mbar was achieved in each subsystem, including both structures.

These structures were vented before assembling the LINAC in the tunnel, which started on November 10th 1989.

After assembling, the system was pumped down by a home made 3 stage cryossorption pump system from atmosphere to $1,0 \times 10^{-4}$ mbar in 2 hours. The ion pumps were started, the roughing system isolated and the pressure reached 10^{-6} mbar in some hours. Spraying He outside the connections and observing the ion pump currents we found only one small leak in one of the RF windows. This was eliminated by simply tightening the flanges.

One day later we had all the gauges and all the ion pump displaying around 1×10^{-7} mbar or less. No "baking in situ" was needed.

The commissioning was started by injecting the electron beam without RF. Then, RF power was injected without the electron beam from 10% to 100% of the klystron power. On December 18th 1989 we had the first electron beam accelerated to around 40 MeV. During this period no serious pressure bump was observed.

A spectrometer to measure the energy dispersion was installed at the LINAC end (fig. 2). It consists of a bending magnet, a quadrupole and a vacuum chamber with 3 ports, one straight ahead, one for the deflected beam and the last with a glass window through which we will try to detect synchrotron radiation.

122 Vacuum Performance of the LNLS Injector Linac

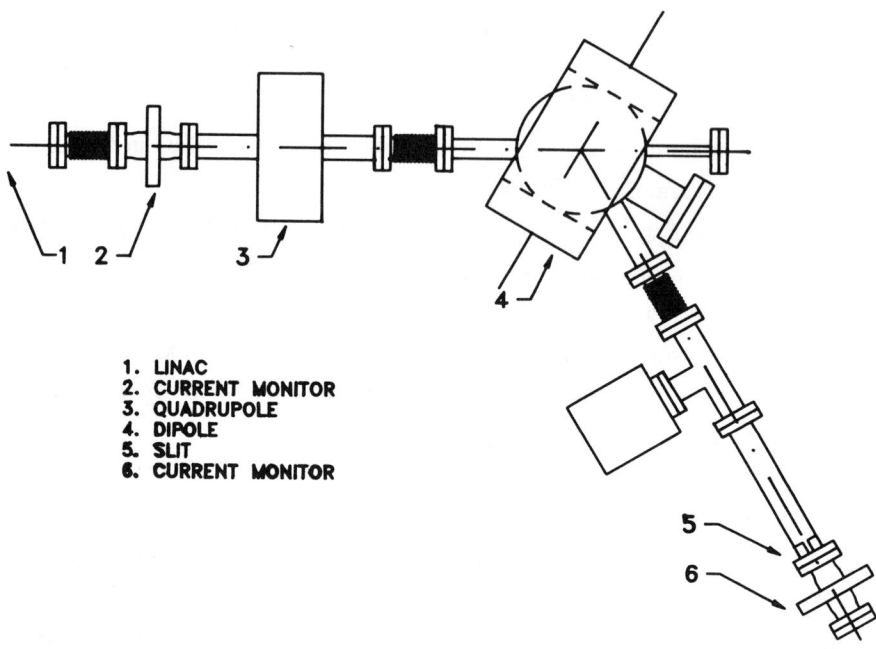

1. LINAC
2. CURRENT MONITOR
3. QUADRUPOLE
4. DIPOLE
5. SLIT
6. CURRENT MONITOR

Fig. 2 - Energy spectrometer.

PRESENT STATUS

The 50 MeV LINAC has been working to specifications routinely since April 1990. Table II shows the vacuum system performance during operation at 50 MeV, 160 mA and a repetition rate of 33 Hz. The position of each gauge can be seen in fig. 1. The pressure between the structures is estimated by current of the Ion Pump 4.

Table II Operating pressures in the LINAC

IG1	IG2	IP4	IG3
2×10^{-8} mbar	3×10^{-8} mbar	8×10^{-8} mbar	1×10^{-7} mbar

RESEARCH AND DEVELOPMENT

The following improvements are planned for the near future:
- increase the energy to 100 MeV by inserting two additional accelerating structures;
- increase the current and the macropulse length through the installation of a new gun under test;
- installation of a better focusing system at the end of the LINAC for experiments with insertion devices.

The LINAC will be used for preliminary studies of fruit irradiation in a joint program with the Food Technology Institute.

Another program will use the LINAC to test room temperature and superconducting micro-undulators under development at LNLS.

CONCLUSIONS

The vacuum system of the LNLS 50 MeV LINAC was presented. The pressures obtained after commissioning compare well with the desired specifications.

The procedure to assemble and commission the LINAC was briefly discussed and the present status shown.

An overview of future improvements and applications of the LINAC was given.

We conclude that the necessary vacuum as well as other techniques involved in the construction of the LINAC are well developed at LNLS.

REFERENCES

1. R. B. Neal, ed, The Stanford Two-Mile Accelerator (W. A. Benjamin, N. Y., 1968).
2. J. W. Wang and G. A. Loew, RF Breakdown Studies in Cooper Electron LINAC Structures. SLAC-PUB-4866, March 1989. (Stanford University, Stanford, CA., 1989).
3. J. Delafosse et G. Mongodin, Les Calculs de la Technique du Vide. Edité par la Société Française des Ingénieurs et Techniciens du Vide. (Paris, France, 1961).

APS STORAGE RING VACUUM CHAMBER FABRICATION

George A. Goeppner
Argonne National Laboratory, Argonne, IL

ABSTRACT

The 1104-m circumference Advanced Photon Source Storage Ring Vacuum System is composed of 240 individual sections, which are fabricated from a combination of aluminum extrusions and machined components. The vacuum chambers will have 3800 weld joints, each subject to strict vacuum requirements, as well as a variety of related design criteria. The vacuum criteria and chamber design are reviewed, including a discussion of the weld joint geometries. The critical fabrication process parameters for meeting the design requirements are discussed. The experiences of the prototype chamber fabrication program are presented. Finally, the required facilities preparation for construction activity is briefly described.

INTRODUCTION

The basic Storage Ring lattice is divided into 40 sectors, each of which is composed of six individual sections, totaling 27.6 meters in length. Except for the insertion device, injection and abort chambers, and rf cavity, the vacuum chamber cross section is identical. Each of the sections is unique; however, being different in length and in the accommodation of ancillary components such as magnets, vacuum pumps, crotch and distributed absorbers, and beam diagnostics. Some of these components require direct access to the interior of the vacuum chamber, resulting in the use of several sizes of vacuum flange. With the varied locations of the flanges on the vacuum chamber, there are several intersection weld joint geometries that must be machined and welded to exacting requirements. In total, there are almost 3800 welded joints, each meeting the design requirements imposed by the operating characteristics of the Storage Ring.

The fabrication of the Storage Ring vacuum chambers requires integration of several manufacturing processes, closely related and well controlled in order to meet the quality requirements of the end product. The critical process steps are extruding, machining, cleaning, welding, inspection and leak checking.

EXTRUDING

The vacuum chamber is fabricated from a 6063T5 aluminum extrusion (see Figure 1). Aluminum was chosen for the vacuum chamber because it can be

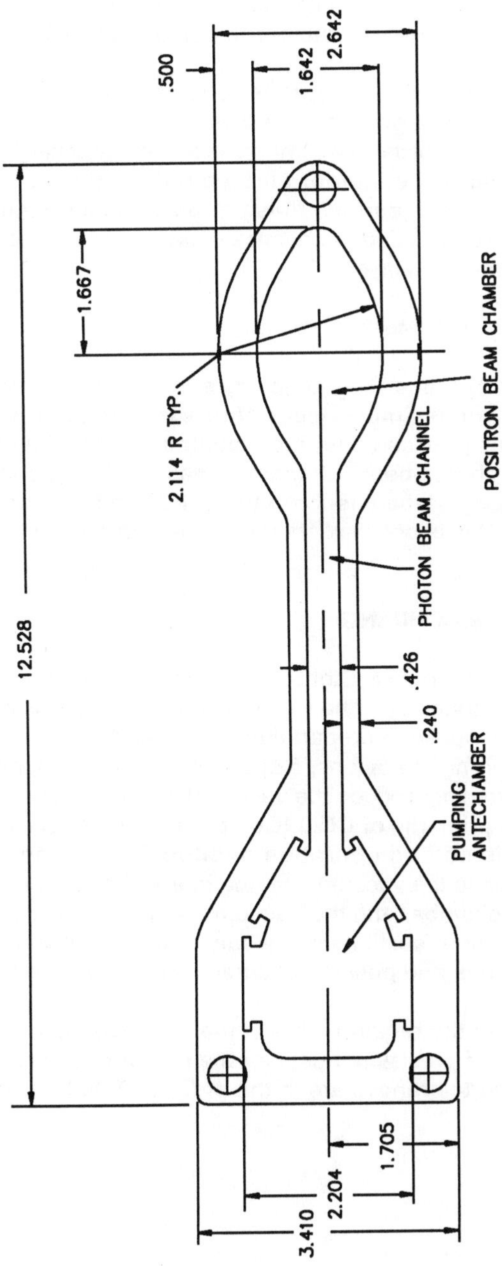

FIGURE 1

STORAGE RING VACUUM CHAMBER EXTRUSION

economically extruded and machined, has good thermal conductivity, low thermal emissivity, low outgassing rate, low residual radioactivity, and is non-magnetic. The 6063 aluminum-silicon-magnesium alloy provides high strength combined with good machining and weldability characteristics.[1] This alloy experiences negligible distortions when subjected to an oven heat treatment to achieve the T5 strength characteristics.

Extruding and heat treatment of 60' lengths is followed by a controlled stretching operation to obtain the straightness required. The stretching also provides for dimensional control of the critical chamber cross section geometry.

The extrusion process provides the interior surface finish needed for the ultrahigh vacuum environment. With a surface finish of 64 μ-inches, the only chamber interior surface preparation needed is chemical cleaning and removal of the magnesium oxide and aluminum oxide layers.

FORMING

There are six sections in the Storage Ring lattice, two of which are bent to a 1533.9" radius. The bending process must occur after extruding and before machining. This process requires precise control, although it can be done using standard shop equipment. The photon beam channel closes up during bending, requiring a subsequent pressurizing of the interior of the extrusion to expand this area to the required dimension. This is accomplished using a high pressure water system.

MACHINING

The machining requirements of the APS Storage Ring can be accommodated using conventional machine tool technology. The vacuum chamber sections have a substantial amount of machining applied in preparation of the weld joints. Ninety-five percent of the machining is milling operations, 62 percent of which is applied to the extrusions. The remaining is turning work on the vacuum Conflat flanges.[2]

The milling activity consists primarily of CNC (Computer Numerical Control) machining. This is a multi-axis cutting operation needed for the contoured intersections of the Conflat flanges to the vacuum chambers, and for the end flange transitions between the vacuum chamber and the vacuum flanges. The sequence of fabrication for the vacuum chamber sections is unusual in that no machining is done after welding; final dimensions and position accuracies must be maintained during the welding operation.

Given this sequence, and with the requirements of the subsequent automatic welding operation, a high degree of precision must be obtained during machining. The mating component machining tolerances are in the order to 0.002" to 0.006".

Due to the large quantity of components, the pieces must be interchangeable with the same degree of accuracy, i.e., no hand-fitting for proper match-up.

An initial constraint was to perform all machining without the use of cutting tool lubricants or coolants. This was thought to be necessary in order to preserve the vacuum qualities of the extrusion interior surfaces. The disadvantages of this approach were reduced cutting speeds and feed rates, rapid tool wear, and lack of proper cutting chip removal necessitating extra care during cleaning. Subsequent studies conducted at Argonne National Laboratory (ANL) and at the University of Texas have concluded that there is little variation in the surface and near surface composition after cleaning between control samples which were dry machined vs those which were machined with water soluble coolants.[3] Future machining operations will therefore allow the use of tool coolant, eliminating the disadvantages previously encountered.

CLEANING

A cleaning step is performed after machining, and just prior to welding. The purpose is to remove all residues of the machining and handling operations, to remove and control the oxide layers resident on the vacuum chamber surfaces, and to prepare the machined surfaces for welding.[1]

Two cleaning stations are required. One is for the 6063 extrusions, the other for the 2219 aluminum flanges and ancillary components.

The degreasers and etchants used are contained within stainless steel tanks. These welded assemblies have circulating pumps and heaters to maintain a constant 65°C temperature, as required. An overhead gantry crane is used to rapidly transfer the extrusion from tank to tank. After final rinse and blow drying, the extrusion is lowered onto a manual material handling cradle for transfer into the welding cell.

The use of ultrasonics is being considered for the extrusion cleaning process. The ultrasonic transducers are placed directly into the cleaning tank, and are activated during the degreasing cycle. Studies have shown that this addition speeds up the cleaning process, and provides a surface that is cosmetically improved over the standard procedure.

WELDING

The performance requirements of the Storage Ring place severe demands on the welds that are necessary to fabricate a chamber assembly. The quantity of ultra-high vacuum welds, and the contoured geometry, requires a process that is repeatable and highly reliable. Aluminum joining is commonly done with TIG (Tungsten Inert Gas) welding; in this application, TIG is combined with the machine tool precision of a standard rectilinear motion robotic welding system. The choice of a well known process applied by a standard machine limits the risk, and reduces the development effort required to establish the weld joint designs, materials selection, tooling, and weld schedule programming. A DC welding power supply was chosen in order to obtain narrow, deep weld cross sections.

The individual welds to be applied to each vacuum chamber section were classified and grouped into six (6) types, Type "A" through Type "F".[4]

This broad classification grouped similar weld joints by the type of joint geometry needed to produce acceptable results. Table I lists the application by type.

Table I Classification of Weld Joints

Type	Application
Type A	End Flange to Vacuum Chamber Extrusion
Type B	6" and 8" Conflat to End Flange 2 3/4", 4 1/2", and 6" Conflat to Vacuum Chamber Extrusion
Type C	Elliptical Beam Tube to End Flange
Type D	8" Conflat to Vacuum Chamber Extrusion 12" Conflat to Vacuum Chamber Extrusion
Type E	Photon Beam Port to Vacuum Chamber Extrusion
Type F	Blank Conflat to Elliptical Beam Tube Mini-Conflat to Blank Flange

All of these welds are to be applied using automatic welding equipment. The water tube extensions from the extrusion water passages were designed to be hand welded. Designs to improve this weld joint are underway.

Given the requirements of UHV vacuum integrity, low rf impedance, and the 150RC chamber bakeout procedure, a set of criteria for weld joint design was established:
- Cracks, incomplete fusion, and cold laps are unacceptable.
- Porosity, inclusions, undercut, craters, and any imperfections that have sharp terminations, are unacceptable.
- Small scattered porosity within the scope allowed by MIL-STD-2219 for Class A welds would be considered acceptable.

The Type A, C, D, and E welds utilized an automatic system having axis motions in x, y, z, and rotary. These motions were programmed as required to accommodate the geometry being welded. The robotic end effector, or weld package, was changed as required. Type B and F welds also used an automatic system, which was an orbital weld head limited to rotary motion with a self contained wire feed mechanism.

Programming of the machine axis motions is a small part of the effort needed to arrive at a suitable welding schedule. The important parameters are arc voltage, current, wire feed speed, and axis velocity. Beginning with set values from experience, the final parameters are arrived at by laboratory weld trials. These are then integrated into the motion program.

The early laboratory trials are directed at simulated weld joints, using the selected materials. For the APS Storage Ring, a thorough weld development program was undertaken, including the development necessary to machine full size extrusions to the exacting tolerances needed for acceptable weld results.

TYPE A WELD JOINT

The end flange to vacuum chamber extrusion weld requires that a portion of the weld joint wrap around the positron beam. Excessive weld underbead in the beam chamber area would increase rf impedance, causing losses. The requirement for this weld was full penetration with an underbead reinforcement not exceeding 0.020" around the beam chamber and into a portion of the photon beam channel.

Critical to this development effort was the selection of material for the end flange. The 6000 series aluminum welds are crack sensitive. 4043 aluminum was attempted, but failed to produce an acceptable surface finish for the UHV environment when machined. The flange material was changed to 2219-T851, requiring the need for filler wire of 4043 aluminum to prevent solidification cracking.

The cross section wall thickness of the extrusion varies, (see Figure 2). This is extremely difficult for the welding process, given the performance requirements of the Storage Ring. Local machining of the extrusion ends inorder to create a butt joint and a weld path geometry suitable for continuous welding from the outer surface was necessary. The geometry was also designed to facilitate the x, y, and r contouring axis of the automatic welding system. The pivot point for the rotary, or r axis, is within the elliptical positron beam port. This locus was chosen to maximize the dynamic accuracy of the velocity along the path, the z-axis position of the tungsten electrode, and the torch attitude with respect to the joint surface in the region where underbead control was essential. The three-axis contouring program established provides constant relative vector velocity between the tip of the torch tungsten electrode and the outer surface of the joint path. It also maintains the axis of the tungsten perpendicular to the joint surface and its distance from that surface constant.

Figure 3 shows the 2219 machined joint relative to the 6063 extrusion machined joint. This design provides a 1/4" ligament of constant thickness on each side of the joint interface which serves to reduce the effect on the weld of the dramatic changes in adjacent extrusion mass. The tongue and groove, when

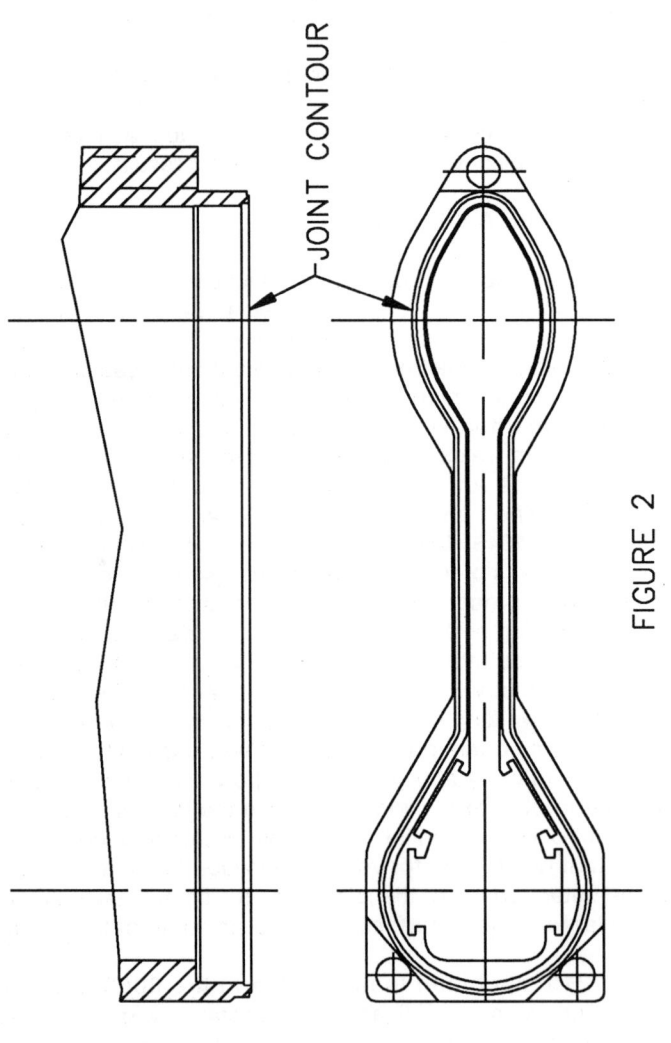

FIGURE 2
CONTOUR OF JOINT MACHINED INTO EACH END OF A STORAGE RING VACUUM CHAMBER SECTION

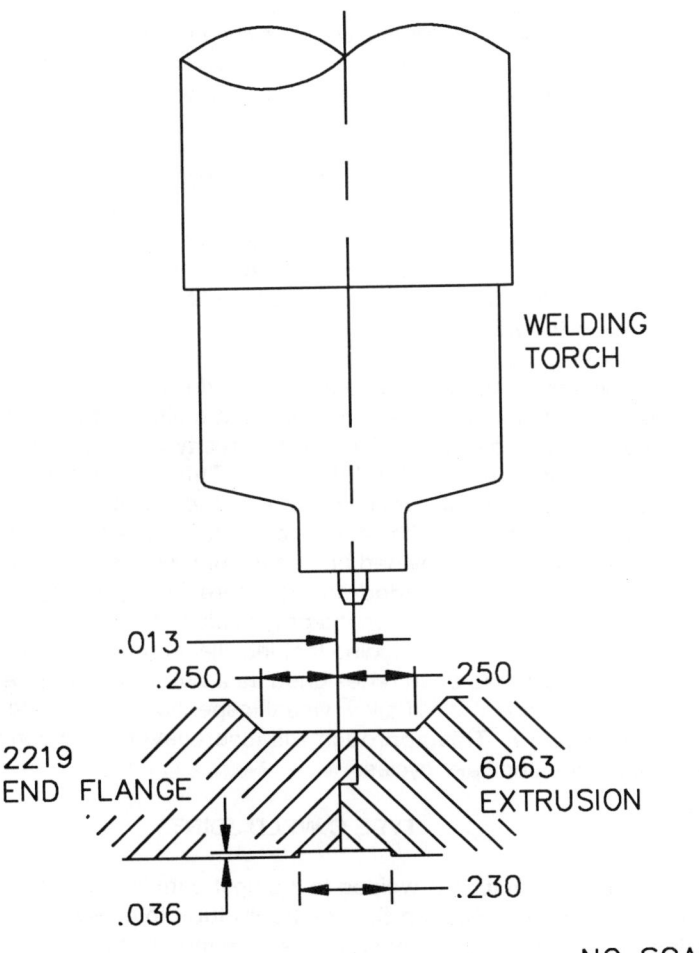

FIGURE 3

CORRECT LOCATION OF TUNGSTEN ELECTRODE FOR WELDING TYPE "A" JOINT

engaged, insures that the half of the contoured joint machined into the end flange is accurately aligned with the mirror image joint machined into the extrusion. The 0.036" deep by 0.230" wide groove on the vacuum side of the butt joint allows an underbead as much as 0.036" to be acceptable from a functional standpoint. The walls of the groove as positioned on both sides of the underbead are very effective heat sinks that inhibit growth of underbead height.

Due to the difference in thermal conductivity between the 6063 and 2219 alloys, the tungsten must be incremented 0.013" toward the 6063 extrusion for the underbead to be centered in the groove. The acceptable position tolerance of this offset is +/- 0.0025".

TYPE B WELD JOINT

The Type B joint accommodated all vacuum Conflat welds not requiring a contoured weld path. They are circular welds, in the same plane. A generalized joint design utilizing fillet welds was initially selected. This approach promised minimum costs for weld joint machining preparation, while also requiring the minimum space.

After an intensive effort, and examination of dozens of cross sections of Conflat fillet welds, it was found that defect-free welds are difficult to obtain. The results of this joint design are weld root defects and porosity which could not be eliminated without edge melting of the Conflat flange. This constitutes an unacceptable condition for vacuum integrity when considering the possibility of fatigue cracking as a result of strains imposed on the vacuum chamber by the 150°C baking cycles.

The fillet weld design provided no convenient means of inspection. Given the quantity of Conflats to be welded on the entire Storage Ring, this presented a dilemma for construction. Vacuum integrity could not be verified immediately after the welding operation; thus concluding that the fillet approach is unacceptable.

The Type B joint has been redesigned as a butt weld. It remains as a circular weld in a single plane, and will allow visual inspection of the underbead as quality verification measure. This approach will also alleviate the concerns of fatigue cracking after repeated bake cycles.

TYPE C WELD JOINT

An elliptical beam tube is welded to the upstream flange of Section 1 and to the downstream flange of Section 5. The beam tube in cross section is an ellipse, retaining the geometry of the positron beam chamber through the opening in the end flange except that it is a closed section. It terminates in a blank Conflat, continuing the elliptical cross section through the thickness of the Conflat.

Two welds are required; one is the Type C, joining the tube to a mating stub machined on the end flange; the other is the Type F. These welds are only 5" apart, so that any distortion of the elliptical tube caused when making one weld will affect the joint fit up for the second weld. A procedure was devised which requires fixturing both weld joints before either weld is made.

The requirements for UHV and underbead control are the same as for the Type A joint. Therefore, a geometry similar to that for Type A was chosen for Type C. The material selected for the beam tube was Al-2219-T851. The axis of the tungsten electrode is positioned directly on the joint interface, with no offset. The weld torch was programmed to follow the elliptical path at a constant vector velocity relative to the outer surface of the joint, and with a constant distance above that surface. This is a three-axis (x, z, r) contoured weld.

By using the Type A joint design, the process development task was limited to establishing a program for varying the welding current as a function of distance along the joint path.

TYPE D WELD JOINT

The 8" Conflats are welded to Sections 3 and 5; the 12" Conflats are welded to Sections 2 and 4. They are welded back-to-back, that is, opposite each other on the same vertical axis (see Figure 4).

The function of these openings is to allow mounting of a distributed absorber (8") or a crotch absorber (12") for intercepting non-experimental synchrotron photon radiation, and ion and NeG pumps to provide local control of desorbed gases. The intersection of the Conflat tube with the extrusion is not in a plane due to the geometry of the extrusion. Hence the need for CNC machining preparation of the weld joint interfaces, and subsequent contour welding.

After experiencing difficulty with the fillet welds on the small conflat joints, it was decided to design a square butt joint for the 8" and 12" Conflats. This design allows the added advantages of providing a full penetration weld, and after-weld visibility of both sides of the joint as a QC measure. The only weld defect to be guarded against is porosity, which can be controlled by careful welding procedures. The full penetration weld provides strength to the interface, as these Conflats are more heavily loaded than the others due to the large vacuum pumps mounted to them.

A physical constraint imposed upon these welds is maintaining parallelism of the Conflat mounting surfaces. This is a problem that must be addressed in machining and welding. The opposing Conflat interfaces to the extrusion are a mirror image, while the vertical projection of the interface forms a plane. These two planes must be machined parallel to the centerline of the extrusion as well as to each other. The maximum gap that can be tolerated by the automatic welding process is 0.005" for this weld, placing demands on the CNC programming and machining for both the extrusion and the Conflats.

In machining to these exacting requirements, control of the photon beam channel dimension of 0.426" required a certain sensitivity. After boring the large

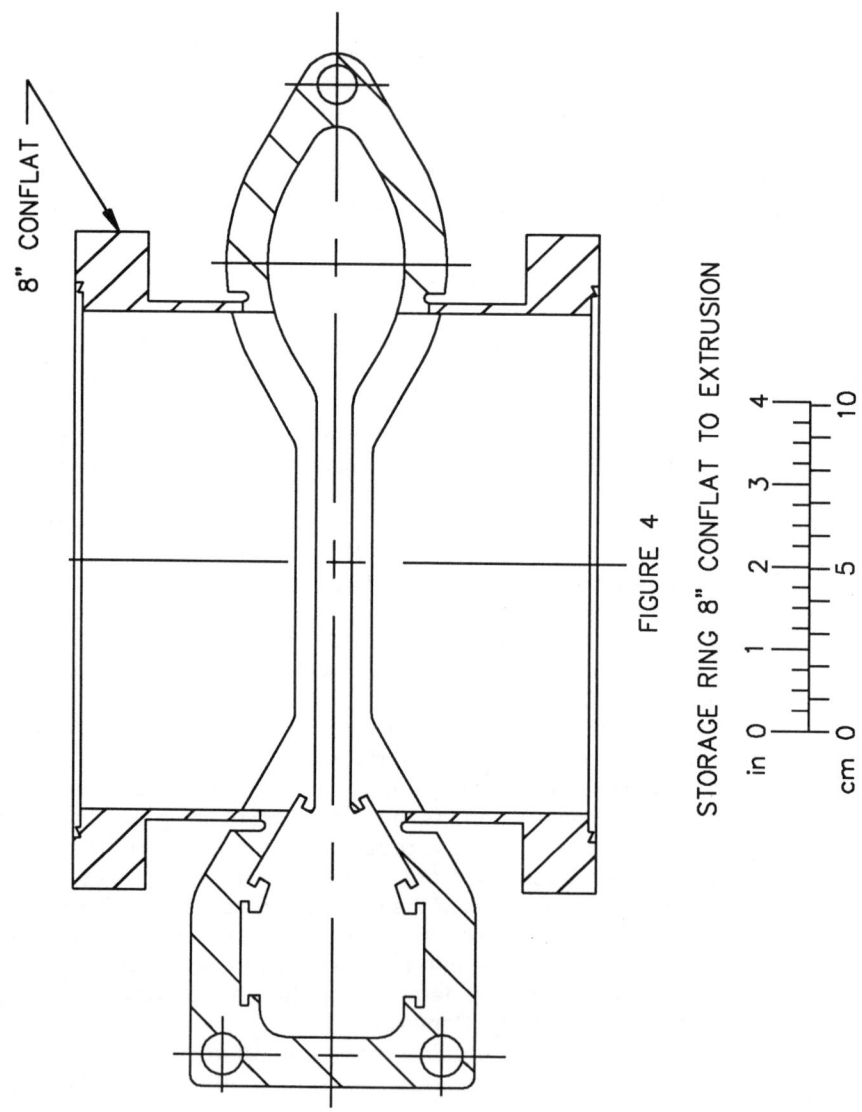

FIGURE 4

STORAGE RING 8" CONFLAT TO EXTRUSION

hole in the extrusion equal to the Conflat inside diameter, the vertical dimension of the photon beam channel decreased.

After initial weld trials, the dimension was found to decrease by 0.017" on the average. The tooling designed for welding restrains the Conflats from translation or rotation, and utilizes four die springs arranged to apply an adjustable clamping force. The compressive load applied by the springs through the opposing Conflats locally deflected the walls of the extrusion inward. This, in combination with the high welding temperatures and the thermo-physical properties of the dissimilar alloys being joined, caused the plastic deformation in the photon beam channel region of the extrusion.

To overcome this problem an adjustable interior support tool was used. The tool is made of stainless steel and it sizes the photon channel, by expanding the gap slightly prior to welding while maintaining Conflat flange surface parallelism. The low thermal conductivity of the stainless steel, with the programming ability for arc current control, ultimately provided a full penetration weld with a uniform underbead for the complete 360 degree 8" Conflat weld. The development efforts, and final implemented solutions, were similar for the 8" and 12" Conflats.

TYPE E WELD JOINT

A photon beam exit port is provided on each of the Sections 3 and 5, allowing the extracted photon beam to enter the experimental facilities. The interior surface of the beam port to extrusion interface is not accessible to the welding torch or tooling. The inner surface contains the underbead, which is in contact with the vacuum environment. There are no requirements for precise underbead height control in this area, but discontinuities that would result in a virtual leak are unacceptable.

Borrowing again from the favorable experience of the Type A joint, a square butt joint geometry providing for alignment of the interior surfaces without the use of back-up tooling, and a balancing of the material mass to compensate for the difference in thermal conductivity between the 6063 extrusion and the 2219 beam port, was successfully arrived at.

The resultant design requires that the torch be tilted at a slight angle instead of being perpendicular to the joint (see Figure 5). Indexing the torch axis toward the base metal of highest thermal conductivity (the 6063 material) also requires the use of a special shielding gas nozzle developed specifically for the Type E joint.

TYPE F WELD JOINT

This joint application is for the elliptical beam tube to blank Conflat at the upstream end of Section 1 and at the downstream end of Section 5, and for the mini-Conflat to blank Conflat for NeG strip electrical feedthroughs. Both welds were

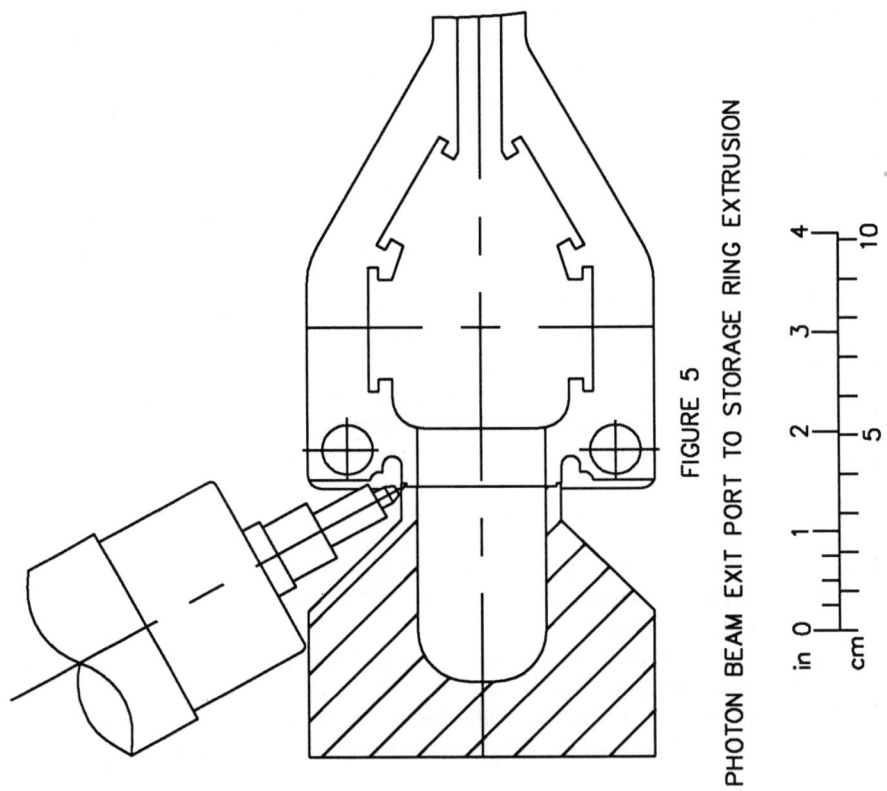

FIGURE 5
PHOTON BEAM EXIT PORT TO STORAGE RING EXTRUSION

designed to be autogenous partial penetration edge butt welds. Suitable procedures were developed to give satisfactory results with a minimum of effort. It was necessary however, to provide a means of venting trapped air from the beam tube to Conflat weld to prevent porosity problems encountered with the initial weld trails.

INSPECTION AND TESTING

Each Storage Ring vacuum chamber section will be inspected for conformance to the drawing dimensions. Conventional inspection equipment and techniques will be used to provide a quality control checkpoint on the set-up and use of the welding fixtures. Vacuum integrity of the welding operation will be verified by leak testing

The vacuum chambers will be assembled individually with all pumping components and valves normally associated with that chamber. An ion gauge and RGA head will also be fitted. A turbopump is used to check for weld leaks at $10E^{-6}$ Torr, after which the bakeout water will bring the chamber to 150°C until the pressure has stabilized (~ 24-36 hours). The NeG strip is activated after the ion pump is flashed and all filaments are outgassed. The chamber is then allowed to cool. The RGA scan is monitored, and the ultimate pressure recorded. This UHV check will also test commercial vacuum components. The chambers are finally vented with dry nitrogen, removed from the test stand, and prepared for a brief storage period.

PROTOTYPE CHAMBER FABRICATION

The developmental work in extruding, machining, cleaning, and welding has been applied to the fabrication of five vacuum chamber sections, using the processes and procedures derived from trials in each of these primary fabrication areas. Substantial data has evolved in support of the final process procedures, and improvements have been identified to provide further efficiency to the construction tasks.

Significant effort has been applied to the design and fabrication of special tooling and heat sinks for the welding operations. The automatic welding system consists of a head and tailstock assembly used with the orthagonal axis motions. Closed loop servo-control is provided for the headstock; the tailstock is mounted on an adjustable slide to accommodate the various lengths of vacuum chambers. Precision tooling is provided as necessary to register weld joint components. An invaluable aid during welding development was a closed circuit TV camera system directed at the end of the wire feed guide tube and the adjacent tungsten electrode. This allowed set-up and adjustment parameters to be established for the wire feed system until the filler wire was reliably fed into the leading edge of the weld pool over the complete path of the Type A joint, for example. Of the five sections that have been fabricated, Sections 1 and 2 have been evaluated for dimensional stability, and Sections 1, 2, 4, and 5 have been evaluated for vacuum performance.

CHAMBER STABILITY DURING BAKEOUT CYCLING

Sections 1 and 2 were mounted to a concrete floor using supports designed for the final installation. They were joined by a connecting bellows. Chamber motion was measured by a combination of optical survey targets and dial indicators. Various locations were targeted for inspection. Several bakeout test cycles were performed between room temperature and 150°C, and back to room temperature. The purpose of the test procedure was to determine if the chamber sections returned to their original position after experiencing bakeout cycling. The section assembly was subjected to 25 bake-cool cycles. The initial tests resulted in a modification to the chamber supports to prevent a loose condition found in the mounting hardware. Most measured variations in the "X" or radial direction were within + or - 0.003", measurements along "Z", in the direction of the positron beam, and "Y" vertical, were within 0.002".

CHAMBER DEFLECTION DURING VACUUM CYCLING

Targets were placed on the outside of the positron beam chamber, and on the photon beam channel area of the extrusion. The beam chamber height deflection averaged approximately 0.010", depending on location. The beam chamber returned to its original location to within + or - .0015". The photon beam channel returned to within 0.001" at three measured locations, after the first two cycles. Deflection of the beam channel under vacuum at the dipole and sextupole areas was 0.027", and 0.021" at the end quadrupole magnet location.[5]

CHAMBER VACUUM EVALUATIONS

Four of the five prototype vacuum chamber sections have been evaluated. Three sections, identified as S1, S2, and S4, were leak tight; Sections S3 and S5 leaked in the 6" Conflat fillet weld area of the end flange assembly. Both leaks have been repaired, and evaluations on S3 and S5 are continuing.

Individual vacuum tests have been completed on S1 and S4. Each chamber was configured with the pumps and instrumentation as designed for use in the Storage Ring. S1 was assembled with two 30 l/s ion pumps, ST 707 NeG strips, and a nude ion gauge. S4 was assembled with a 220 l/s ion pump, NeG strips, and an ion gauge. Additionally, a quadrupole mass spectrometer (QMS) was added for diagnostic purposes.

The chambers were roughed out with a 360 l/s turbo pump, while baking at 150°C with heat tapes. The pressure stabilized after 25 to 36 hours, at which point the instrumentation was outgassed before cooldown commenced.

The ultimate pressure for S1 was $4E^{-11}$ Torr; S4 ultimate pressure was $3E^{-11}$ Torr. The pressure of S4 was confirmed by use of an extractor gauge. After bakeout, the QMS indicated a water peak almost as large as the carbon monoxide peak. This could be the result of a relaxation of the flange to gasket seal, which was observed during cooldown, just prior to reaching room temperature. The Storage Ring vacuum Conflat bolt holes are threaded; this design has presented additional sealing problems, which are currently under consideration.[6]

FABRICATION FACILITY

An area of approximately 9600 square feet is being prepared to clean, weld, assemble, and test the approximately 240 vacuum chamber sections which will be installed in the Storage Ring (see Figure 6). The inter-dependance of these primary process steps, with the attendant quality control requirements, is best managed under one roof. The area is assigned according to functional process operations.

Of utmost importance is facility cleanliness and humidity control. The interior has been thoroughly cleaned and painted, including sealing of the concrete floor. Electrical and water service is being installed for each station, as required. The welding cell will be enclosed and air conditioned, and provides for local weld fume removal from each of two automatic welding systems.

The vacuum chamber assembly and leak check stations will also be provided with local air conditioned environments. Movement into and out of the fabrication area of materials and personnel will be routed through an adjoining wall to another area, thus providing a form of air lock to the outside. Material handling will be accomplished with a combination of gantry cranes, floor mounted jib cranes, and specialized material handling carts.

Two small component storage areas are located within the building, one of which is a controlled access area for tools, gauges, and instruments. A mobile office is adjacent to the building, providing a location for a Production Management Center, building construction personnel, technical documents, and a master drawing set.

ACKNOWLEDGEMENTS

The author is grateful to T. Gill, R. Nielsen, R. Niemann, B. Roop, and R. Wehrle for their assistance and direction; to W. Farrell for his significant contribution to the welding development effort and support in preparation of this manuscript; to A. Salzbrunn for the preparation of the manuscript; and to Y. Amer for editing.

140 APS Storage Ring Vacuum Chamber Fabrication

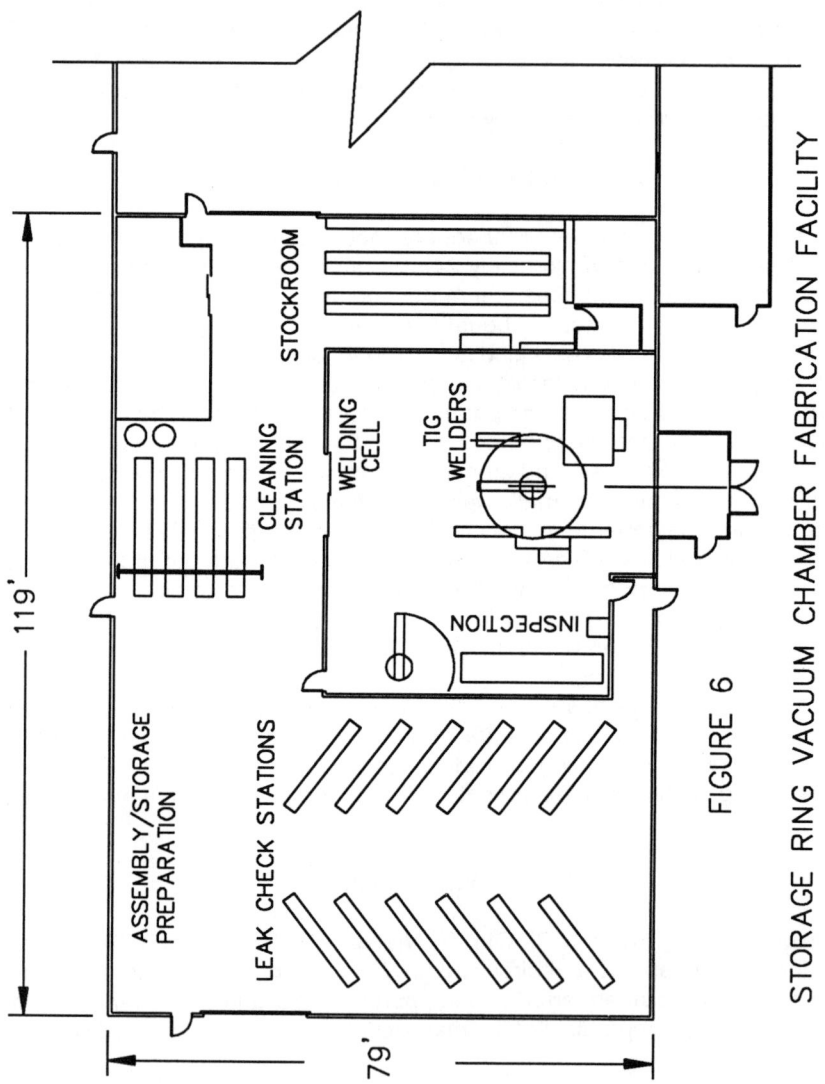

FIGURE 6
STORAGE RING VACUUM CHAMBER FABRICATION FACILITY

REFERENCES

1. R. B. Wehrle and R. W. Nielsen, "Design for APS 7 GeV Storage Ring Vacuum System at ANL": American Institute of Physics Conference Proceedings No. 171 American Vacuum Society Series 5, Upton, NY, 1988.
2. R. Campbell, "Manufacturing Engineering Study for Advanced Photon Source Storage Ring Vacuum Chamber Assemblies," private communication.
3. B. Roop, "Cleaning of Aluminum after Machining with Coolants," to be published.
4. W. Farrell, "Development of Weldments for the Vacuum Chambers of The 7 GeV Advanced Photon Source Storage Ring," private communication.
5. R. B. Wehrle, "APS Storage Ring Vacuum Chamber Tests for Dimensional Stability after Bakeout Cycling While under Vacuum," private communication.
6. B. Roop, "Sections 1-5 Storage Ring Vacuum Chamber Evaluations," private communication.

The vacuum system of IHI's compact SOR "LUNA"

M.Oishi, M.Uesaka, S.Mandai, T.Nishidono, T.Nakashizu and Y.Hoshi

Ishikawajima-Harima Heavy Industries Co.,Ltd.(IHI)
1-6-2 Marunouchi,Chiyoda-ku,Tokyo,100,Japan

ABSTRACT

The construction of the 800MeV compact synchrotron light source "LUNA" was completed in April 1989 at Tsuchiura Facility of IHI. 20 mA beam have been successfully accelerated up to 800 MeV and accumulated with the lifetime of more than 30 min. Now, the machine study and modifications are being done in order to get more accumulated beam currents and longer lifetime.

In most of vacuum sections, the static vacuum pressure of less than $1 \times 10E-7$ Pa has been obtained and in the presence of synchrotron radiation, the vacuum pressure is measured as less than $1 \times 10E-6$ Pa.

The SUS316L vacuum chambers were baked at 200 C For 70 hours in total. The surfaces of the vacuum chambers were treated with buff polishing, acid and freon flashing cleaning. In the 23 meters long vacuum system, ten 400 liter/sec ion pumps and 15 cartridge type nonevaporable getter (NEG) pumps are installed and the total vacuum pumping capacity is 6660 liter/sec.

Additionally, the beam channels for X-ray lithography and other experiments are going to be constructed in 1991.

INTRODUCTION

Synchrotron radiation is expected to be used for various industrial needs, especially for the X-ray lithography for LSI production. IHI has developed LUNA in order to apply synchrotron radiation to various research. [1]

The injector of LUNA is a 45 MeV linear accelerator. Synchrotron radiation is extracted from normal bending magnet of the 800 MeV synchrotron (storage ring). We designed the beam energy and magnetic field so as to set the peak wavelength of SR around 10 A. Basic parameters are summarized in Table 1.

The lattice of LUNA is very simple. It has four normal cells. Each cell consists of a 90 degree sector bending magnet, a horizontal focusing quadrupole magnet and a vertical focusing quadrupole magnet. Also, we set 4 sextupole magnets for chromaticity correction, 2 skew quadrupole magnets for coupling correction and 8 horizontal and 4 vertical dipole steering magnets for closed orbit distortion correction. The layout of synchrotron is shown in Fig.1.

The installation of LUNA was completed in April 1989. Synchrotron radiation at 800 MeV was observed in December 1989. In April 1990, 20 mA beam current at 800 MeV was obtained, and its lifetime was more than 30 minutes.

Table 1 Basic parameters of synchrotron

Energy	800 MeV
Critical wave length	21.8 A (peak 9.2A)
Beam current	20 mA
Beam lifetime	30 min
Circumference	23.5 m
Vacuum pressure	10E-7 Pa
Bending magnet	90 degree sector
	1.33 Tesla
RF frequency	178.5 MHz
Injector	45 MeV Linear Accelerator

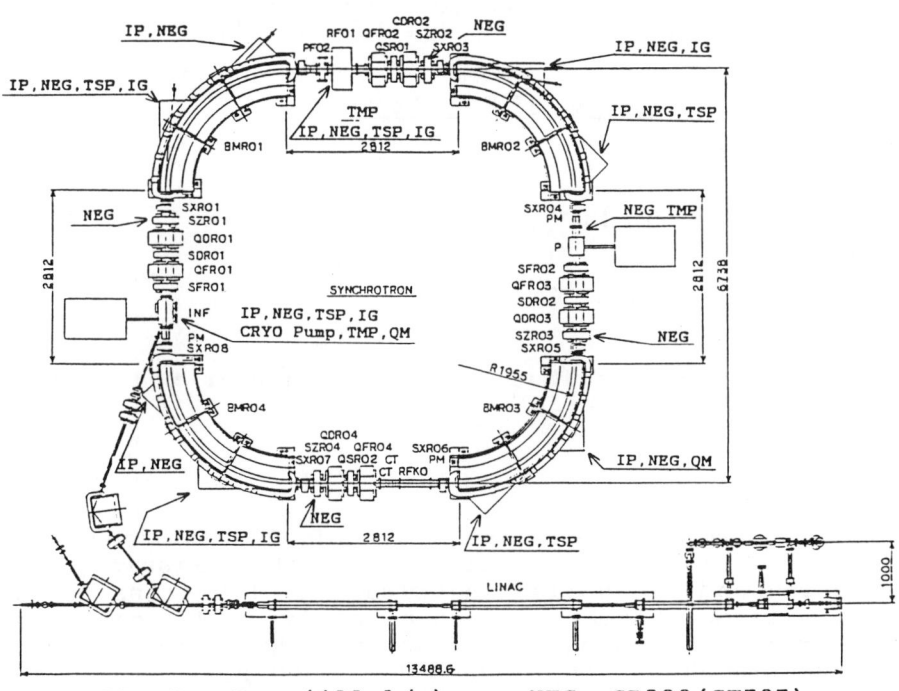

IP: Ion Pump(400 l/s)   NEG: GP200(ST707)

Fig.1 Layout of IHI's compact SOR "LUNA"

VACUUM SYSTEM

Pressure and Lifetime

The lifetime of electron beam is neccesary to be more than ten hours for the lithography experiment, and is neccesary to be more than ten minutes for experiment at the injection energy. These require that the average pressure is less than 10E-7 Pa when we take into account the residual gas scattering. The main component of the residual gases are assumed to be CO.

Outgassing and Pressure distribution

The thermal outgassing rate is assumed to be up to 8x10E-10 Pa liter/cm2 sec conservatively where we consider not only SUS316L but also Ceramics, viton sealed valve, even the leak rate which is estimated through the previous experiment and experience. Photo stimulated desorption (PSD) is estimated by the following equations[2],[3].

$$\frac{d^3 N}{d\epsilon\, dt\, di} = 1.51 \times 10^{14} \frac{\rho}{E^2} \left(\frac{\epsilon}{\epsilon_c}\right)^{-2/3} \quad (1)$$

$$\epsilon_c = \frac{2.2 \times 10^3 \, E^3 \, (GeV)}{\rho \, (m)} \quad (2)$$

$$M\gamma = 5 \, 10^{-8} \, D^{-2/3} \quad (3)$$

$$q\,th = 2n \, M\gamma K \quad (4)$$

The pressure distribution of the quarter model of the ring is shown in Fig.2.

Vacuum Chambers

All vacuum chambers are made of SUS316L and their inner surfaces are treated by buff polishing, acid cleaning and Freon flashing cleaning. The cross section of a straight chamber is shown in Fig.3. The shape is a racetrack, its height is 32 mm and the width is 92 mm. Every straight chamber has a NEG pump port as shown in Fig.3. The structure of a bending chamber is shown in Fig.4. The shape is a rectangular and the height and width are 30 mm and 92 mm, respectively.

Every bending chamber has an ionization gauge port, two evacuation ports and there is space for 5 beam line ports at most and DIP (Distributed Ion Pump). The outside wall has a cooling water channel.

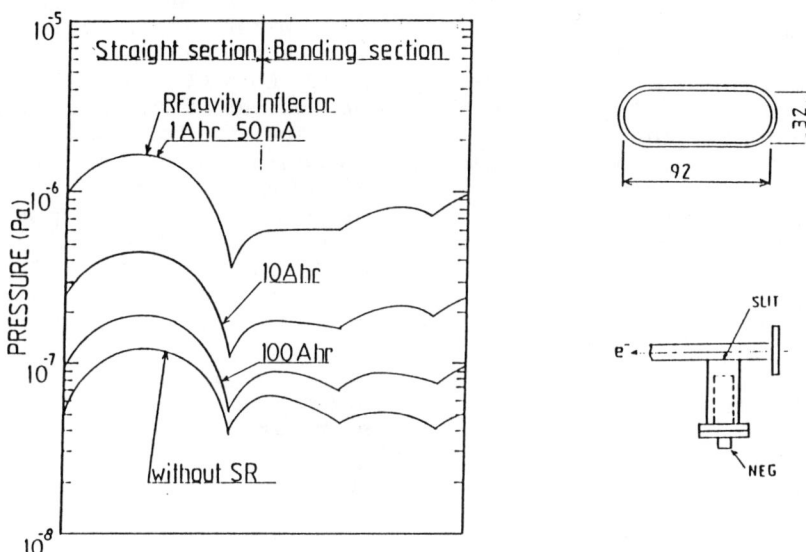

Fig.2 Pressure distribution of quarter model of the ring

Fig.3 The cross section structure of straight chamber

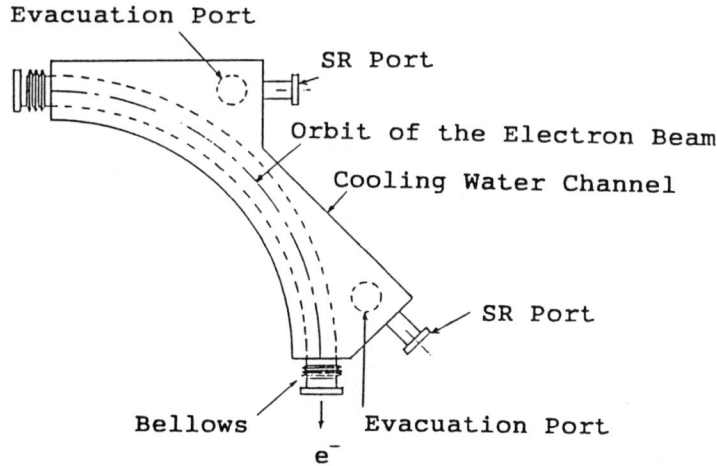

Fig.4. Structure of bending chamber

Pump

The main pumps are 10 triode sputter ion pumps of 400 liter/sec, 15 NEG cartridge pumps (Saes getters ST707 GP200) and 6 titanium getter pumps are installed as sub-pumps as shown in Fig.1. The structure of an evacuate port at a bending section is shown in Fig.5. The triode sputter ion pumps were chosen as main pumps because of the ability to evacuate inactive gases. The NEG pump has a large pumping speed for CO and H2. We chose the cartridge type NEG pumps because of the easy maintenance and large pumping speed for CO. We confirmed by the preliminary experiment that they worked well at the specified pumping speed. The titanium getter pumps are only used when the ion pumps start before baking at the pressure of 10E-5 Pa. The ion pumps can't work at the pressure of more than 10E-4 Pa, because the compact power supplies don't have large capacities of ion pump currents in that range of pressure.

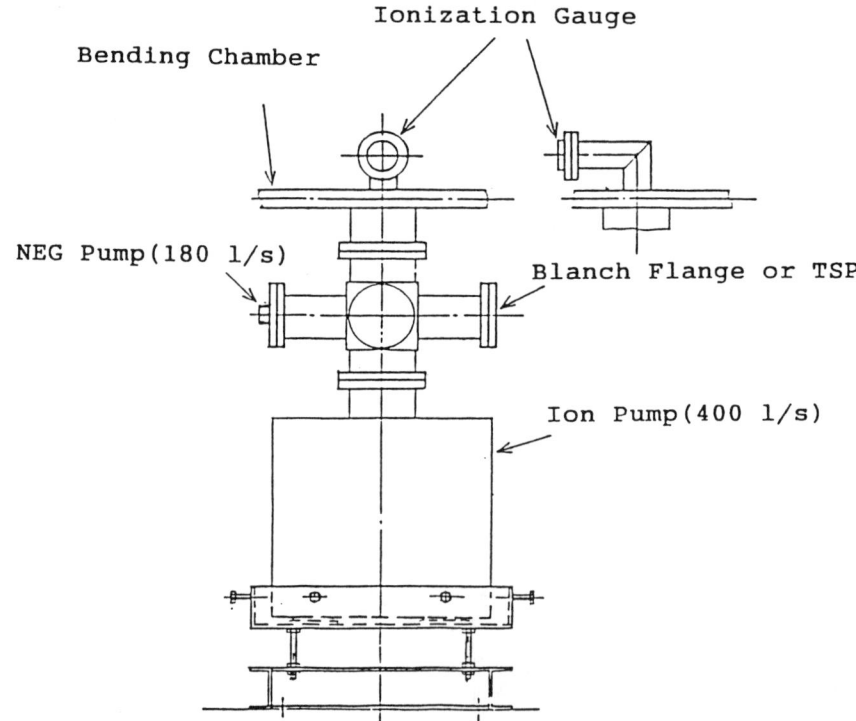

Fig.5 Structure of evacuation port.

Vacuum Control

Hot filament ionization gauges are used for the pressure measurement. Vacuum pressures at every section are displayed and diagnosed at the display board in the control room. 2 quadrupole mass filters are used for the residual gas analysis. All valves and beam shutters are controlled and their status is displayed in the control room. All pumps are controlled locally, but the status of several operation signals of the ion and turbo-molecular pumps are displayed and used for the vacuum interlock.

Baking and Conditioning

The vacuum chambers are baked at 200 C for 70 hours by electrical heaters. The heaters are controlled automatically. When the chambers have cooled down to about 100 C, the NEG pump heaters, the ion pumps, the titanium getter pumps and the ionization gauges are degassed. After the baking is finished, the NEG pumps are activated. The NEG activation is done till partial pressure of H2 decrease.

VACUUM STATUS

Pressure

Measured values of pressures at the bending sections have satisfied the specification very well. The history of the pressure at the bending section (BMR02 as shown in Fig.1) is shown in Fig.6. In the initial 700 hours the SR ring was filled with atmospheric air several times. In Fig.6, in period A, the SR ring was not baked and NEG pumps were not activated. The pressure was between 10E-6 and 10E-7 Pa. Between period A and period B, the SR ring was baked at 200 C for 70 hours and NEG pumps were activated. Then, we achieved the pressure of 2x10E-8 Pa. After baking, the SR ring was filled with atmospheric air twice for maintenance. After the maintenance,the pressures were reduced to between 10E-7 and 10E-8 Pa immediately (about a couple of days) without baking. The presssure at the straight sections (INF, RF as shown in Fig.1) is 1x10E-7 Pa after baking, and the pressure at the pertubater (P as shown in Fig.1) is up to 4x10E-8 Pa.

148 The Vacuum System of IHI's Compact SOR "LUNA"

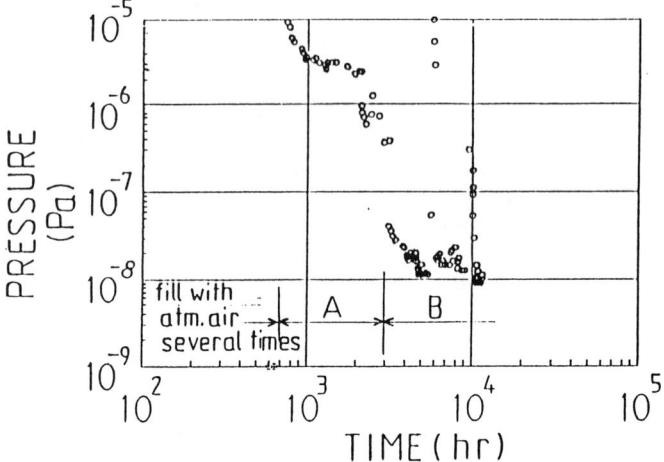

Fig.6. History of the pressure at bending section

Thermal Desorption

The thermal outgassing rate of LUNA is shown in Fig.7. It is estimated approximately from the measured pressures that all pumps are working at the specified pumping speed. In addition, the calculated pressure distribution as shown in Fig.2 was confirmed measurement. It's about 10E-8 Pa l/cm2 sec without baking. After baking, it become 10E-10 Pa l/cm2 sec. Note that both of them are less than the design values.

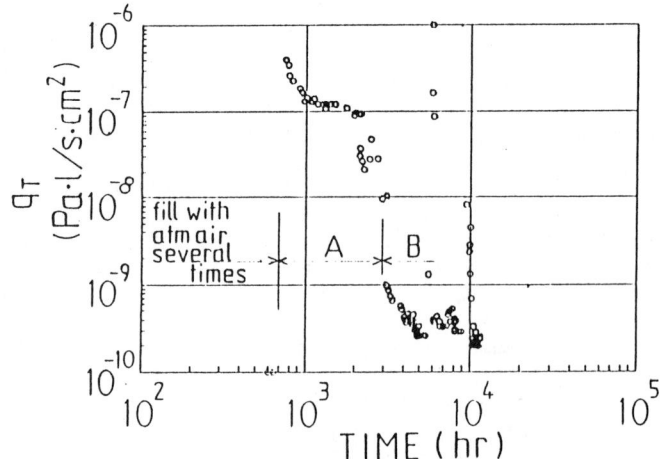

Fig.7. History of the thermal desorption.

Photo Stimulated Desorption (PSD)

The pressure at the bending section (BMR02 as shown in Fig.1) under the Synchrotron Radiation is shown in Fig.8. We think these measurements are nearly equal to the actual pressures in the electron beam channel, because the conductance between the spaces where the vacuum gauges and the beam stay, respectively, is designed to be high.

We plotted the pressure divided by the beam current as a function of time-integrated beam current. The circles and the line represent measured results and calculated ones, obtained by use of eqs.(1) through (4), respectively. The measured values almost agree with the calculated ones.

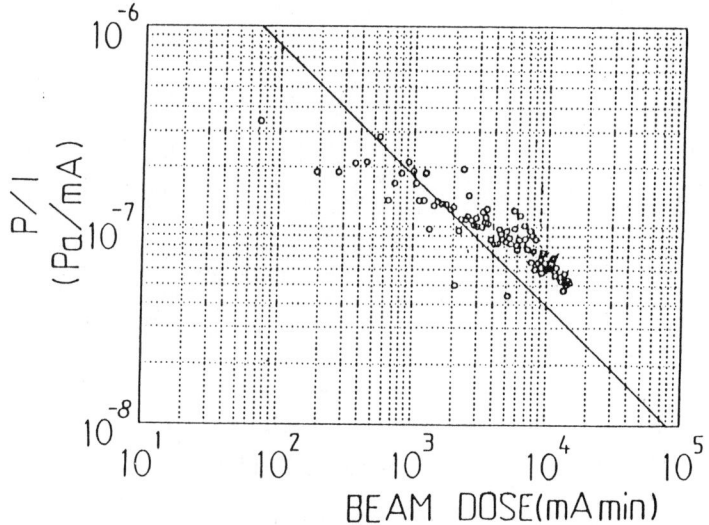

Fig.8. Pressure vs. time-integrated beam current at the bending section (BMR02 as shown in Fig.1)

We try to estimate the PSD approximately in the same way as thermal desorption. The PSD history is shown in Fig.9. We plotted PSD divided by beam current as a function of time-integrated beam current. The circles and line represent the estimated results and calculated ones, obtained by use of eqs.(1) through (4), respectively. The discrepancy between the estimated and calculated results, shown in Fig.9, indicates that more considerations about the estimation of the PSD and thermal desorption are needed. We calculated PSD as a product of measured pressure and estimated pumping speed. When we estimate the pumping speed, we have to consider the pressure and sorption quantity dependencies of the pumping speed.

We considered them for the ion pumps, but not for the NEG pumps. Namely, we assume that the pumping speed of the NEG pumps is constant. We of course know that it is not true because we have observed the pressure and sorption quantity dependencies in the preliminary experiment. However, we could not get enough data to determine the pressure dependency curve quantitatively. We think that this fact is the main reason for the above discrepancy. We plan to carry out the additional experiment in order to understand the behavior of the NEG pumps more precisely in next year.

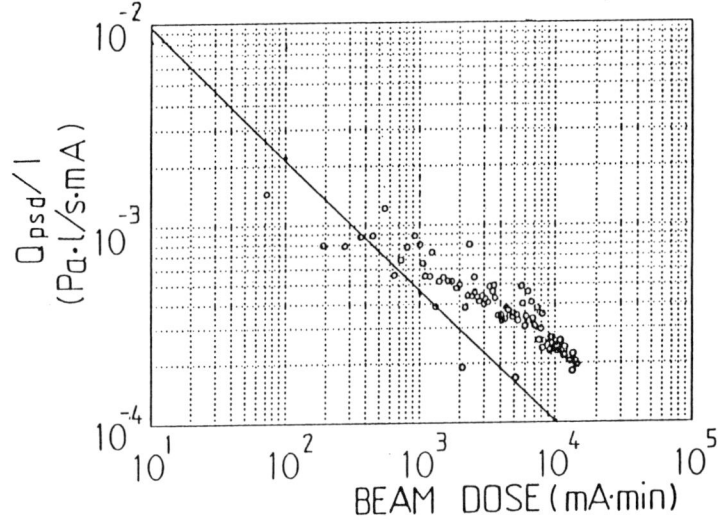

Fig.9. PSD due to beam vs. time-integrated beam current

CONCLUDING REMARKS

We have successfully established the ultra high vacuum system of the compact synchrotron. The average pressure on the whole is 1x10E-7 Pa. we adopted the conventional devices such as the triode sputter ion, NEG cartridges and titanium getter pumps because we emphasized simplicity of the maintenance. However, we also understand the necessity of better vacuum pressure below 10E-8 Pa and no baking system. We are accumulating the data concerned with vacuum and then we are going to develop the low desorption treatment of materials in near future.

REFERENCES

[1] S.Mandai,et al.,"Development of Compact Synchrotron light Source for X-ray Lithography", The 3rd International Conference on Synchrotrom Radiation Instrumentation, Tsukuba, Japan, 1988.
[2] ANL: 7GeV Advanced photon Source Conceptual Design Report,1987.
[3] ESRF: Foundation Phase Report,1987.

Submitted to the conference on Vacuum Design of Synchrotron Light Sources,
Argonne National Laboratory, November 13-15, 1990.

VACUUM CONTROL AND INTERLOCK SYSTEM FOR VUV PHOTOCHEMISTRY BEAMLINE OF HESYRL

Xilin Xu, Weimin Xu, Chanrong Yao
Hefei National Synchrotron Radiation Laboratory
University of Science and Technology of China
Hefei, Anhui 230026, P.R.China

Abstract

The paper describes the vacuum interlock system for the VUV photo-chemistry beamline. The interlock system is designed to limit the extent of damage in the vacuum failure of a beamline component. In particular it must protect the storage ring and other beamlines from the contamination, rising pressure, or catastrophic failure of one beamline. The principle and operation of the vacuum control and interlock system are presented.

1. Introduction

Hefei Synchrotron Radiation Laboratory (HESYRL) is a facility for dedicated use of synchrotron radiation which comes from a fully dedicated 800 Mev electron storage ring. It launched the first radiation light on April 26, 1989. Meanwhile photon beamlines and experimental stations have been under construction in order to keep pace with the light source. The VUV photochemistry beamline BL-U10A is installed. Fig.1 shows the side view of the beamline BL-U10A (excepting the front end) and photochemistry experimental station.

Synchrotron radiation from the storage ring is introduced through the front end of the beamline, via downstream segment to the experimental area.. Experimental beamline shares a common vacuum with the storage ring. It is necessary to protect the storage ring vacuum from contamination, or rising pressure, or catastrophic failure. A vacuum control and interlock system is designed and has been implemented for the beamline.

2. Outline of the beamline vacuum control and interlock system

The vacuum control and interlock system should have the following functions:
1) Minimize the vulnerability of the stroage ring vacuum system from the synchrotron radiation beamlines.
2) Allow individual branches of the beamline to function independently of one another.
3) Provide an electrical interface for opening & closing of the radi-

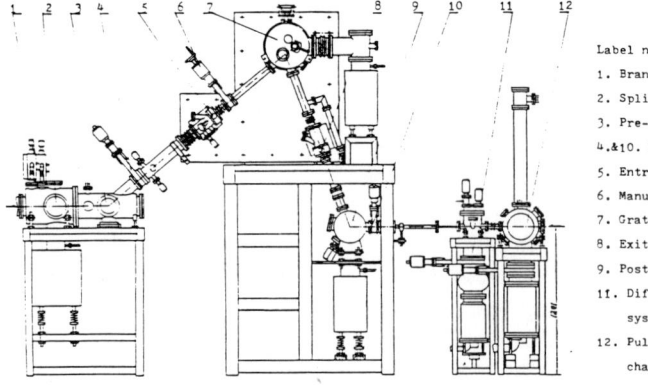

Label notes:
1. Branch beam shutter
2. Splitting mirror chamber
3. Pre-mirror chamber
4.&10. Pneumatic valve
5. Entrance slit
6. Manual valve
7. Grating chamber
8. Exit slit
9. Post-mirror chamber
11. Differential pumping system
12. Pulse molecular beam chamber

Fig. 1 Side view of the beamline BL-U10A and photochemistry station

ation protection system.
4) Display vacuum status for beamline segments and all valves states.
5) Issue warning signals to both control room and experimental station by flash light and audio alarm.

A block diagram of the vacuum control system for the beam line BL-U10A is shown in Fig.2. There are 5 vacuum guages being arranged along the beamline. These devices are monitored by a vacuum control and interlock system which causes valves to be closed automatically in the event of the vacuum problems.

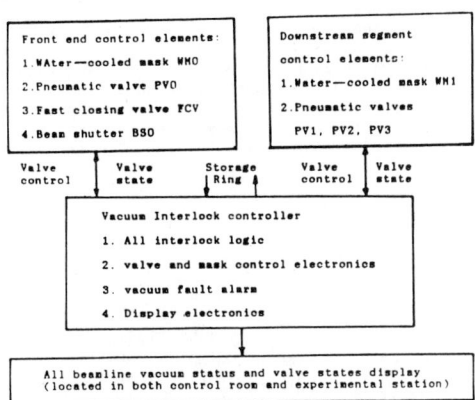

Fig.2 Block Diagram of Vacuum Control System

3. Operation of the vacuum control and interlock system

Fig.3 shows the configuration of the vacuum control and interlock system for the VUV beam line BL-U10A. At the front end, a fast closing valve FCV, which is designed by our laboratory is installed. This valve, along with its associated electronic circuit, has a closure time of 6 milliseconds. The valve does not form a perfect vacuum seal, but does minimize the gas load into the storage ring in the case of a vacuum accident. For better efficiency, a fast closing valve controller is placed as close as possible to the site of where the FCV is installed.

If vacuum failure occures and the pressure rises to a prescribed limit($1*10^{-4}$ torr or $5*10^{-5}$ torr, selected by a toggle switch on the front panel of the FCV controller) in the downstream of the beamline, a Type DL-7 vacuum gauge unit 2 (G U2) as a pressure sensor will issue an interlock signal to control logic circuit 2 (CL2), then triggles the fast closing valve FCV and closes the pneumatic gate valve PV0 as well as the water-cooled mask WM0 simultaneously.

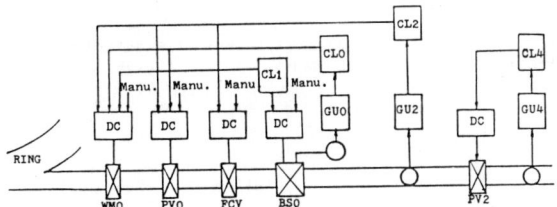

Fig. 3 Configuration of the vacuum control and interlock system
DC: Driving Circuit; CLn: Control Logic; GUn: Gauge Unit.

When the pressure in the front end area exceeds $5*10^{-8}$ torr, the cold cathode vacuum gauge unit 0 (GU0) mounted on the beam shutter chamber will give an interlock signal which is sent to control logic circuit 0 (CL0) to close the pneumatic gate valve PV0 and the water-cooled mask WM0. It is necessary to delay the interlock signal for 5 seconds for protecting PV0 from operating frequently due to vacuum instantaneous disturbance.

In the downstream of the beamline, it has a Type DL-7 vacuum gauge unit 4(GU 4) to monitor the pressure of vacuum segment for local vacuum protection.

The vacuum control and interlock system also provides to close or open water-cooled movable masks WM0 and WM1, the pneumatic gate valves PV0, PV1 and PV2, a fast closing valve FCV, and the beam shutter BS0 independently by using the push-button on the front panel of the vacuum interlock controller at the experimental station. If only the beamline pressure is equal to or lower than $5*10^{-9}$ torr the controller permits user to open valves at the front end. The controller is protected with specific lock key which is carefully maintained by the staff and users who supervise the experiment station. The vacuum control and interlock system can be actuated only if this key is inserted in the panel of the controller.

The experimental station is equipped with a display panel that indicates the position of the valves & masks and vacuum status of different vacuum segments of beamline. Fig. 4 shows the panel of the vacuum control and status display for BL-U10A photochemistry beamline.

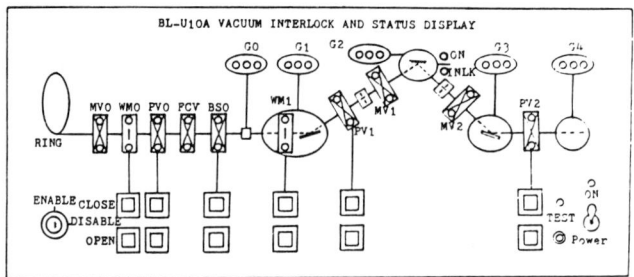

Fig. 4 The panel of BL-U10A vacuum control and status display

4. Prospect

HESYRL has four beamlines at the first stage. Each beamline owns its independent vacuum control and interlock system. However, consideration is now being given to a centralized vacuum control system for both existing beamlines and those being built in near future. The system being studied would have each beamline controlled by a microcomputer (such as PDP-11/73 or IBM-PC/AT) which would be linked to the main computer in the control room.

5. Acknowledgement

The authors would like to thank Dr. John Yang of Stanford Synchrotron Radiation Laboratory (SSRL) and Dr. Noriichi Kanaya of the Photon Factory (KEK) for their helpful discussions.

References:

[1] Y.W. Zhang et al., A photochemistry station at HESYRL, International Conference on SR Application, May 9, 1989, Hefei, Anhui, China.
[2] C.Y. Xu, X.L. Xu, Y.W. Zhang et al., Front End of Beamline for HESYRL, International Conference on SR Application, May 9, 1989, Hefei, Anhui, China.
[3] X.L. Xu, The Vacuum Interlock System for the Front End of HESYRL Beamlines, HESYRL Technical Memo. TM-0104, Jan. 5, 1987.

VACUUM SYSTEM DESIGN OF SRS INDUS-I

S.S.Ramamurthi,M.G.Karmarkar,R.J.Patel
CENTRE FOR ADVANCED TECHNOLOGY,INDORE,-452012 INDIA.

ABSTRACT

The low energy,450 MeV VUV ring INDUS-I, under construction will have a 20 MeV Microtron pre-injector, a 450/700 MeV Booster synchrotron and 450 MeV <u>Storage Ring Indus-I</u>. Vacuum system is designed to maintain a beam-on operating pressure of 1×10^{-9} mbar. Vacuum envelope is an all metal bakable system fabricated out of stainless steel (S.S) 304L material. Fabricated components are elaborately preconditioned to get hydrocarbon free vacuum. Cryo-sorption pumps and Turbomolecular pumps are used for roughing and bakeout, whereas triode ion pumps (TIP) and titanium Sublimation-pumps (Ti-SP) are used for maintaining UHV. Distributed ion pumps (DIP) built in the bending magnet (BM) chamber will pump gas molecules produced by photon induced desorption. To minimize distortion in magnetic field due to eddy current set up in dipole chamber wall, during field ramping in the <u>Booster Synchrotron</u>,specially designed thin wall corrugated chambers are used. The <u>Microtron</u> uses 2000 l/s cryopump to take care of desorption of water vapour from Microtron magnet polefaces, which form a part of vacuum chamber.

Studies were conducted (i) To investigate hydrocarbon free pumping with Turbomolecular pumps (TMP) and (ii) Electron induced desorption to estimate the gas load during electron storage. Design details of vacuum system for storage ring INDUS-I, Booster Synchrotron and Microtron are presented.

INTRODUCTION

Synchrotron Radiation Source, Indus-1 consists of a 20 MeV Microtron as pre-injector, 450/700 MeV Booster Synchrotron and 450 MeV Electron Storage Ring. Their layout and design parameters are shown in figure 1. For stored beam lifetime of more than 10 hrs,(due to gas scattering at 450 MeV and 100 mA) , the vacuum system of Storage Ring is designed to maintain pressure less than

10^{-9} mbar (1), whereas the pressure less than 10^{-7} mbar is adequate in injection system viz, Microtron, Booster and Transferlines. The vacuum chambers are fabricated out of stainless steel (S.S) material. Flange joints with copper gaskets are used for interconnection of chambers. The chambers and components forming part of UHV envelope are preconditioned by (i) ultrasonic, alkaline and acid cleaning (ii) electropolishing followed by high temperature vacuum degassing, and (iii) argon glow discharge cleaning (GDC). This will ensure hydrocarbon free clean UHV conditions in vacuum system (2,3). GDC will reduce beam conditioning time during operation of the machine, whereas high temperature degassing will help to reduce pump down time after vacuum chambers are exposed to air. Due to large linear size (about 86 meters from microtron to storage ring) and small cross section of vacuum chamber, a large number of small capacity (140 and 270 l/s) TIPs are used. The dominant gas load produced during operation of electron storage ring is the synchrotron radiation induced photo-desorption from the walls of bending magnet vacuum chambers. This is taken care by installing distributed ion pumps in BM chambers of Indus-I. Forced cooling of storage ring vacuum chamber is not envisaged since linear power density is rather small (60 w/m) . Penning gauge and BA gauges are used for measurement and control of pressure, whereas Residual Gas Analyzer (RGA) and compact helium leak detector cells mounted at some locations will help in monitoring the composition of residual gases and leak hunting during operation of machine. Ion clearing electrodes are mounted inside the vacuum chamber of storage ring to mitigate ion trapping problems.

This paper describes the basic design of vacuum system of SRS facility and design and construction features of dipole vacuum chambers . In addition results of (i) clean pumping studies with TMP and (ii) electron induced desorption studies are also presented.

VACUUM SYSTEM DESIGN

<u>INDUS-I.</u>
450 MeV 100 mA dedicated synchrotron radiation source Indus-1 has circumference of 18.96m with 4 BM, each having 2 beam ports. To obtain the required low pressure ($<10^{-9}$ mbar), desorption due to Syn.radiation and thermal outgassing rate must be below certain

definite value (4). [average number of molecules desorbed by photon D_E = 2.5 x 10^{-6} and specific thermal outgassing 5 x 10^{-12} torr l/s /cm^2]

To minimize the thermal outgassing the vacuum system is designed as all metal bakable system (up to 250° C). It is reported (5), that the gas scattering beam life time depends on atomic mass number (Z) of the residual gas and with higher Z, life time reduces exponentially. Hence all attempts are made in vacuum chamber construction, pre-treatment and pumping system design so that high Z gases (CO, CO_2, N_2) are minimized in the chamber.

In order to find the pumping speed for maintaining the pressure 1×10^{-9} torr at 100 mA, the total outgassing rate is to be estimated. Outgassing is due to both thermal release of molecules from the surface of vacuum chambers and desorption of molecules induced by synchrotron radiation. The release of molecules /sec. can be computed using empirical relations (6). For Indus-1 it works out to be 7.1956 x 10^{13} mol/sec. which can be expressed as 2.248 x 10^{-6} torr l/s.

Thermal outgassing from stainless steel type 304L baked to 250°C for 24 hrs is assumed to be 5 x 10^{-12} torr l/s/cm^2 at room temperature. Since the total surface area in Indus-I is 12.6×10^4 cm^2 the rate of outgassing from the system is

Q_{TH} = 6.3 x 10^{-7} torr l/s

Thus total outgassing rate at 100 mA works out to

Q_T = Q_{SR} + Q_{TH}

= 2.248 x 10^{-6} + 6.3 x 10^{-7}

= 2.88 x 10^{-6} torr l/s

In order to realize pressure of 1×10^{-9} torr in indus-I, the total pumping speed required at beam current of 100 mA, works out to be 2880 l/s.

Much of the gas load due to the Synchrotron radiation occurs in the bending magnet (B.M) chambers and as such distributed ion pump (pump using fringing magnetic field of B.M) installed along the whole length of each of B.M vacuum chamber would be highly advantageous. However attainable pumping speed of DIP is governed by space and conductance limitations. Available space in the dipole chamber will limit DIP speed to about 250 l/s at 10^{-9} torr. Fig-2 shows DIP in vacuum chamber in B.M and one module respectively. The anode cell diameter

10.0 mm and length 10 mm is chosen. It is of conventional diode design and based on semi-empirical relation (7).

The space available for mounting of triode sputter ion pumps is very limited in the ring. Pumping system consists of Triode Sputter Ion pumps and Titanium Sublimation pumps installed at 9 locations (atleast 2 in each straight sections). The total installed capacity of TIP, Ti-SP & DIP together gives pumping speed of 8200 l/s. The pump locations and conductance limitations will give effective pumping speed of 2880 l/s. .

BOOSTER SYNCHROTRON

The Booster Synchrotron is designed to accelerate electrons to energy up to 450 MeV for Indus-1 from 20 MeV. The injection into the synchrotron will be carried out with 20 MeV Microtron. The Booster consists of 6 dipoles and 6 pairs of Quadrupole Magnets. Because of the short stay of the beam in the synchrotron a vacuum of the order of 10^{-6} mbar will be adequate. The total internal surface area of Booster vacuum chamber is 24 m sq. To estimate the pumping speed it is necessary to take into account the low value of conductance, specially in bends. It has been designed to install total pumping capacity of 4050 l/s distributed over 12 pumping ports (including RF cavity, Injection and Extraction septum chambers, kickers).

MICROTRON

A 20 MeV Microtron is used as pre-injector (8). The vacuum envelope of microtron is made up of carbon steel magnet poles (surface area : 14176 cm^2) The total volume of the microtron chamber is approximately 86 liters. Apart from these, RF cavity, waveguide, emitter, etc. are also housed in this vacuum envelope, which degas and add to the gas load. The ultimate pressure of 10^{-7} torr is to be maintained with the electron beam, so that the cavity life is prolonged . Considering outgassing rates and relative areas of materials exposed to vacuum, the carbon steel magnet poles govern the design of the vacuum system. A system for mild bake out of magnet poles is planned to reduce pump down time. The major load of a mild baked system is that of water vapour. Hence, it is beneficial to have cryopump and turbomolecular pump combination for pumping, with an isolating gate valve for protection against power failure. To attain ultimate pressure of 10^{-7} torr, close cycle helium cryopump of 2000 l/s capacity and one TMP of 500 l/s capacity are provided.

DESIGN AND FABRICATION OF VACUUM CHAMBERS

Vacuum chambers are fabricated out of stainless steel (AISI 304L). Indus 1 bending magnet chamber is a box type design with two tangential ports for tapping Synchrotron Radiation (9). The chamber has two compartments, one for the beam and other for DIPs. Fig 3 shows the chamber. It is fabricated by machining 20 mm thick S.S plates in two halves,and welded by TIG process. Seamless S.S.Tubes of 69 mm ID form vacuum chamber in the straight sections. S.S.bellows provided in straight sections take care of thermal expansions during bakeout and offer flexibility in assembly of chambers.

The most difficult part of the booster vacuum system was to design and fabricate bending magnet vacuum chamber. A thin wall chamber is required to minimize distortion of magnetic field due to eddy currents set up during field ramping. In addition, oblong cross section of chamber is necessary to accommodate it in the pole gap of C-shaped bending magnet. The chamber has to be rigid against external atmospheric pressure.

A stainless steel bellows type chamber (with 0.3 mm wall thickness) having race track cross section bent to 1800 mm radius has been designed and developed (10). Fig.3 shows the chamber. Number of designs were considered viz (i) External rib type structure (ii) corrugated chamber with unequal depth of corrugations on inner and outer radii (iii) close pitch bellows type chamber. Sample length of each type was fabricated. Due to uncertainty in the quality of spot welding /furnace brazing, rib type construction was discarded. Second type corrugated chamber produced with unequal depth denting, resulted in a single piece construction (1800 R and $60°$ sector), however the mechanical rigidity under external pressure was not satisfactory (chamber deflection under vacuum was of the order of 3mm). The third type ie, closely pitched bellows type chamber was fabricated and with three joints a full bent sector (1800 R and $60°$) was obtained. The close pitch corrugations were generated in two stages, first wide spaced corrugations using rubber expanding tools were generated in a 0.3 mm thick S.S tube and then these were compressed with external metallic dies to get closely pitched bellows type configuration .Three such pieces were welded with their flared ends tilted such that after welding a 60° section is formed. All 6 chambers have been successfully fabricated and tested.

EXPERIMENTAL INVESTIGATIONS

STUDY FOR CLEAN PUMPING WITH TURBO MOLECULAR PUMP [TMP]

The TMP, in principle can be switched on directly from atmospheric pressure. However, the design of the TMP is such that at higher inlet pressure the rotor runs at low rpm (thereby reducing the compression ratio and pumping speed) and it takes in our system about 3 to 5 minutes to attain full rated rpm. (volume is about 10 times the pumping speed) During this reduced rpm run, there is always a possibility of back diffusion of oil vapours from TMP bearings and back up pump (11). (usually an oil filled rotary pump) resulting in the oil contamination of the system.

The simpler method during the start-up is to independently pump down the chamber to 10^{-2} torr through cryosorption pumps and then connect the TMP rotating at its full speed, to the chamber. An experiment was conducted using 170 l/s TMP which was operated with two conditions viz.

1. TMP directly connected to the system at atmospheric pressure
2. TMP connected to the system only after the attainment of full speed and the system pumped to 10^{-2} mbar pressure (this initial pumping carried out by cryosorption pumps)

The experiments were conducted with a system baked at 200°C for 4 hrs. Partial pressure of residual gases in the system were measured with Quadrupole Mass Analyzer. The results are tabulated below.

Experimental results :
Condition -1.
(TMP started directly from At.pressure)

Mass No	Without Bakeout (partial pressure	with bakeout (200°C/4hrs) in mbar)
H_2	1.74×10^{-8}	0.88×10^{-8}
C	2.23×10^{-9}	6.31×10^{-10}
CH_4	2.14×10^{-8}	1.78×10^{-9}
H_2O	1.13×10^{-6}	2.43×10^{-8}
CO/N_2	2.55×10^{-8}	2.57×10^{-9}
O_2	4.53×10^{-9}	0.64×10^{-10}
CO_2	2.66×10^{-8}	1.71×10^{-10}
Total pressure	9×10^{-7}	4.5×10^{-8}

Condition -2 (full speed TMP connected to a system pre roughed to 10^{-2} mbar by cryosorption pump).

Residual gas	Without bakeout [Partial	With bakeout 200°C/4 hrs pressure in mbar]
H_2	3.53×10^{-8}	0.88×10^{-8}
C	3.94×10^{-9}	5.83×10^{-10}
CH_4	2.73×10^{-8}	1.65×10^{-9}
H_2O	1.13×10^{-6}	2.24×10^{-8}
CO/N_2	5.5×10^{-8}	2.11×10^{-9}
O_2	1.07×10^{-9}	0.71×10^{-10}
CO_2	2.99×10^{-8}	0.98×10^{-10}
Total pressure	9×10^{-7}	4.25×10^{-8}

From the data obtained on partial pressures it is seen that the partial pressure with and without sorption pump ie, starting TMP at atmosphere and 10^{-2} mbar is not appreciably different up to 10^{-8} mbar pressure. However if one critically examines the data, one may conclude that the difference, though marginal, the effect on beam lifetime would be appreciable.

The results can be summarized as follows :

Residual gas	Starting from atm. [partial pressure in mbar]	starting at 10^{-2} mbar	difference
CH_4	1.78×10^{-9}	1.65×10^{-9}	0.13×10^{-9}
CO/N_2	2.57×10^{-9}	2.11×10^{-9}	0.46×10^{-9}
CO_2	1.71×10^{-10}	0.98×10^{-10}	0.73×10^{-10}
Total pressure	4.5×10^{-8}	4.25×10^{-8}	0.25×10^{-8}

A large number of partial pressure readings (with experiment repeated for about 5 times) were taken and it can be concluded that for applications upto 10^{-9} mbar regions. (for transport lines, Booster etc) there is no appreciable problem even if TMP is operated directly at atmospheric pressure. However for applications in 10^{-10} mbar range (storage ring) where even slightly higher partial pressure of mass 44 and above is not tolerated, the system may be first roughed to 10^{-2} mbar with oil free pumps (Cryosorption pumps etc) and then a full speed TMP can be connected for further pump down.

ELECTRON INDUCED DESORPTION STUDIES_TO OPTIMIZE PRECONDITIONING OF VACUUM CHAMBERS

Difficulty in maintaining the vacuum in a storage ring, arises due to the various interactions of synchrotron radiation with the walls of the chamber. The main processes are.
i) The photons striking the walls, increase the wall temperature, causing thermal outgassing.
ii) Photons induce desorption of gases from the wall.
iii) Secondary electrons produced by the photons, desorb gases from the walls.

The first one of the above, can be reduced by baking the chamber to sufficiently high temperature before operation. The second effect is found negligible. The third effect namely the effect of photo electron induced desorption of gases from the chamber walls is most dominant (12).

The gas load due to photoelectron induced desorption depends on surface condition of the chamber. To optimize the preconditioning requirement and to confirm that the number of molecules desorbed per electron is in the range of 10^{-6}, an experiment was performed. A chamber with suitable pumping system was used for this experiment. The Tungsten wire, biased at 200 Volts is used as the source of thermionic electrons. The maximum energy of the secondary electrons produced by the Synchrotron radiation in indus-1, impinging on the walls of the chamber is 200 eV. The filament is flashed for a known time and the pressure is noted. The experiment was conducted with three surface conditions of the chamber; viz,
  i) Chamber given only a detergent cleaning and wiped with acetone.
  ii) Chamber is chemically cleaned (alkaline bath and pickling) and electropolished.
  iii) Chamber cleaned as in (ii) and subjected to argon glow discharge.

The rough estimate of the desorption rate is worked out : Condition I: 10^{-1} mol/ele. Condition II: 6×10^{-3} mol/ele. and Condition III: 5×10^{-4} mol/ele.

Further Investigations are being carried out to bring down the desorption rate to the desired value of 10^{-6} adopting high temperature argon glow discharge cleaning.

PRESENT STATUS

Fabrication of various components of Indus-I, Booster Synchrotron, Microtron and transferlines is at various stages of completion. Booster vacuum envelope with all 6 bellows type chambers has been assembled, tested and vacuum runs are in progress. 2 numbers of dipole chambers for Indus-I have been fabricated and a quadrant with one dipole chamber has been assembled and UHV of the order of 5×10^{-10} is realized. A prototype distributed ion pump module has been fabricated. This is assembled in one of the dipole vacuum chambers of Indus-I with magnetic field provided by ceramic magnets, the DIP alone creates a vacuum of the order of 10^{-9} in the SRS chamber.

CONCLUSION

The vacuum rings are being made operational prior to the commissioning of the synchrotron and Storage Ring to gain working experience in UHV region. However extensive experimental studies and careful construction of vacuum envelope will be essential to understand the photon induced desorption problems, typical with SRS installations.

REFERENCES

1. ESRF Vacuum workshop Rep.1987, BP220 Grenoble.
2. Cleaning and surface Analysis of S.S UHV chambers by Argon Glow Discharge CERN/ESR/VA/77-59, Sept.77
3. EPA Initial vacuum performance and Beam life time with electrons . CERN /PS/87/42MC, A.Poncet.
4. Ultra high vacuum technology for Synchrotron radiation.F.Fisher, IEEE NSc Vol.NS24 No.3 1977 June.
5. Design study of SRS Daresburry,UK DL/SRS/R-2/ 1975
6. SOR-Ring INS -TH -108 Construction and performance, June.76.
7. H Hartwing, J.S.Kouptsidis, JVST2 1154 (1974)
8. Internal report on 20 MeV Microtron H.C Soni et al
9. Internal report on design and fab.of BM vacuum chamber for storage ring Indus-I,M.G.Karmarkar et al
10. Internal report on Mech.design and fabrication of BM vacuum chamber for Booster synchrotron P.K.Nema M.G.karmarkar et al.
11. TMP & their limitations ,J Henning , Advanced Syn.light source conference, BNL Proc. May.1988.
12. A.G.Mathewson, CERN, X Italian National Congress on vacuum science and technology, Italy,1987.

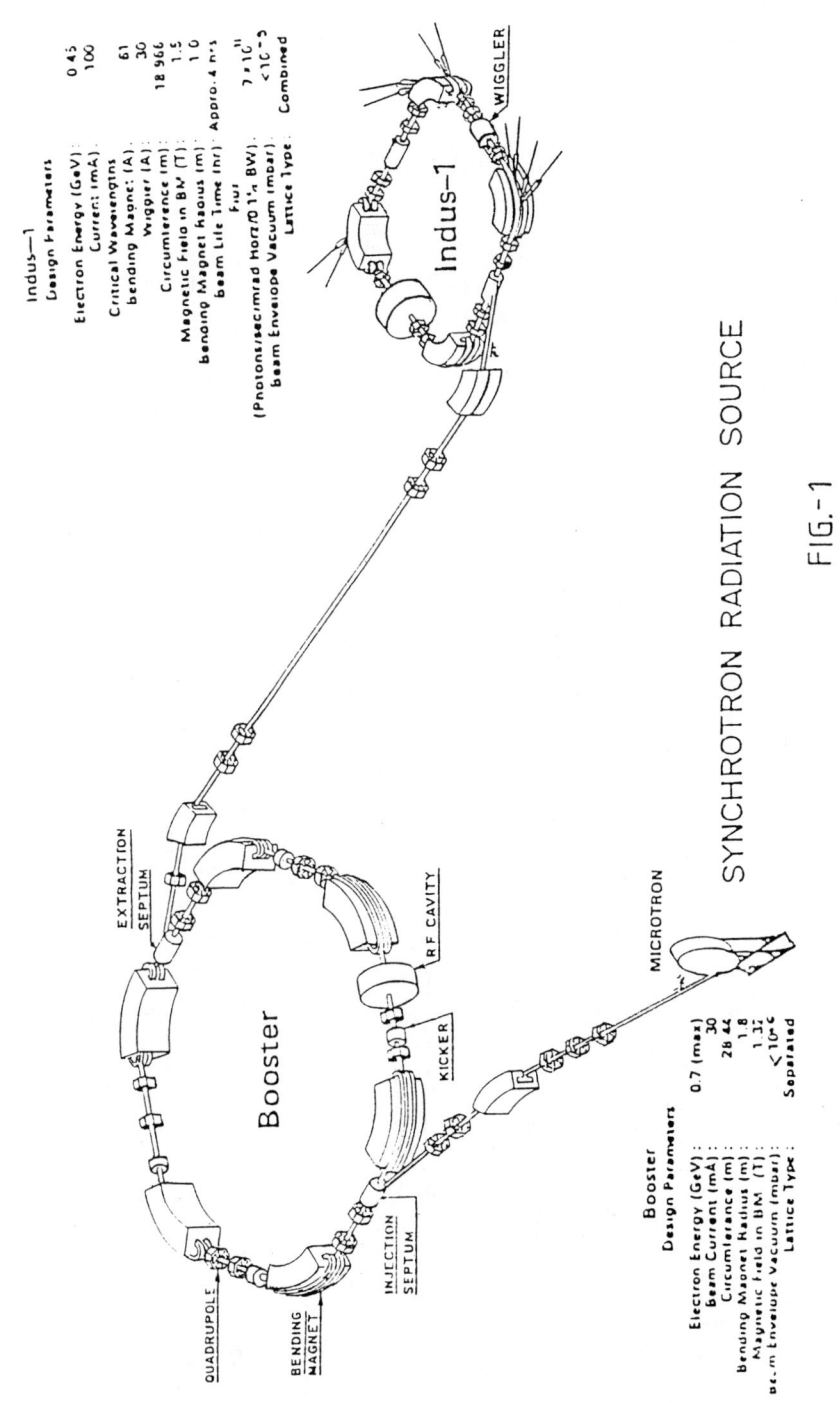

FIG.-1 SYNCHROTRON RADIATION SOURCE

166 Vacuum System Design of SRS Indus-I

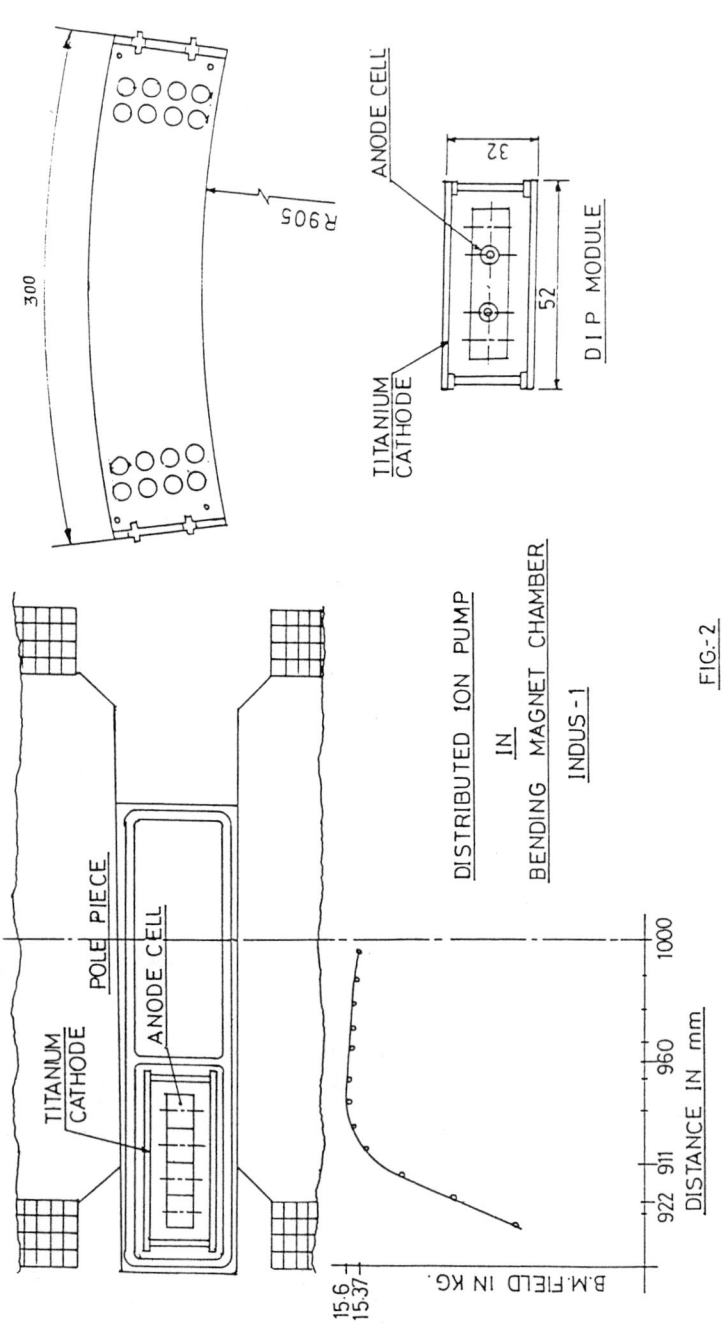

FIG.-2

All dimensions are in mm

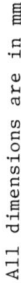

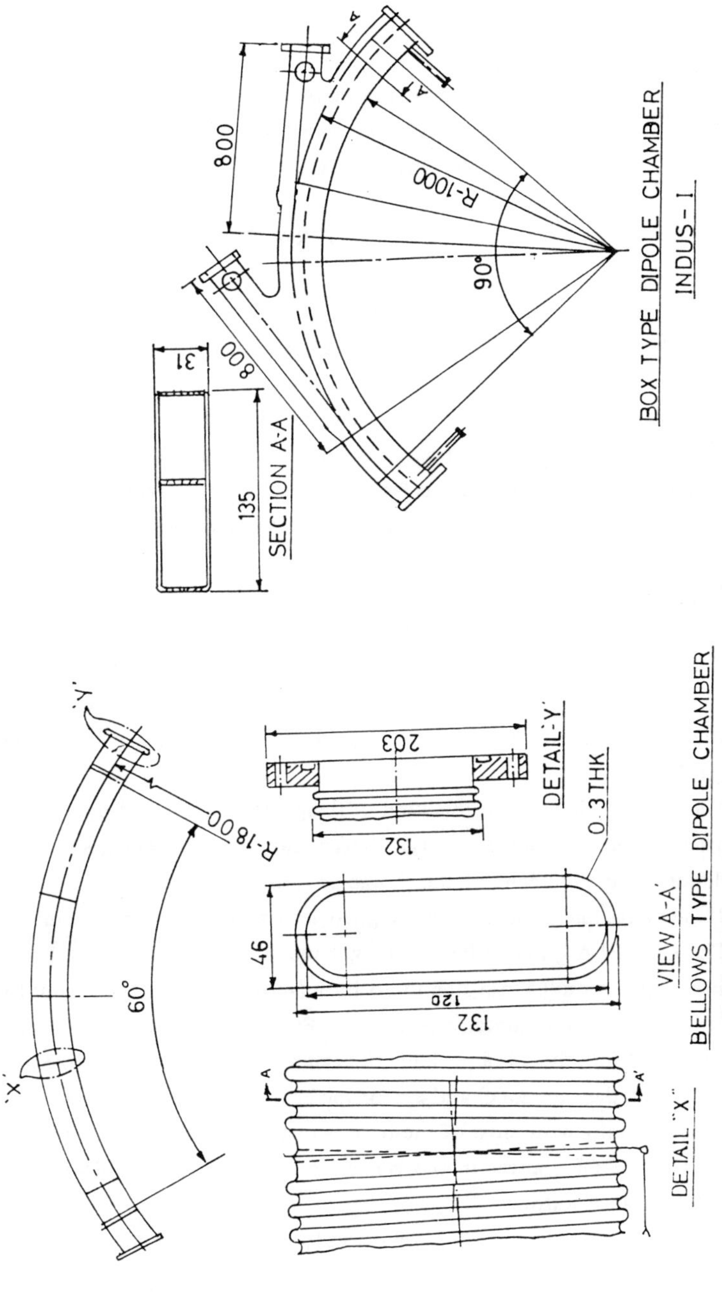
BOX TYPE DIPOLE CHAMBER
INDUS-I
800
800
R-1000
90°
A
A
31
135
SECTION A-A
BELLOWS TYPE DIPOLE CHAMBER
BOOSTER SYNCHROTRON
'X'
'Y'
60°
R-1800
DETAIL 'X'
A
A'
VIEW A-A'
46
120
132
0.3THK
132
203
DETAIL-'Y'

FIG.-3

COPPER WASHER POSITIONER FOR CONFLAT FLANGE

Guihe LI

Hefei National Synchrotron Radiation Lab.
University of Science and Technology of China
Hefei, Anhui, 230029, China

ABSTRACT

This paper introduces the function of copper washer positioner for CF flange. Vacuum chambers for particle accelerators are usually arranged in the horizontal plane and their flanges are usually installed vertically. It is difficult to hold copper sealing washer accurately in position, which can influence the vacuum sealing function. The copper washer positioner performs a double function. It can set a limit to the position of the copper washer and also check the flange for sealing before fastening the bolts. We have raised the effectiveness of installation and reduced the rejection rate of copper washers down to zero.

1. INTRODUCTION

The CONFLAT seal applys the cutting edge principle to obtain a reliable vacuum seal. It is an all-metal sealing flange (which is called CF flange) made by the Varian Company for ultra-high vacuum systems. The flange is widely used, can be baked, and provides a reliable seal.

The sealing reliability of this kind of flange depends on not only the design of its sealing structure, but also on the material chosen and processing craft, and on the quality of the installation. The alignment of the flange components determines the effectiveness of the seal. If the flange and washer are misaligned, a reliable seal can not be made.

The CONFLAT flange has very tight manufacture tolerances. The spacer of 2 mm thickness depends on 0.5 mm step to position (shown in Fig1.(c)). It is not very difficult for the flanges to use horizontally (shown in Fig1.(a)). But when they are used vertically (shown in Fig1.(c)), there is a problem in keeping the spacers and flanges in the proper position. However, the vacuum chambers for the beam current in particle accelerators are all installed horizontally with most of the flanges installed vertically. Ensuring the successful installation of the CF flanges in the vertical orientation is a problem in the vacuum engineering of the accelerator.

Until now, there were only a few ways to hold the copper washer in the position, but they are all not satisfactory or perfectly reliable.

For the need of the accelerator engineering, we have made a new copper washer positioner.

2. THE PRINCIPLE and MANUFACTURE of the COPPER WASHER POSITIONER

It has been realized that the misalignment of the flange is a result of the misalignment of the washer. So if the problem of the washer in proper position is solved, the reliability of flange can be made. The aligned flange obeys the axial symmetry principle, and the washer and flange must be an ideal coaxial body.

Based on this, a special copper washer was designed. It consists of two concentric arc steps. The inner radius equals the copper washer external diameter and the height of the step is 1 mm. The outer radius equals the flange external diameter and the height of the step is 2 mm. The structure is illustrated in Fig2.

The copper washer positioner has the distinctive feature and is used conveniently and reliably. It has two functions.

(1). It is used to set a limit to the position of the washer accurately.

(2). It is used to check the proper position of the flange and washer.

3. APPLYING RULE of the COPPER WASHER POSITIONER

Before the copper washer is placed, two bolts are penetrated in the proper position. The position can be found based on the following table. The central angle between the centers of these two bolt holes can not be less than 90°. For example, four bolt holes must be kept vacant between these two holes through which the two bolts are passed for CF150 flange. The central angle between the centers of the two bolt holes is $18 \times (4+1) = 90°$.

Table 1.

Flange mod.	Central angle between the centers of adjacent two bolt holes(deg.)	Vacant bolt holes' number
CF35	60	1
CF63	45	1
CF100	22.5	3
CF150	18	4
CF200	15	5

The steps of installing the flanges are as follows.

(1). The two flanges are laid oppositely, then two screw bolts are penetrated at the lower parts of the flanges according to the above table. The gap between the lower parts of the flanges is then adjusted to 2–3 mm. (Fig3.(a).(b)).

(2). The copper washer is placed between these two flanges from above and held on the screw bolts (shown in Fig3.(b).(c)).

(3). The copper washer is held up with the positioner and its outer arc mated with the lower edge of one of the flanges (as shown in Fig3.(d)). The other flange is adjusted until it mates with the washer (as shown in Fig3.(e)) and the copper washer is accurately positioned.

(4). Continuing the process, the two bolts are screwed by hand, and another bolt is penetrated into the higher parts of the flanges and screwed by hand.

Through the steps as above, the proper positioning of the washer and flanges has been established. The copper washer positioner must be used again to examine the entire seal for reliability.

(1). When the copper washer positioner is entered into the gap between the two flanges from three directions, and the positioner's outer step can mate with the edge of any one of the flanges, then the flanges are accurately coaxial with the copper washer and there is not any mistake in the installation.

(2). Another function of the copper washer positioner is as a thickness gauge. When the copper washer is positioned properly, the gap between the two flanges is 1 mm and the copper washer positioner can be inserted into the gap. If there are some difference in the gap's dimensions between the three directions, the copper washer is not properly positioned. If the copper washer positioner can not be inserted into the gap between the two flanges, the thickness of the copper is less than 2 mm or the height of the flange's cutting edge is more than 0.5 mm, then the copper washer must be replaced.

The copper washer positioner performs a double function, which is to set a limit to the position of the copper washer and flange, and to check for the sealing function of the copper washer and flange. So it should be realized that the hidden troubles can be found before the screw bolts are tightly fastened.

4. CONCLUSION

The copper washer positioner is easily made, is convenient to use, and is a necessary instrument for the installation of CF flanges. If it is used, the efficiency of the installation can be increased and the reliability of the vacuum seal can be assured.

The copper washer positioner was made by the writer on Oct. 11 1985, and was used in the installation of the vacuum system for the KEK–PF in Japan for the first time. It was very successful in this installation.

After more than three years, the copper washer positioner was used again in the installation of the vacuum system for HESYRL. We achieved a very reliable seal in the construction of the HESYRL vacuum system in a short time.

ACKNOWLEDGENENT

Thanks to Mr. M.Kobayashi in KEK–PF, Mr. X.Q.Wang and Mr. J.P.Wang in HESYRL for their helpful assistance.

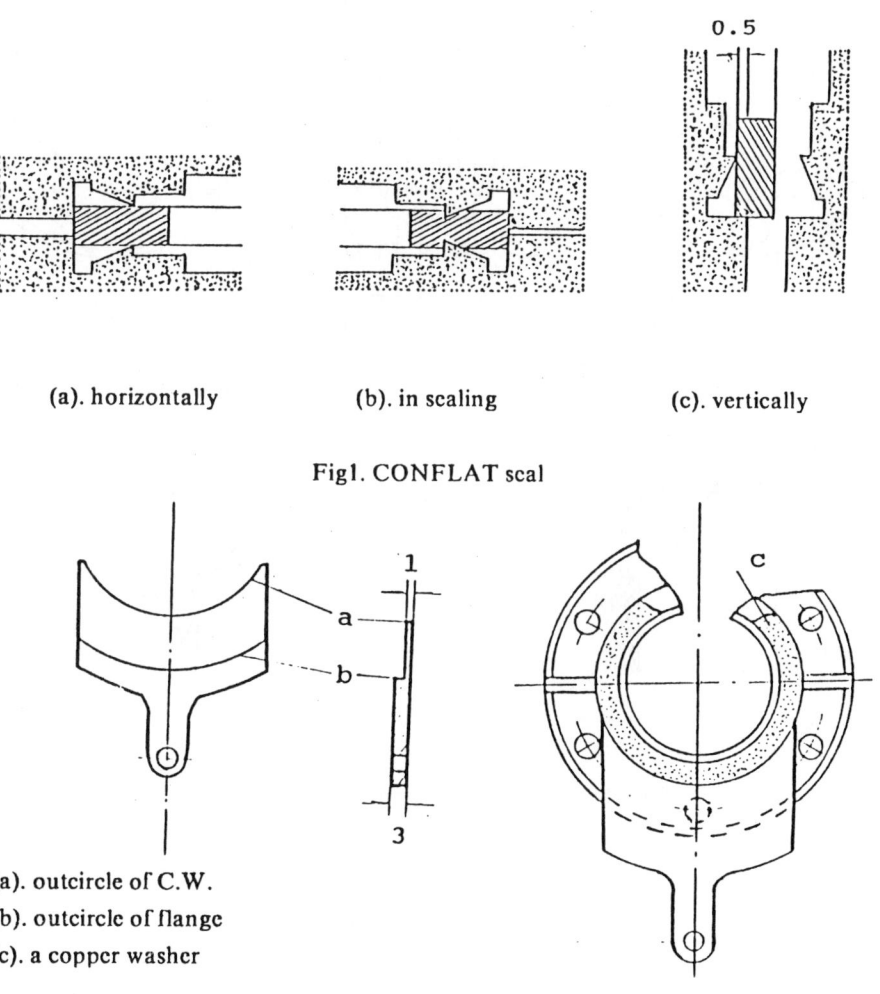

(a). horizontally (b). in sealing (c). vertically

Fig1. CONFLAT seal

(a). outcircle of C.W.
(b). outcircle of flange
(c). a copper washer

Fig2. Scheme for copper washer positioner

172 Copper Washer Positioner for Conflat Flange

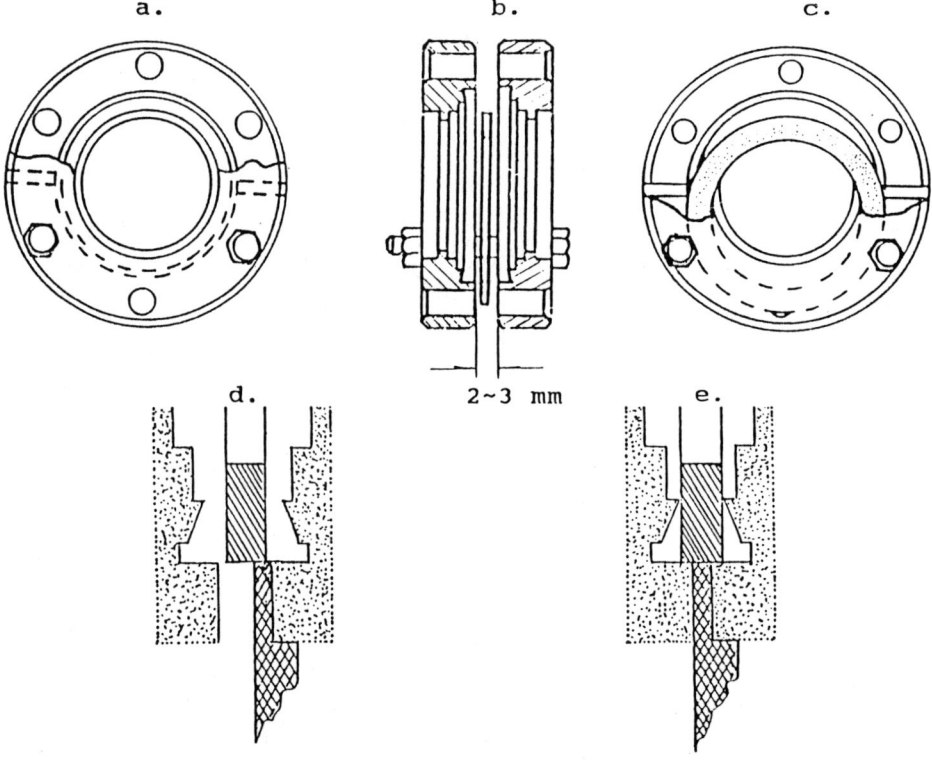

Fig3. Scheme for the steps of installing flanges

SRRC VACUUM SYSTEM

Y.C. Liu[a] and J.R. Chen[b]
Synchrotron Radiation Research Center, No.1 R&D VI, Hsinchu Science
Based Industrial Park, Hsinchu, Taiwan 30077, Republic of China

ABSTRACT

The design, fabrication and primary testings of the SRRC vacuum system are described. Aluminum-alloy is chosen as the vacuum chamber material due to its high thermal conductivity, easy fabrication, low outgassing rate and other benefits. A computer simulation for the outgassing rate and pressure distribution was made to determine the locations and pumping speeds of the pumps. For evacuation, oil-less turbomolecular pump systems, ion pumps and getter pumps are used. A "concentrated+DIP" pumping method is applied in the bending chamber to get efficient pumping. An oil-less machining process and a special extrusion method are applied to fabricate the bending chamber and the straight chamber, respectively. Surface analysis method showed that the surface made by these processes are less contaminated. A static vacuum of $<1 \times 10^{-10}$ Torr was reached in an one-sixth ring vacuum system after a bakeout at $\sim 130\,°C$ for 24 hours. In order to minimize deformation, careful welding processes are performed. Smooth beam duct cross section is designed for reducing RF impedance.

INTRODUCTION

In order to meet the high-quality requirement of a synchrotron light source, every component of the facility should have the proper function. It is essential that the vacuum system offer clean and comfortable space for the circulating electron beam. This is an important factor that must be considered carefully in the design, construction and the operation stages. First of all, a good vacuum system must reduce the gas molecule density in order to minimize the collision events between the electron beam and gas molecules, which would shorten the beam life time as well as degrade the beam quality.[1-3] The beam duct should not be changed abruptly in order to minimize the RF impedance of the vacuum chamber. For an electron storage ring, the vacuum chamber will suffer a high intensity synchrotron radiation. The chamber material and cooling method must be chosen carefully to avoid the melting down or deformation of the vacuum chamber.[4]

Due to many benefits of aluminum alloy, e.g. high thermal conductivity, easy fabrication, low outgassing rate etc.,[5-7] the material is chosen as the vacuum chamber material of the 1.3GeV electron ring of the Synchrotron Radiation Research Center (SRRC). A concept of an oil-less

a) Department of Physics, National Tsing-Hua University, Hsinchu, Taiwan 30043, Republic of China.
b) also Institute of Nuclear Science, National Tsing-Hua University, Hsinchu, Taiwan 30043, Republic of China.

system, an oil-less machining process and oil-less pumping systems, was adopted. In addition to that, the requirements on less deformation and less displacement after welding or during in-situ bakeout process were also considered in the system design. After prototype testings,[8,9] a sector (one-sixth ring, 20 m) of the SRRC ring vacuum system has been fabricated and tested. The design, fabrication and testing results of this system are described in this paper.

DESIGN AND FABRICATION

A. Beam life time and average pressure

A major goal of the design of the ring vacuum system is to offer low pressure which will meet the 10 hour beam life time requirement of the SRRC 1.3 GeV storage ring. Under this criterion on beam life time, we have calculated that an average pressure of $<1 \times 10^{-9}$ Torr has to be reached with beam on. In order to determine the location, quantity and pumping speeds requirements on pumps, a computer program has been established in this lab to calculate the outgassing rate and pressure distribution around the ring.[8,10] Figure 1 shows schematically the pump locations

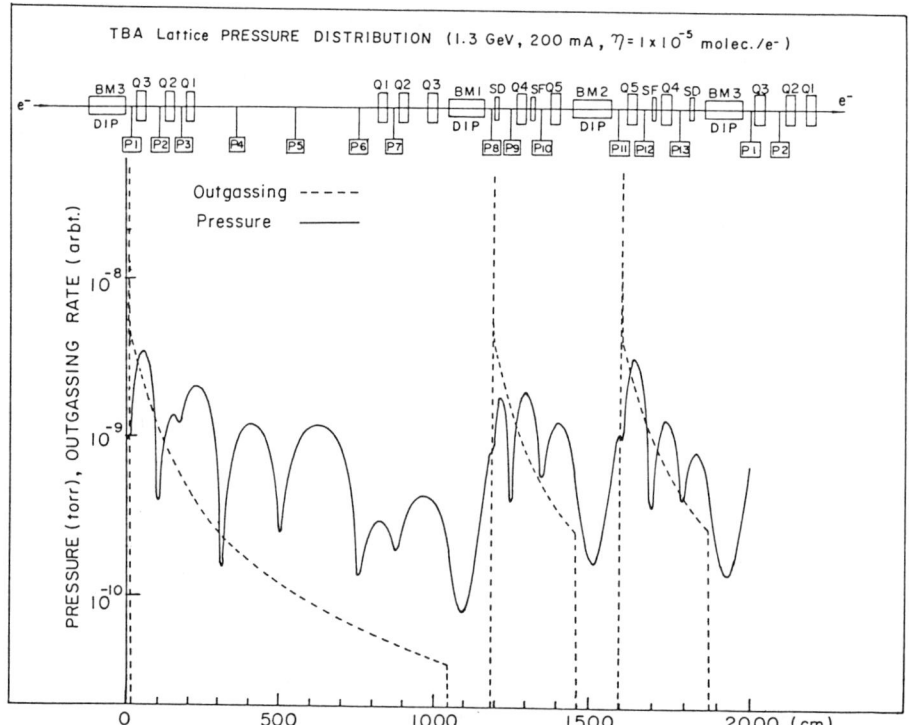

Fig.1 Pressure distribution curve. The average pressure is 1×10^{-9} Torr. P1-P13 : pumping stations.

and the results of outgassing rate and pressure distribution. Under this pumping configuration, the average pressure can reach $\sim 1 \times 10^{-9}$ Torr after an accumulated dosage of ~ 100 Ah (a photon desorption coefficient of 1×10^{-5} molec./e$^-$), which will meet the requirement of 10 hour beam life time.

B. Vacuum components

There are six long straight sections in the SRRC 1.3 GeV storage ring. Two sections are used for injection and RF acceleration purposes, the remaining four sections will be used for insertion devices after the machine is commissioned. In this report, the vacuum system is related to the phase-I machine that has no insertion device in the long straight section.

There are many components in the one-sixth ring vacuum system. The major components are vacuum chambers : the bending vacuum chamber (B-chamber, $\times$ 3) and the straight vacuum chamber (S-chamber, $\times$ 7). Figure 2 and figure 3 show the schematics of a B-chamber and a S-chamber, respectively. Four kinds of pump are adopted, they are ion pumps (IP, lumped and distributed), getter pumps (TSP and NEG), oil-less turbomolecular pumps (TMP) and sorption pumps. The first two are used in the stage of normal operation and the last two are used in the stage of

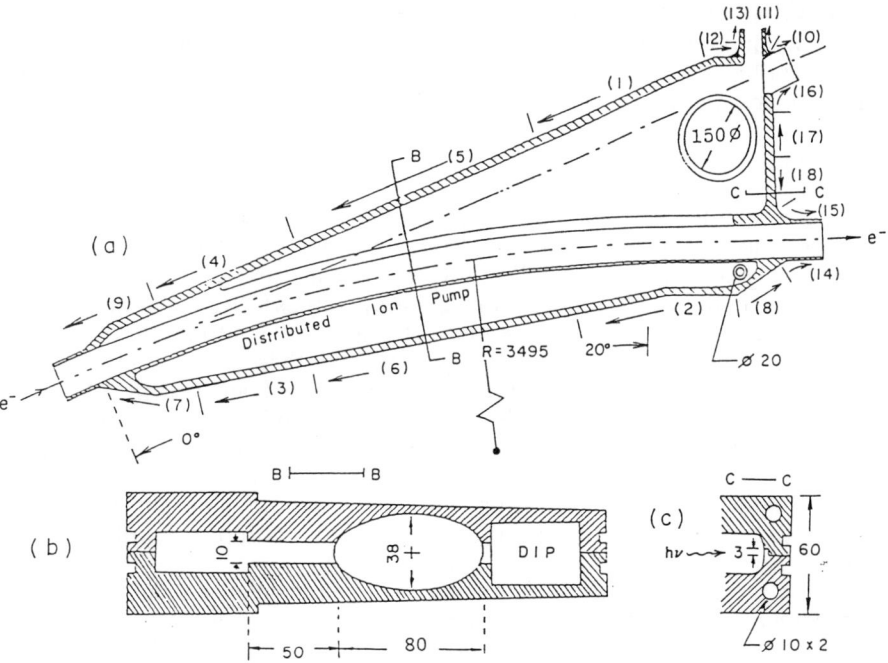

Fig.2 Schematic drawings of a B-chamber, top view (a) and cross section views (b)&(c). The numbers (1) to (18) are the welding procedures for the circumference welding.

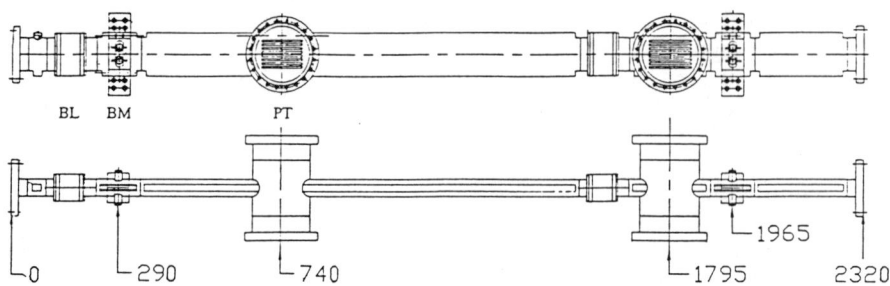

Fig.3 Schematics of a typical S-chamber. In this chamber, there are bellows (BL), beam position monitors (BM), and pumping ports (PT).

roughing. During in-situ bakeout only turbomolcular pumps are used for the evacuation. There are three kinds of valve in the system : 100 CF all metal gate valve with RF contact ($\times$ 2) for dividing the ring into sections, 150 CF O-ring gate valve ($\times$ 4) for isolating turbomolecular pumps from the ring vacuum, and a 35 CF all metal angle valve for isolating sorption pumps ($\times$ 2) after the roughing stage. Two kinds of vacuum sensor, ionization gauge and quadrupole mass spectrometer, are used in the system for measuring total pressure and partial pressure, respectively.

In addition to the previous items, beam position monitors ($\times$ 8), aluminum bellows (with spring-finger RF contact, $\times$ 14), supporting frame and other components are also included in the system.

C. Vacuum chamber

In order to remove the synchrotron radiation heat from the vacuum chamber easily and to match the dipole magnet gap, the SRRC 1.3 GeV storage ring uses aluminum alloy (A6061-T6) as the bending chamber material. The bending chamber is made by an oil-less NC machining process to improve the outgassing rate performance and contour precision, then the upper and lower halves are joined together by the TIG welding method.[11] From the results of the Auger Electron Spectroscopy (AES) surface analysis, the surface made by this oil-less machining process is less carbon contaminated (Fig.4). For the cooling system design, a cooling channel is drilled along the end wall of the chamber (Fig.2) to avoid the water channel from penetrating into the vacuum side. In addition, a "concentrated" pumping with a distributed ion pump is applied to the chamber.[8]

For the beam ducts of S-chamber, aluminum alloy (A6063-T4) is used. They are made by the extrusion method. There are two cooling channels (one is dummy) on the opposite sides of the beam duct. The inner cross section is 38 mm high $\times$ 80 mm wide with an elliptical shape the same as that of the electron beam path in the bending chamber.

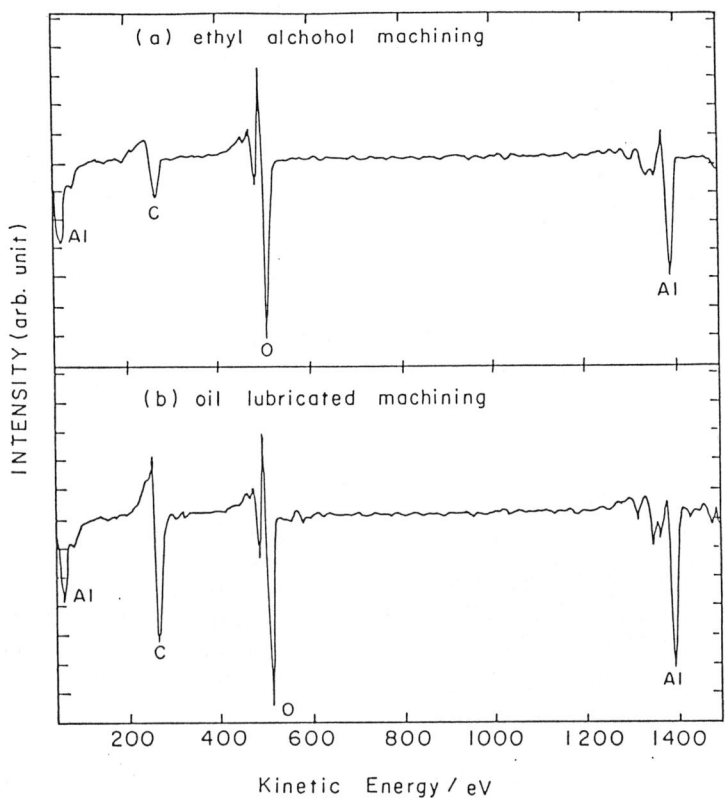

Fig.4 The AES results of the ethyl alchohol machining process (a), and of the conventional oil-lubricated machining process (b).

The pumping port of the S-chamber is an I.D. 150 mm pipe penetrated by a piece of straight beam duct, as shown in Fig.5. In order to reduce the electron beam instability and to improve the conductance of the pumping port, the straight beam duct chamber within the I.D. 150 mm pipe is machined with many short slots (5mm × 25mm).

D. Joining

The TIG welding and the ConFlat (CF) flange sealing methods are used as the major joining methods in the SRRC vaccuum system. The two halves of the bending chamber are joined together by the AC-TIG welding method. In order to prevent distortion, the welding current is kept below 180 Amperes, and the welding processes were performed carefully. The present results show that the maximum deformation of the bending chamber is < 0.3 mm per meter length.[12]

In the S-chamber, the joinings between aluminum flanges, bellows,

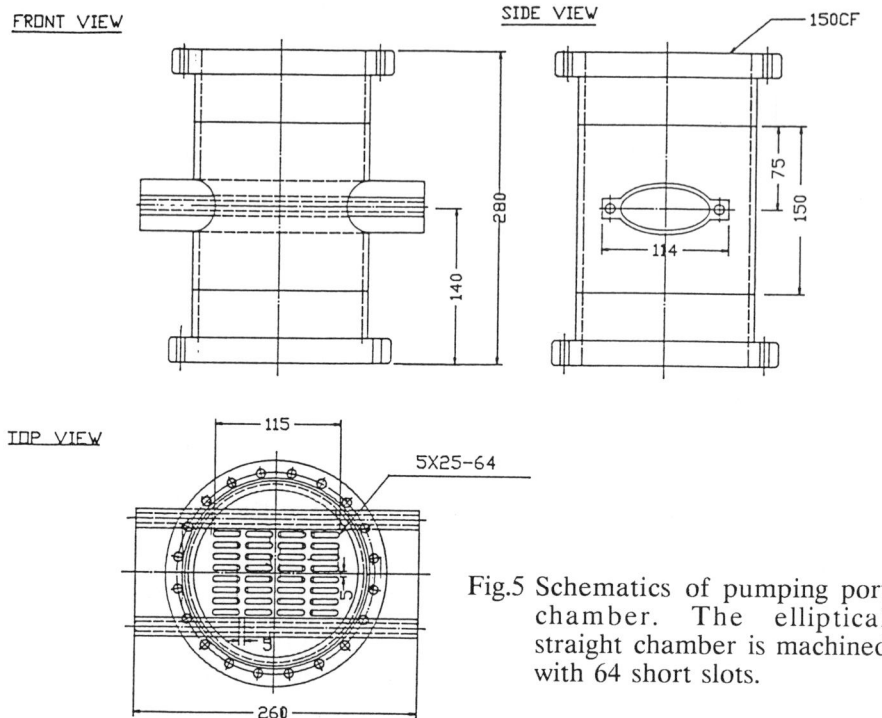

Fig.5 Schematics of pumping port chamber. The elliptical straight chamber is machined with 64 short slots.

beam position monitors (four electrodes were e-beam welded to a piece of chamber to form a complete set, Fig.6) and pumping ports are made by the AC-TIG welding method, too. The welding deformation of straight section chamber is controlled to be ≤ 1 mm per meter length.

The aluminum CF flange sealing method (aluminum gasket, bolts and nuts) is applied to the joining between two aluminum flanges on the beam duct. Aluminum gasket, aluminum bolts and nuts are also used in the joining between an aluminum flange and a stainless steel flange. This joining method can maintain a leak rate of $<1 \times 10^{-10}$ Torr l s^{-1} after bakeout. In the case of two stainless steel flanges, they are joined together by using copper gasket, stainless steel bolts and nuts.

E. Supporting frame

Two types of supports are used in this ring vaccuum system. One is fixed support, the other is floating support. Besides, some limitation-frames are designed to avoid large transverse displacement of bending chambers during evacuation and in-situ bakeout. Beam position monitor and a point near to the exit beam port of the bending chamber are in fixed positions. The fixed type support is mounted on the rigid magnet stand. Even in the case of baking, the positions of the fixed point can not be moved. Between two fixed points, a bellows is installed in order to

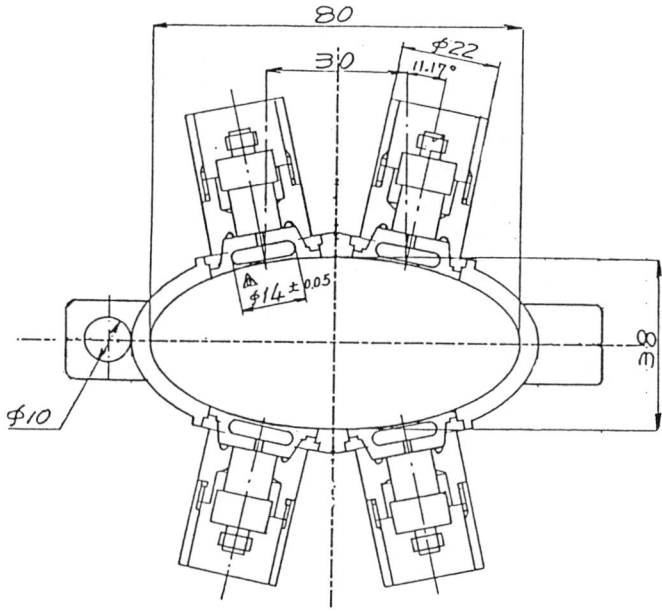

Fig.6 Schematic of the beam position monitor. Four electrodes are e-beam welded to the elliptical beam pipe.

make assembly easily and to absorb the deformation during bakeout. For the floating supports, two kinds of supports are adopted. The first kind is springs, with an effective spring constant $\sim$ 5 kg/mm, which support the heavier components, e.g. 400 l/s IP. The second kind is a thin stainless steel plate (1.5mm) mounted vertically on the magnet stand to support the lighter vacuum chamber. For easy installation and replacement of vacuum chambers and for matching the design of magnets, the ring vacuum system is divided into several sections. Between two sections, CF flanges are used as the joining method.

SYSTEM TEST

A. Vacuum test

The system was evacuated initially by turbomolecular pump systems. In order to reduce the gas load on the TMP system, two sorption pumps had been applied to help the evacuation for a few minutes. The sorption pumps were later isolated from the system by a metal valve. After a leakage was stopped at $\sim$ 3 h, the slope of the pump down curve changed abruptly at that time. After 44 h, the system was baked to $\sim$ 130 °C for one day. A pressure of $< 1 \times 10^{-10}$ Torr was obtained quickly as the system was cooled down to room temperature.

Figure 7a and 7b show the mass spectra of the residual gases before and after in-situ bakeout, respectively. The major residual gases before bakeout were H_2O, H_2, CO/N_2, CO_2 and O_2 gases, in sequence of intensity. After bakeout, they are H_2, CO/N_2, H_2O and CO_2 gases. Among these peaks, O_2 gas was oringially thought to be from air leak. However, it decreased with time and was removed significantly after bakeout. Therefore, that O_2 gases were probably adsorbed on the surface of aluminum chamber and NEG pump, where the porosity is high, during air exposure. The origin of N_2 gases is thought to be same as that of O_2 gases, because that the system had been vented to dry N_2 gases and exposed to the air. During bakeout, the adsorbed O_2 and N_2 gases released and a peak with mass number 30 appeared. This peak might be from the combination of nitrogen and oxygen. After bakeout, however, the residual gas of mass number 28 is

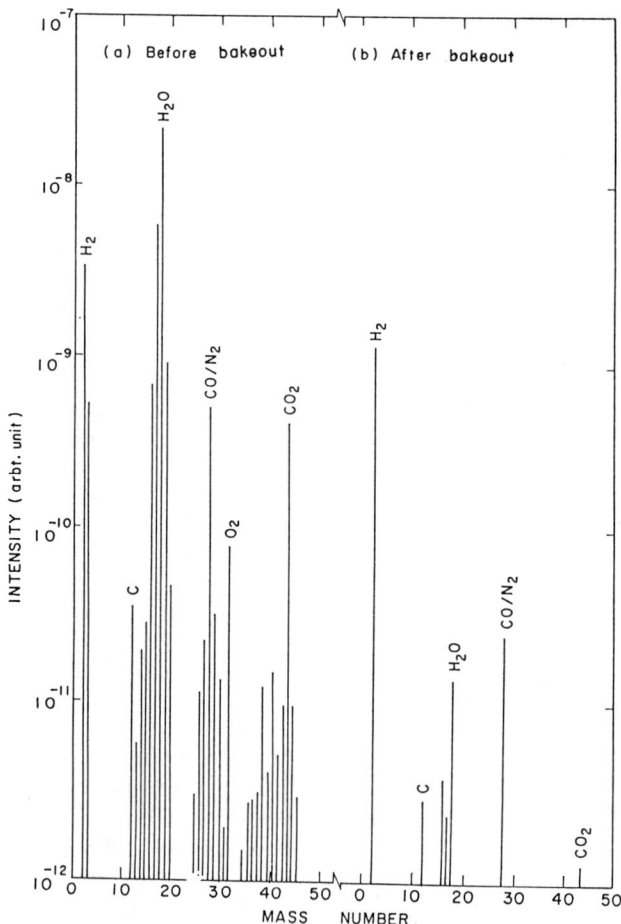

Fig.7 The mass spectra of the residual gases in the system; before bakeout (a), and after bakeout (b).

mainly CO, no evident N_2 cracking pattern was observed.

Due to baking temperature was not high enough and was not uniform in the system, H_2O peak is significant even after bakeout. In addition, the cracking patterns of freon and hydrocarbon had been observed during bakeout. The freon gas might be from the degreasing process of B-chamber which was by rinsing in a freon bath; the hydrocarbon resulted from the contamination during machining, assembling, or the residues of solvent. The cracking patterns of freon and hydrocarbon disappeared after baking.

B. Deformation and displacement

Deformation and displacemnt of the system during in-situ bakeout were measured by dial gauges. The results show that the maximum longitudinal displacement of the "fixed-points" was ~ 0.4 mm at $\sim 130°C$, and recovered to < 0.1 mm after cooled down to room temperature. The transverse dispacement is about two times lower.

In addition to fixed-points, the displacement of some flexible points were also measured. The maximum elongation of long chambers (~ 2 m) was ~ 4 mm, and recovered to < 0.1 mm after to cool down to room temperature. A significant transverse displacement (> 1mm) at the upstream end of the B-chamber was observed during bakeout. The displacement at this point recovered to < 0.4 mm after bakeout.

All the above-mentioned results are satisfactory to the SRRC ring vacuum system. An improvement is being considered to get less dispacement of the fixed-points.

SUMMARY

The vacuum system of the SRRC 1.3GeV synchrotron light source was designed, and a section vacuum system of one-sixth ring has been fabricated and tested. The results show that the system has small deformation and good vacuum performance after bakeout. The design, fabrication and testing results are summarized as the following :

1) Aluminum alloy is chosen as the chamber material due to its many benefits, especially for its high thermal conductivity, easy machining and low outgassing rate.
2) A computer simulation for the outgassing rate and pressure distribution was made to determine the locations and pumping speeds of pumps.
3) An oil-less NC machining and a careful welding process were adopted in the B-chamber fabrication. Less carbon contamination and small deformation (<0.3mm per meter length) of the chamber were obtained.
4) The beam duct cross section is kept constant in this one-sixth ring vacuum system. Bellows and sector gate valves of the system are equipped with RF contact.
5) The straight section chamber is a combination of aluminum beam ducts, flanges, bellows, beam position monitors and pumping ports. The welding process for the S-chamber was controlled so that a deviation

(transverse or longitudinal) of <1mm per meter length was obtained.
6) The conflat flange joining method between aluminum and stainless steel flanges was adopted by using aluminum gaskets, aluminum bolts and nuts. No air leak was observed after an in-situ bakeout at ~130℃.
7) The fixed and the floating support design in this one-sixth ring vacuum system are satisfactory. The maximum displacement (or defomation) is <0.5mm after bakeout.
8) A static vacuum of $<1 \times 10^{-10}$ Torr was quickly obtained after a bakeout at ~130℃ for one day.

ACKNOWLEDGEMENT

The authors would like to thank their colleagues of the SRRC vacuum group for their contribution in this work.

REFERENCES

1. J. Kouptsidis and A.G. Mathewson, DESY Report, DESY 76/49, (1976).
2. A.G. Mathewson and G. Horikoshi, KEK Report, KEK-78-9, (1978).
3. J. Le Duff, Nucl. Inst. Meth. A239, 83 (1985).
4. S. Krinsky, BNL Report, BNL 23749, (1977).
5. H. Ishimaru, J. Vac. Sci. Technol. A2(2), 1170 (1984).
6. J.R. Chen, K. Narushima, and H. Ishimaru, J. Vac. Sci. Technol. A3(6), 2188 (1985).
7. M. Suemitsu, T. Kaneko and N. Miyamoto, J. Vac. Sci. Technol. A5(1), 37 (1986).
8. J.R. Chen, G.Y. Hsiung, D.C. Chen, D.J. Wang, G.S. Chen, and Y.C. Liu, in Proceedings of the Topical Conference on Vacuum Design of Advanced and Compact Synchrotron Light Sources, (BNL, New York), MAY 16-18, 1988.
9. J.R. Chen, G.S. Chen, D.J. Wang, G.Y. Hsiung and Y.C. Liu, Vacuum, Sept./Oct., (1990).
10. D.C. Chen, G.Y. Hsiung and J.R. Chen, J. Vac. Soc. R.O.C. 1(1), 24 (1987).
11. D.J. Wang, J.R. Chen, S.N. Hsu, G.Y. Hsiung, G.S. Chen, S.Y. Perng, H. S. Tzeng and Y.C. Liu, Proceedings of the 2nd National Conference on Welding Technology, (Kaoshiung, Taiwan), 1989, p.129.

Vacuum System Experience at the Daresbury SRS

R.J. Reid, S.F. Hill and P.A. Crank
SERC Daresbury Laboratory, Warrington, WA4 4AD, U.K.

ABSTRACT

The performance of the vacuum system of the Daresbury SRS over the past two years of operation is reviewed. Beam lifetimes in excess of 30 hours with 200mA of stored beam at 2GeV are regularly obtained. It has proved possible to achieve this performance in a reasonable time-scale without *in situ* bakeout being required after the system has been let up to atmosphere, because of the memory effect for stainless steel and copper of long periods of beam scrubbing. The importance of the use of distributed pumping in the dipole chambers will be discussed, and their effect on beam lifetime will be demonstrated.

An important technique in obtaining rapid machine recovery after insertion of a new beam chamber is the use of careful pre-cleaning of the vacuum component, including glow discharge cleaning. A description will be given of experience using various gasses or gas mixtures (Argon/Oxygen; Oxygen; Hydrogen; Helium/Oxygen) for such cleaning, with a discussion of the attractions and possible problems associated with each of them.

INTRODUCTION

The initial vacuum performance of SRS2, following the upgrade to high brilliance operation has been described elsewhere.[1] This paper surveys its continuing performance.

Over the past two years of operation, the machine has achieved an excellent reputation for the performance of the vacuum system. In particular, its beam lifetime at high current is well in excess of specification, allowing the machine to operate with one fill per day.

The machine and its vacuum system have been fully described elsewhere,[2,3] and, in brief, comprise an electron storage ring of 2GeV energy, approximately 100m in circumference of stainless steel vacuum chamber construction. In the bending magnet vessels, however, the beam space is largely surrounded by OFHC copper, as indicated in Figure 1. The pumping system uses conventional lumped triode sputter ion pumps, titanium sublimation getter pumps and integrated diode sputter ion pumps utilising the fringe fields of the bending magnets. In all, the available pumping speed at 10^{-9} Torr is estimated to be about 20,000 litre sec^{-1}.

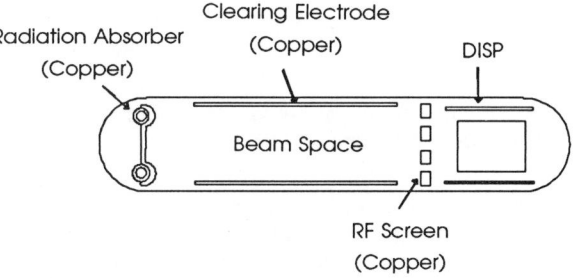

Fig 1 - Cross section (schematic) of bending magnet vessel.

The machine was designed for *in situ* bakeout at 250°C, but since the major rebuild in 1987, has not, in fact, been baked in its entirety.

VACUUM PERFORMANCE

An important operational parameter of a synchrotron radiation source for a user is the beam lifetime and its recovery after atmospheric excursion of the vacuum vessel. At the end of 1989, about half of the SRS vacuum envelope was let up to dry nitrogen for some modifications to be made, and some new components were inserted. Figure 2 shows the

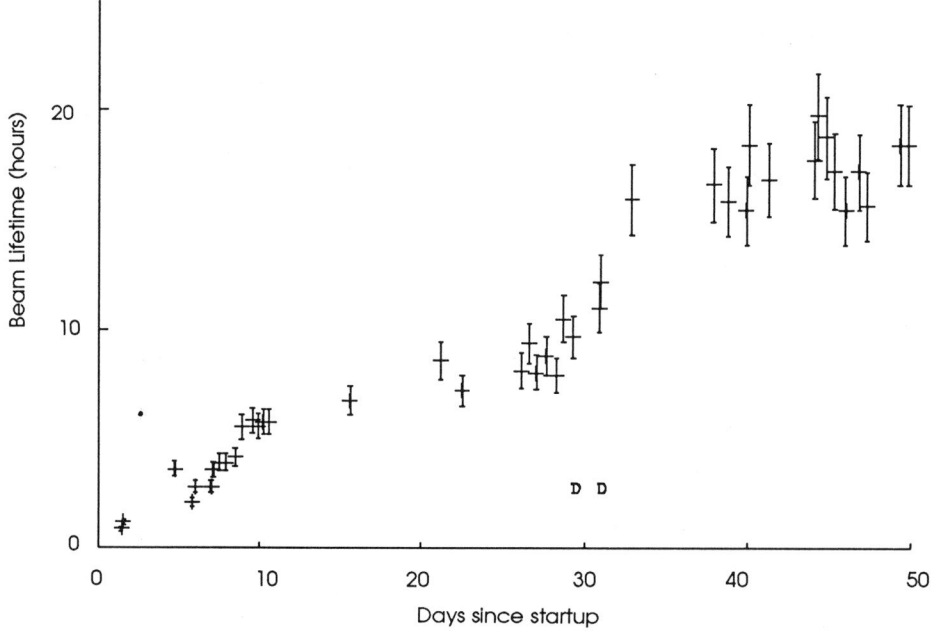

Fig 2 - Beam Lifetime at 100mA beam Current after startup. "D" indicates when a group of DISPs were switched on (see text)

recovery of the beam lifetime with 100mA stored beam current over the first 50 days of beaming. Although there is the inevitable scatter in data, it can be seen that lifetimes close to 10 hours (the machine specification) were achieved after about 30 days. At this time, it was found that some of the sixteen installed integrated sputter pumps (DISPS) had not been reconnected. Seven of them were switched on over the next two days, and it will be seen that a rapid increase in lifetime to 15 hours was achieved.

Figure 3 shows the machine average pressure (determined as an arithmetic average of the readings of sixteen Bayard-Alpert gauges spaced around the ring, one on each straight section). Taking into account the scatter in the readings, switching on the DISPs gives a marginal improvement in the average gauge reading, despite the good improvement in lifetime. In other words, the beam is a much more sensitive indicator of true average

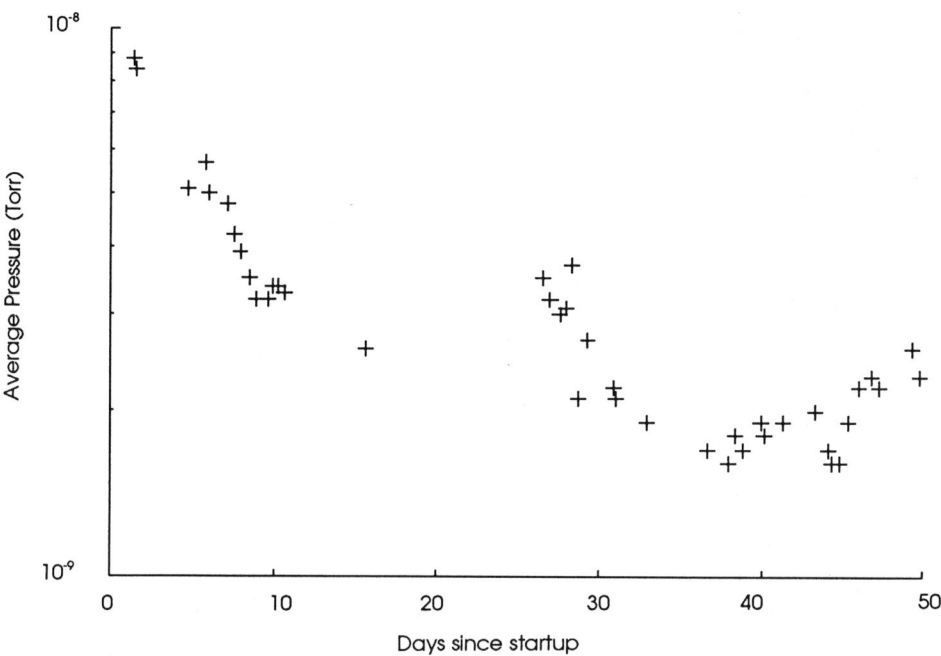

Fig 3 - Average ring pressure with 100mA beam in the machine after startup.

pressure than the gauges. This is not surprising, as the majority of beam desorption occurs in the dipole vessels, well away from the measuring points.

Figure 4 plots the 200mA beam lifetime over a longer period of time. It will be observed that at about day 54, a further 4 DISPs were switched on, giving the full complement of 16. The 200mA lifetime quickly doubled to about 24 hours, and subsequently increased to being regularly in excess of 30 hours. Although extracting desorption coefficients from machine data is difficult, since some of the parameters (e.g. actual pumping speeds available) are not well defined, the indications are that η_{tot} lies in the range 5.10^{-7} - 1.10^{-6} molecules per photon.

Summarising, with only 5 DISPs operating, a 10 hour lifetime was achieved at 100mA in 30 days and at 200 mA shortly thereafter with 12 DISPs operating, without bakeout. If all DISPs had been on from the start, it is likely that the machine would have achieved the 10 hour figure within about 20 days of startup. This is comparable to the time which would be required to bake half of the machine.

This good performance may be attributed to two reasons, viz., the long beam scrubbing of much of the machine (many parts have been exposed to synchrotron radiation in the machine for 10 years) and the careful preparation of new components before installation, which includes glow discharge cleaning. Both of these effects are only important because of the memory effect of stainless steel and copper for desorption.[4]

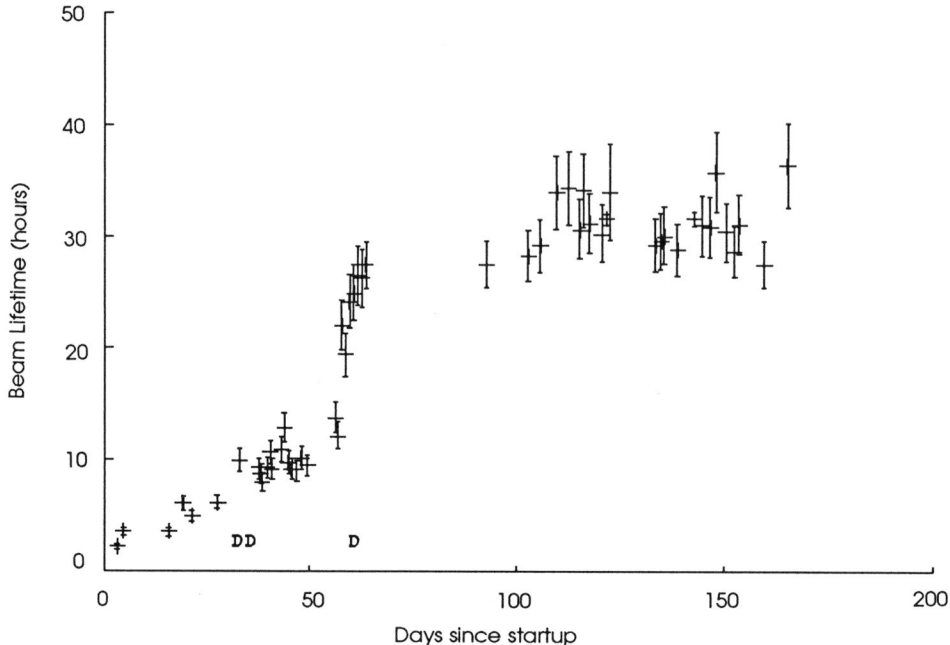

Fig 4 - Beam lifetime with 200mA beam current over an extended period of time.

GLOW DISCHARGE CLEANING

DC glow discharge cleaning has been used for many years for obtaining low gas desorption in accelerators.[5] Traditionally, a gas mixture of Argon/10% Oxygen has been used,[6] but this leads to noticeable amounts of Argon in the residual gas atmosphere for long time periods. For this reason, because of the possible effect on beam lifetime, pure oxygen discharges were used at the SRS for a time to obviate Argon hangup. However, as mentioned earlier, much of the inside of the machine is copper, and glow discharge in oxygen leads to oxidation of the copper surfaces. Such surfaces look dirty, although from a vacuum point of view they are clean and perfectly satisfactory. For many years the fusion community has used hydrogen as the discharge medium.[7] This was tried and yielded good cleaning. However, excessive sputtering of copper was obtained, leading to electrical breakdown of insulators and feedthroughs. Satisfactory results, as regards cleaning, desorption and residual gas, have been obtained in our Laboratory by the use of Helium/10% Oxygen gas mix as the discharge medium. This work has been more fully reported elsewhere.[8]

ACKNOWLEDGEMENTS

The contribution of many of our colleagues to this work is gratefully acknowledged.

REFERENCES

1. R.J. Reid, AIP Conference Proceedings **171**, 46 (1988)

2. D.J. Thomson and V.P. Suller, Rev Sci Instrum **60**, 1377 (1989)
3. R.J. Reid, Proc EPAC 1, 1988 (World Scientific Publishing Co Pte Ltd, Singapore, 1989), 1337
4. R.S. Vaughan Watkins and E.M. Williams, Vacuum, **28**, 459 (1978)
5. R.S. Calder, Vacuum, **24**, 437 (1974)
6. R.S. Calder *et al*, Proc 7th IVC, Vienna, 1977 (F. Berger & Sons, Vienna) 231
7. H.F. Dylla, AIP Conference Proceedings **171**, 144 (1988)
8. R.J. Reid and P.A. Crank, Proc 2nd European Vacuum Conference, Il Nuovo Cimento, *in press*

ELETTRA VACUUM SYSTEM

M. Bernardini

Sincrotrone Trieste, Padriciano 99-34012 Trieste, Italy

ABSTRACT

A status report of the vacuum system of ELETTRA, the 2 GeV, 400 mA light source under construction in Trieste, will be described.

The Vacuum project, presented at "Synchrotron Radiation Vacuum Workshop" at Riken (Japan 22-24 March 1990) and more recently at EVC-2, the European Vacuum Conference at Trieste (Italy 21-26 May 1990), is now in the phase of testing a protoype sector, which is 1/24 of the ring circumference. Details and some technological aspects of the fabrication will be reviewed together with the vacuum performances.

Results of laboratory experiments on components, standard or not, allowed us to finalize the main choices in light of the general philosophy of the project and will be properly summarized.

VACUUM CHAMBER

ELETTRA is a highly brilliant synchrotron light source presently under construction at Trieste (Italy) and scheduled for initial operation in 1993. The essential parameters are summarized in table I.

Table I.

Beam Energy	[GeV]	2
Magnetic Bending Radius	[m]	5.5
Circumference	[m]	259.2
Beam Current	[mA]	400
Critical Energy	[KeV]	3.23
Pressure with Beam	[Torr]	2×10^{-9}
Bending Magnet Synchrotron Power	[KW]	103
Insertion Device Synchrotron Power	[KW]	40

The vacuum system forms approximately a 260 m diameter ring and consists of two different shaped 316 LN kinds of stainless steel chambers, one to be used in the straight sections and the second to be used in the bending magnets. Together with some aluminium alloys, AISI 316 LN-SS shows ultimate outgassing rate of the order of $10^{-13} \div 10^{-14}$ Torr l/s cm^2, after proper thermal and chemical treatments, and can be considered one of the best materials for vacuum chambers of synchrotron light sources.

Compared to aluminium it presents a lower thermal conductivity, but this disadvantage is clearly overcome by several advantages like easy weldability, low thermal espansion coefficient, easy coupling with commercial vacuum components such as bellows, pumps and feedthroughs. In addition to that, it can be baked at higher temperature (350 °C ÷ 400 °C) and its desorption yield (molecules/photon) is about two orders of magnitude lower, at least in the starting period [1].

Figure 1 shows the cross-sectional view of the straight section chamber which covers more than 200 m of the ring circumference. The rhomboidal shape has been chosen to fit all the magnetic device apertures, namely bending magnets, quadrupoles, sextupoles and steering magnets, maintaining everywhere the same size and, at least, 5mm of space from the poles for bake-out operation in situ.

Around the 24 bending magnets, each of it covering a 15 degrees arc, the vacuum chamber presents a different cross-section (see Figure 2) to allow the synchrotron radiation to pass through a 10 mm slot into another connected chamber (antechamber) which contains the port of one beam line. In this way, the electron beam in the bendings circulates in a chamber similar to that of the previous straight sections, except for the small slot on the external side.

Figure 3 shows a period (achromat) out of 12 of the complete ring with two straight sections, two dipoles and two beam lines.

Some machines under construction, like the 2 GeV ALS at Berkeley and the 7 GeV APS at Argonne [2], foresee a vacuum chamber presenting the outer antechamber all along the ring. At present, we do not know if an antechamber will become a resonant radiofrequency structure because of the continuous slot communicating with the beam.

Our philosophy for ELETTRA was to limit this danger, which would cause instabilities to the electron beam, to those portions of chamber where the beam lines have to be anyway located to extract the radiation.

All along the machine, about 100 monitors have to be placed to check the position and the characteristics of the electron beam, and each of them is connected with the rest of the chamber by flat flanges with a VAT ultra-high-vacuum sealing method (VATSEAL). The tightness is realized by means of a silver plated copper gasket which matches almost exactly the rhomboidal shape of the chamber and it has been chosen because it does not create dangerous cavities as the standard UHV Conflat seals. It is worthwhile to recall that any cavity-like change of the chamber cross-section introduces impedances and thus instabilities. For the same reason, each bellow presents an internal RF contact to avoid as much as possible any changes from the standard chamber.

SYNCHROTRON RADIATION INDUCED DESORPTION

Photon absorbers and specific pumping systems allocated in the antechamber sections deal properly with the consequent gas desorption far away from the electron beam channel. However, for a prudent approach to the vacuum project, a non negligible fraction of this desorption has to be considered to influence also the vacuum of the electron channel and thus potentially the electron beam lifetime.

In addition, the ELETTRA vacuum chamber design, described in the previous section, allows the photons of the radiation distribution tail to impinge the chamber walls downstream each bending magnet.

Without entering into the details of the topic, discussed in ref.3, it is worthwhile to outline the proceeding of the evaluation that, finally, gives a reasonable estimate of the pressure profile and the proper pumping system, necessary to obtain a lifetime of the electron beam longer than 10 hours under the planned operating conditions. The number of the molecules, desorbed by the expected photons in ELETTRA at 2 GeV and 400 mA [4] is:

$$N_M = \eta \; 6.4 \times 10^{20} \; [\text{molecule/s}] \qquad (1)$$

where η represents the desorption efficiency (molecules/photon) which varies with the beam dose according to a power - law:

$$\eta(D) = A \, D^{-2/3} + B \qquad (2)$$

with A and B depending on chamber material, cleaning treatments, geometry etc. The beam dose D measured in ampere hours (Ah) is a sort of 'conditioning' due to the beam itself and can be scaled from one machine to another [5].

For ELETTRA we used the data of the SRS at Daresbury [6], and in particular the value of $\eta=10^{-6}$ [mol/ph] (for typical UHV gas composition) obtainable after a dose of 50.2 [Ah] operation.

By taking into account the variation of the angle between incident photon and chamber wall and the gas desorption, the pressure distribution along an achromat can be obtained by using a finite element analysis. Figure 4 shows the pressure profile, the number and the position of 120 l/s ion pumps.

An additional 400 l/s ion pump has to be installed just underneath each photon absorber with the function of trapping the major part of the relative desorption, 90% in the present computation.

The average pressure, after 50 [Ah] operation, results 2.5×10^{-9} [torr] (N_2 equivalent) sufficient to match the required value of the electron beam lifetime. In fact bremsstrahlung is the main effective beam-residual gas interaction which affects the lifetime τ in ELETTRA at 2 Gev, 400 mA:

$$\tau_B = \frac{40}{\rho \text{ (n Torr)}} \text{ hours.}$$

VACUUM PLANT

The final vacuum system consists of 24 400 l/s and 108 120 l/s ion pumps, all of them with a NEG module installed inside the body to roughly double the pumping speed for H_2, CO and CO_2.

In addition to that wafer shaped NEG ribbons will be placed in the bending antechamber to increase the desorption trap capability of the 400 l/s ion pumps with their supplementary H_2 pumping speed of 500 l/s each.

Two all metal gate valves are foreseen for each (11) insertion device straight section so that the whole ring vacuum chamber can be separated in sectors and the in situ bake-out phase performed in steps.

The roughing groups, a combination of magnetic bearing turbomolecular pump with oil free forepump, are movable and valved off as soon as the operation is finished.

The longest chamber module is about 4 m and after an accurate chemical treatment all the modules will be vacuum degassed at 400 ° C, 10^{-6} Torr in a dedicated furnace.

After assembling, every bake-out in situ will be done at 150 °C by means of a premounted microprocessor-controlled heating system.

STATUS OF THE PROJECT

The manufacture of a vacuum chamber prototype was completed at the De Pretto factory and the delivery is scheduled for next December 1990. It refers to a half achromat, out of 24 semiachromats of the ring, and includes a dipole vessel with the light exit.

To ensure the required tolerances, the antechamber is fabricated starting from a 60 mm SS sheet and obtaining the final shape by milling cavities for vacuum and the slot for the light. This solution came out from our specifications on stresses (< 100 N/mm^2) and slot deformation (< 0.1 mm) as derived from the load due to vacuum at the bake-out temperature in situ (150 °C).

The standard vacuum chamber with the rhomboidal cross-section is obtained by cold drawing, starting from a round pipe. The portion to be coupled to the antechamber is properly calendered to the right curving radius and welded along the slot after a proper cut preparation. Particular care was devoted to this welding, operated from inside according to the best rules for UHV.

The prototype will undergo to the same treatments as the final chamber modules. The special, 6 m long, furnace under vacuum for the first bake-out is already operating in the laboratory and achieved the pressure of $\approx 10^{-9}$ Torr at 400 °C. The pumping system includes three oil free vacuum groups similar to those to be used on the ring for the following 150 °C bake-out operations in situ, according to the principle of maximum cleanliness.

Photon absorber and photon shutter design s fig. 5, 6 have been completed taking into account the power density associated with the dipole magnets, namely 140 W/mm^2.

This figure will be reduced by inclining the surface with the respect to the incident radiation; however to avoid any direct connection of water and vacuum in case of leaks in copper solderings, the cooling channels, where possible, are machined into the copper blocks. Figure 6 shows a cross-sectional view of the photon absorber directly obtained by machining from solid, confining the only brazing to an SS flange far away from the radiation.

Prototypes of flanges with VAT seals and bellows with internal RF contact have been tested in the laboratory together with a model of beam position monitor. The goal of introducing the minimum possible impedance in the chamber would be satisfactorily achieved by our mechanical choices. In fact, some estimates, based on approximate formulae, indicate that the total value of the low frequency coupling broad band impedance of our ring is reasonably low and pratically due to the contribution of RF cavities, provided that the bellows are properly shielded.

Concerning the reliability, all the components were submitted to several mechanical and thermal cycles without any leakage and/or drawback.

REFERENCES

1. M. Andritschky et al. CERN, Technical Note (LEP-VA 1988).
2. 'Vacuum Design of Advanced and Compact Synchrotron Light Sources' (Brookhaven May 16-18 1988), Proceeding on AVS series 5 (Upton NY 1988).
3. M. Bernardini and R. Kersevan, 'Synchrotron Radiation Induced Outgassing Profiles and Pressure Curves in the ELETTRA Vacuum Chamber', 'Presented at European Vacuum Conference' (Trieste May 21-25, 1990).
4. O. Gröbner et al., Vacuum 33, 397, (1983).
5. O. Gröbner, CERN/ Technical note ISR-VA/OG/sm/tn-4 (1982).
6. A. Poncet, CERN/PS/87-42 (ML) (1987).

194 ELETTRA Vacuum System

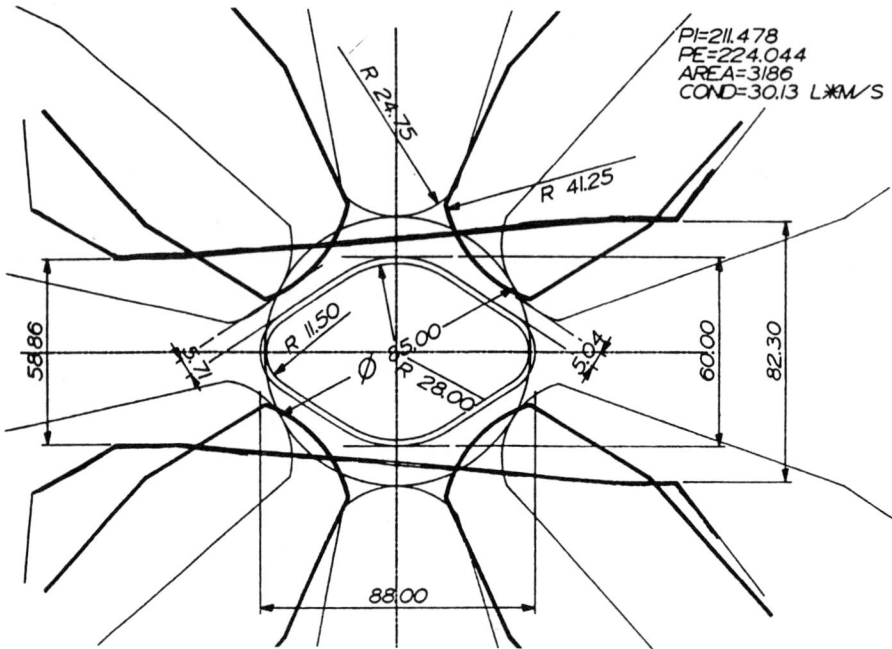

Fig.1. Vacuum chamber profile in the straight sections.

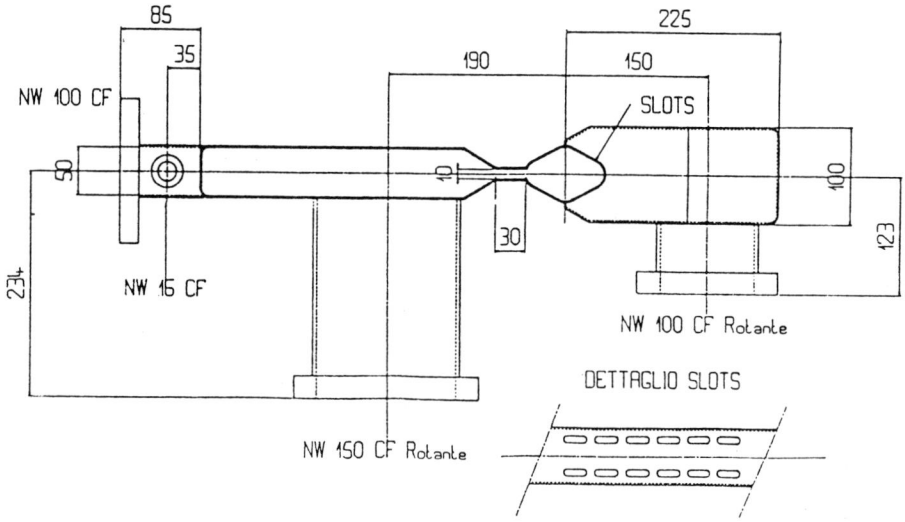

Fig.2. Bending magnet vacuum chamber cross-section.

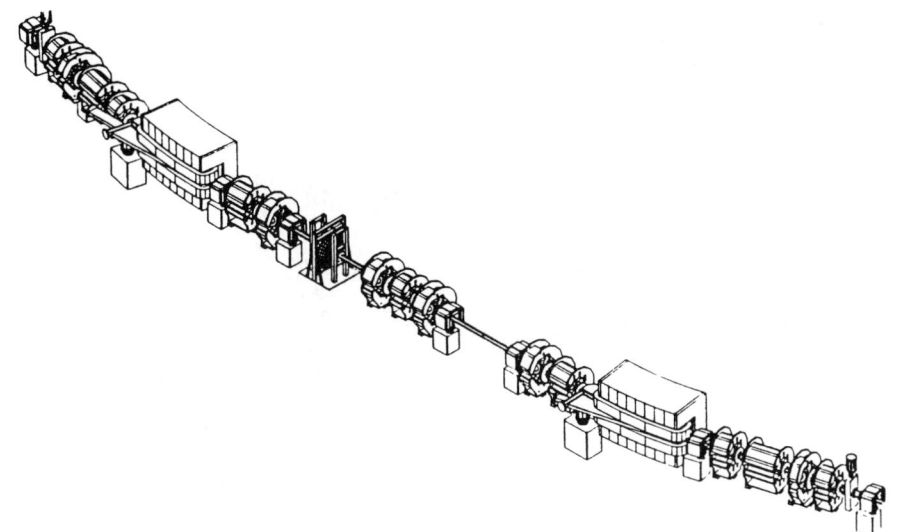

Fig.3. View of one achromat (1/12-th of the ring). ID vacuum chamber is not shown.

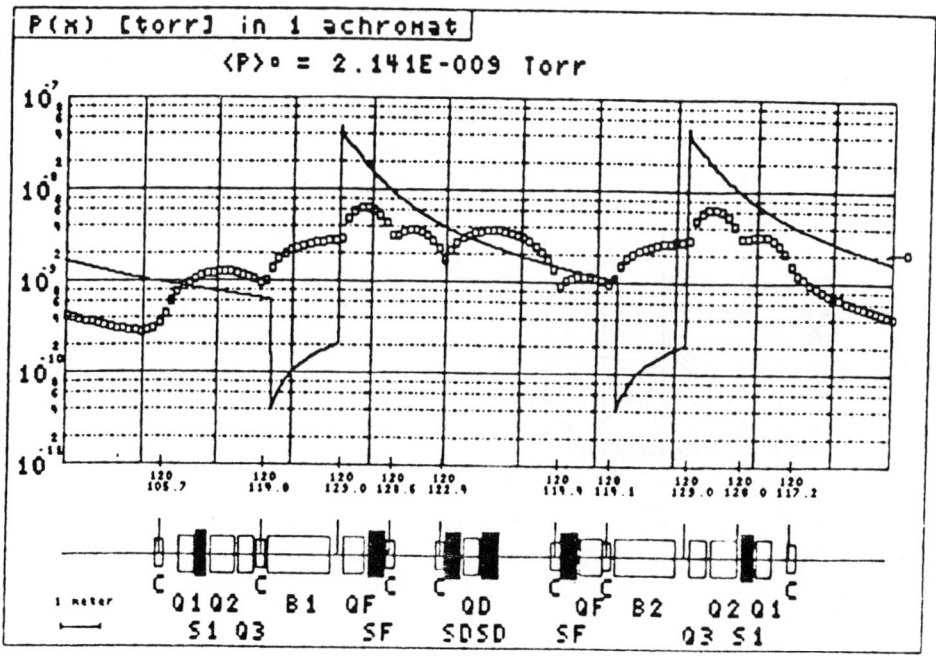

Fig.4. Calculated pressure profile along one achromat with $\eta = 10^{-6}$ [mol/ph]. $E = 2$ GeV, $I = 400$ mA. Solid line = outgassing profile.

196 ELETTRA Vacuum System

Fig.5. (Right) Drawing of the photon beam-stopper.

Fig.6. (Below) Cross-section of the photon absorber.

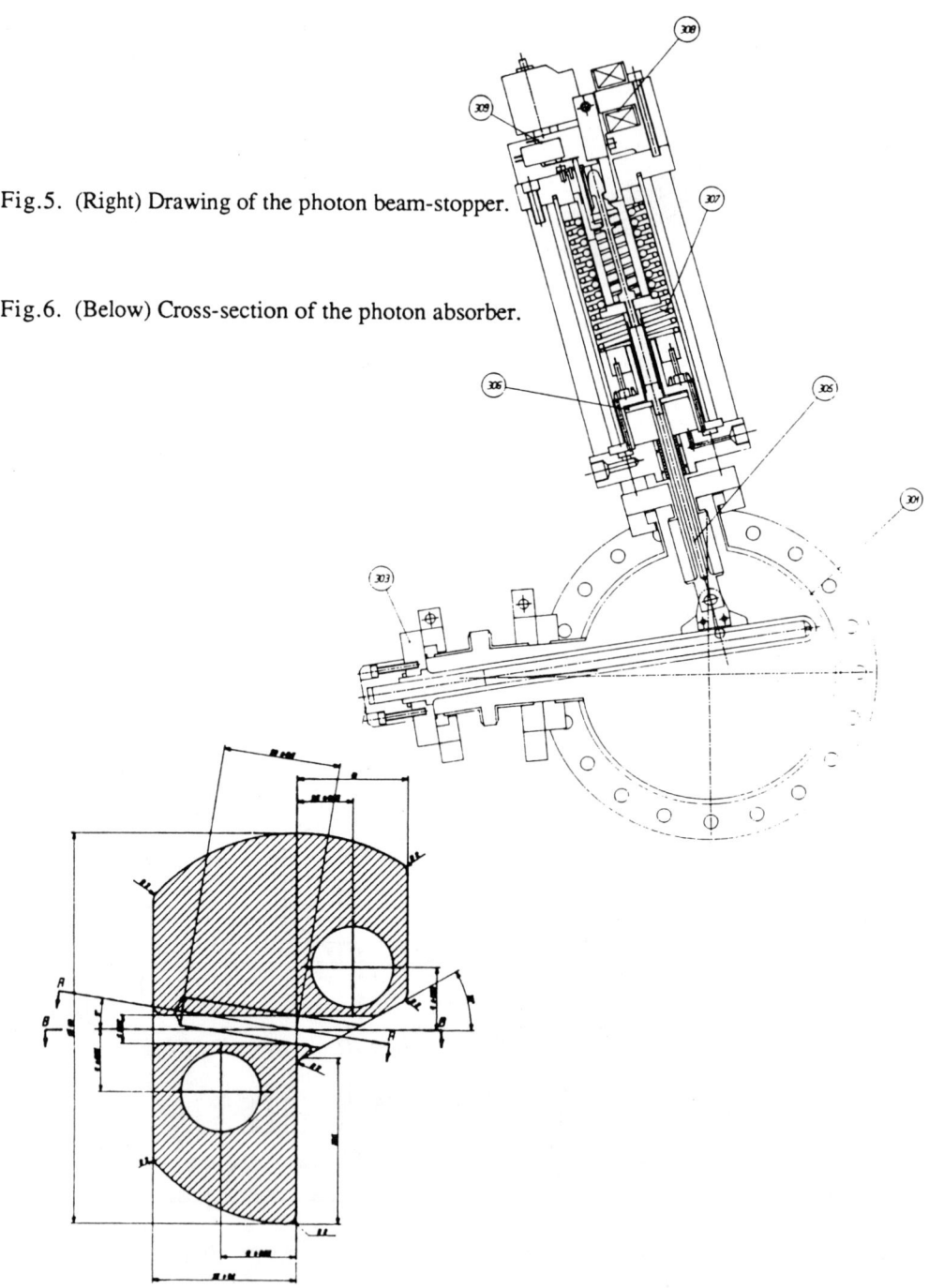

MANUFACTURE OF THE ALS STORAGE RING VACUUM SYSTEM*

Kurt Kennedy
Lawrence Berkeley Laboratory,
University of California, Berkeley, CA 94720

ABSTRACT

The Advanced Light Source (ALS) storage ring has a 4.9 meter magnetic radius and an antechamber type vacuum chamber. These two requirements makes conventional bent tube manufacturing techniques difficult. The ALS sector vacuum chambers have been made by machining two halves out of aluminum plate and welding at the mid plane. Each of these chambers have over 50 penetrations with metal sealed flanges and seven metal sealed poppet valves which use the chamber wall as the valve seat. The sector chambers are 10 meter long and some features in the chambers must be located to .25 mm. This paper describes how and how successfully these features have been achieved.

INTRODUCTION

The ALS is a synchrotron light source with an electron storage ring that requires a vacuum system with an small electron channel and very low pressure. An antechamber[1] type vacuum system was chosen because it:
1) Improves longitudinal conductance
2) Lowers impedance
3) Permits adsorption of photons away from the electron beam and close to discrete pumping
4) Gives freedom to make vacuum penetrations for photon lines, diagnostic ports, pumps, etc
Fig.1 shows the cross section of such a vacuum vessel with an antechamber.

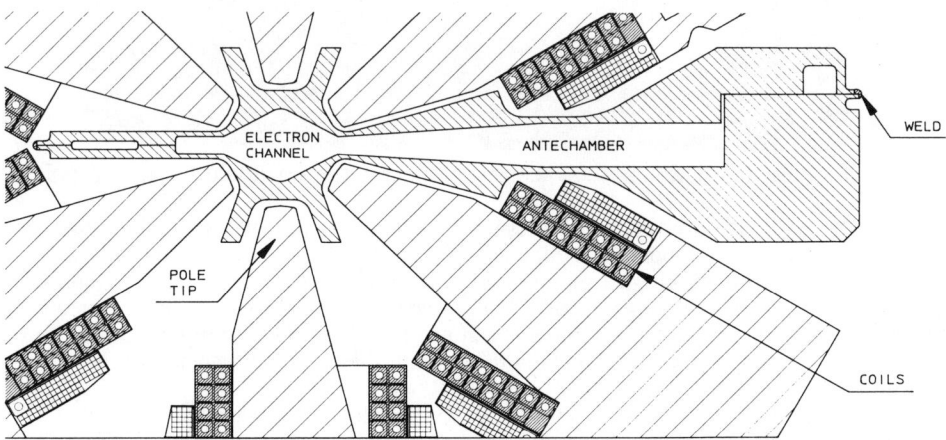

Fig. 1 Section of the vacuum vessel with antechamber at the sextupole

* This work was supported by the Director, Office of Energy Research, Office of Basic Energy Sciences Division, of the U.S. Department of Energy under Contract No. DE-AC03-76SF00098

The benefits of an antechamber are not without a price. A lot more surface, with attendant outgassing, is exposed the vacuum and the span of metal subjected to vacuum load is much larger. The thickness of metal required to limit the vacuum deflection to 0.1 mm is over 20 mm for aluminum. The traditional manufacturing method of extrusion and bending seemed impossible because of the relatively small magnetic radius and the added width that the antechamber required. A machined aluminum chamber was pursued.

DESIGN OF A MACHINED VACUUM CHAMBER

Vacuum chambers machined in two halves and welded on the mid plane offer design freedoms that off set some of the inherent problems of this technique. Fig.2 shows how integral ribbing between magnet pole tips can support the vacuum load and keep the vacuum wall as thin as 1 mm. Flanges can be machined directly into the chamber. Complex shapes can be created, such as electron channels that follow the beam stay clear.

Fig. 2 Picture of a section of the vacuum chamber at the sextupole and quadrupole

The ALS vacuum system divides up in such a way that, it is logical to machine the 10 meter sector chambers as a unit. The dimensional accuracy required for such a chamber is 0.25 mm, making Computer Aided Design (CAD) a near necessity and the complexity is such that 8 megabits of CAD memory is needed for two dimensional drafting on half of a complete chamber. Features that must register top to bottom must have adjustability equal to about one ten thousandth the largest dimension of the chamber.

The selection of the aluminum alloy was based on distortion and ease of welding. The alloy need only be strong enough to resist plastic deformation from the metal gasket material used in the flanges. The alloy selected was 5083, which is a solid solution alloy, strengthened primarily with magnesium and has a minimum yield of 215 MPa (31 ksi) with a H321 temper. This alloy is a nonheat treatable alloy free of the residual stress associated with quenching and with good welding and machining proprieties. This alloy is copper free and does not have to be nitric acid rinsed after mild caustic cleaning. This permits the cleaning process to be drained to a sewer.

FLANGES

The design of flanges that could seal directly to the vacuum chamber had to take account of the relative softness of 5083 H321 aluminum. A conflat type design requires much harder aluminum than 5083. Metal O rings requires a very smooth surface for sealing and it was felt that the flange surfaces machined into the vacuum chamber would offer too great a scratch hazard . The sealing system selected was one derived from a high pressure Mott seal[2]. A cross section of the seal is shown in Fig. 3. The sealing surfaces are out of harms way and oriented such that the tool marks of an end mill lay perpendicular to the leak path. The flange thickness and bolt schedule are the same as standard conflats. Almost 1000 of these seals are used in the ALS storage ring.

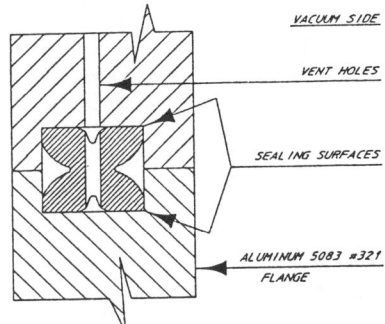

Fig. 3 Metal seal system

CHAMBER MACHINING

Numerically controlled milling machines capable of making parts four by thirty meters have developed to serve the needs of the aircraft industry Fig 4.. These machines have three spindles, each in effect have five axis of motion and three parts can be made simultaneously. These machines are well suited for making vacuum chambers except that their usual mission use aircraft rather than accelerator accuracy. Every 1°C change in the ALS chamber temperature changes it's length by 0.25 mm. This large thermal dilation occurs in the relatively innocuous longitudinal direction. The part must, however, be clamped during machining so that thermal expansion and contraction originate from a single point near the center of the part and when the part returns to starting temperature the part is in the exact starting place.

The machining procedure consisted of: blank the plate to within 6 mm of the perimeter of the part, rough machining all surfaces to with in 2.5 mm of finish size, release all restraints for 24 hours, reclamp and finish machine. The time to machine three complete chambers was 750 hours. Hand finishing added another 100 hours.

Fig. 4 Numerically controlled mill machining three vacuum chambers

CHAMBER CLEANING AND ASSEMBLY

The size of the vacuum chamber made a non-immersion cleaning process necessary. The parts had to be cleaned hanging from a crane using men with brushes. The cleaning process had to be compatible with the workmen and the floor drain, that went to the sanitary sewer. The procedure used was: steam clean for gross oil removal and heating, scrub the surface with detergent, scrub the surface with a proprietary[3] mild alkaline cleaner, rinse with copious amounts the hot de-ionized water and blow dry with liquid nitrogen boil off.

The perimeter was sealed with a filet weld using a hand held TIG torch and added 5183 alloy. The potential virtual leak at the mid plane of the chamber was vented by a channel just inside the weld and this channel was flooded with argon during welding. Features that required good top to bottom registry, such the electron channel and valves parts, had special alignment tooling to insure optimal distribution of errors. The assembled vacuum is shown in Fig. 5

The combined machining and distortion error in the vertical direction was negligible small and the chamber lay on a granite flat without measurable gap. In the transverse direction, the maximum combined error was about 3 mm and this error could be reduced to 0.3 mm with modest lateral forces applied by the chamber support system. The longitudinal error is dominated thermal dilation and tends to be less important then errors in the other axes.

Fig. 5 Picture of an assembled ALS sector chamber

VACUUM PERFORMANCE

The vacuum chamber was vacuum baked to 150°C for 24 hours using 20 permanently mounted strip heaters with a total capacity of 24 KW. The heavy aluminum cross section of chamber permitted the use of modest insulation and still maintain acceptable temperature gradients. The chamber achieved 3×10^{-10} Torr in 48 hours using only ion pumps, whose cumulative speed was 420 l/s. Based on the geometric area, this would indicate mediocre but adequate outgassing rate of 10^{-12}

Tl/s/cm^2. This relatively high outgassing rate is the result of the machining techniques used to minimize fabrication cost. Much of the metal removal was done with convex end mills whose center moves very slowly and leaves a rough surface.

DISCUSSION

Machining techniques produced very complex ALS sector vacuum chambers with good dimensional accuracy and acceptable vacuum properties but no discussion of fabrication method would be complete without mentioning economics. The cost of the aluminum and machining for one chamber was $75K not including tooling and programming. If these fixed costs are distributed over 12 chambers, the unit price goes to $80K. An additional 600 hours of cleaning, welded and assembly were needed to complete each chamber. The total cost per meter, not including pumps, of an ALS sector chamber is about $10K.

$10K / meter is a very large cost if compared to a simple extrusion, however, there are ameliorating factors. Each meter of chamber will have, on average, 5 all metal flanges. Large aluminum pieces were salvaged which will supply all the metal for 12 insertion device chambers. Without machining, magnet gaps would had to be increase 10 mm (gradient) to 30 mm (sextupole). This increase translates to between $12K and $31K per meter increase in magnet cost depending on how cost scales with aperture. This analysis would imply that if magnet apertures can be reduced by about 10% by machining vacuum chambers, you can't afford not to.

REFERENCES

1. Kurt Kennedy, Topical Conference on: Vacuum Design of Advanced and Compact Synchrotron Light Sources, 52, (1988)

2. Fred Middleton, IEEE Transactions on Nuclear Science, Vol. NS-28, No. 3, 3298, (1981)

3. Mirachem 100, The Mirachem Corp., 2107 East 5th. St., Tempe, AZ. 85281

VACUUM SYSTEM DESIGN FOR THE PLS STORAGE RING

C. D. Park, K. H. Kil, C. K. Kim and S. M. Chung
Pohang Accelerator Laboratory, Pohang Institute of Science and Technology,
P. O. Box 125, Pohang 790-600, Korea

ABSTRACT

The 2 GeV Pohang Light Source (PLS) is being built at Pohang Accelerator Laboratory (PAL). The vacuum system for the storage ring designed to maintain an operating pressure of 1 ntorr is described. Vacuum chambers will be made of 316 LN stainless steel. The major source of gas in an operating storage ring is the interaction between the fan of photons coming from the dipole magnets and the surfaces of the vacuum chamber. In the PLS vacuum system this interaction is performed in a controlled way to a large extent at the discrete photon absorbers with concentrated high capacity pumps. Finite element analysis of the pressure gradients indicates that such a vacuum system design will be adequate in its vacuum performance.

1. Introduction

The PLS is a dedicated synchrotron radiation facility being built at PAL with the aim of commissioning the machine in late 1994.

The proposed machine is optimized to generate synchrotron radiation extending from the vacuum ultraviolet to the soft x-ray region of the spectrum with good tunability over this range and the capability to accommodate a large number of insertion devices.

Fig.1 shows the layout of the facility. The accelerator complex consists of a 2 GeV electron Linac as a full energy injector, a 2 GeV electron storage ring with a circumference of 280 m as a dedicated light source, and experimental hall for exploiting synchrotron radiation. The Linac is connected to the storage ring through a beam transport line underground. A full energy injection scheme has been chosen due to the high requirements on orbit stability and reproducibility. Full energy injection with high injection rates will lead, in turn, to fast vacuum chamber clean-up[1].

For the magnet lattice, a Triple-Bend-Achromat (TBA) structure has been chosen for the PLS storage ring. The storage ring is made up of 36 dipole-magnet vacuum chambers, each about 1.5 m long, interconnected with straight sections. It also incorporates 12 long straight sections. Ten sections will be used to accommodate insertion devices (IDs) such as undulators and wigglers, which will meet the needs of users for hard x-rays of high brilliance. The remaining two will be occupied by the beam injection apparatus and RF accelerating cavities. The storage ring will be capable, ultimately, of accepting 400 mA

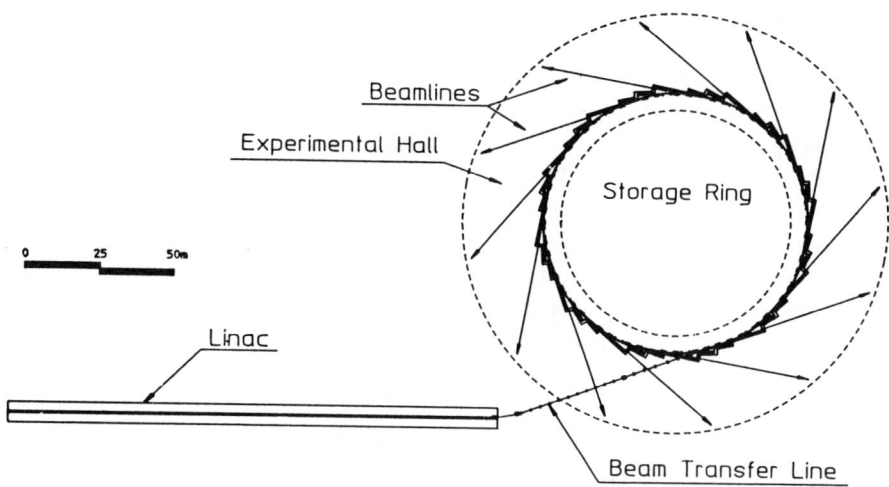

Figure 1. Layout of the facility.

of beam current at 2 GeV. The principal machine parameters relevant to vacuum considerations are listed in Table 1. The scope of this paper will be limited to describing the storage ring vacuum system.

From the user's point of view, a long beam lifetime is an important characteristic of a synchrotron radiation source. It will assure a slow decay of the amount of available photons for the research work with synchrotron light and the possibility of experimental data taking over an uninterrupted period of many hours. To achieve such lifetimes, it is imperative that the particle loss due to interactions with residual gas molecules is minimal. In the PLS, the vacuum system is designed to maintain a beam-on operating pressure of around 1 ntorr for a beam lifetime due to gas scattering of about 10 hours. Achieving such a low pressure is no mean task when one considers the low conductance of the vacuum duct and the high synchrotron radiation induced gas load.

Table 1. Storage ring parameters.

Energy	2.0	GeV
Beam current	100	mA
Bending radius	6.30	m
Bending field	1.06	T
Bending length	1.1	m
Critical energy	2.82	KeV
Circumference	280.56	m
Superperiod	12	
Syn. power radiated	22.5	KW
Total photon flux	1.6×10^{20}	ph/s

2. Vacuum Chamber

Fig. 2 shows a plan view of a sector which is 1/12 of the ring circumference. It shows three dipole chambers with three photon beam ports, one per dipole, straight sections, and one long straight section. The dipole magnet photon beam ports provide for a horizontal opening angle of a 28 and 26 mrad, respectively, while zero degree ID beam ports are designed for 14 mrad. In addition, locations of sputter ion pumps (SIPs) and photon absorbers are also shown. One RF-shorted gate valve will be installed at each end of the sector, so that any sector can be isolated from the rest of the machine. Each dipole chamber is joined to its adjacent chambers with Conflat-style flanged joints as standard. Hydro-formed bellows are located upstream of each dipole chamber and at each end of the long straight section to allow for thermal expansions and facilitate installation.

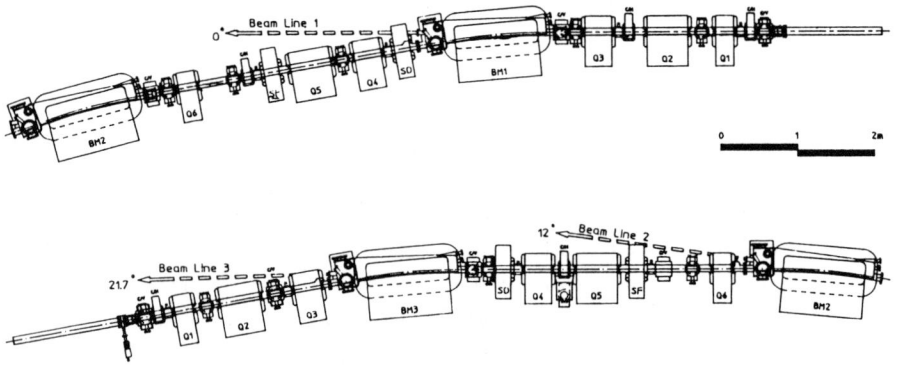

Figure 2. Plan view of a sector.

A cross section of a typical straight section chamber is shown in Fig.3. The inner chamber aperture is compatible with the beam stay clear requirements set forth by beam dynamics. The minimum aperture is therefore 78 mm in horizontal plane and 41 mm in the vertical one. This aperture cross section will be kept constant wherever possible all around the ring.

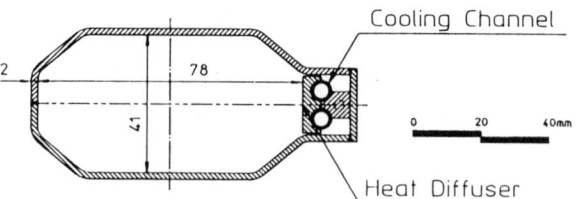

Figure 3. Cross section of a typical straight section chamber.

The maximum power density immediately downstream of the photon absorber is around 3 W/mm^2 and this does not present a heavy thermal load to deal with. However, austenitic stainless steel is particularly sensitive to thermal fatigue because of its low thermal conductivity and high thermal expansion. The water-cooled distributed strip absorber is thus inserted inside the vessel to protect the chamber walls.

Taking the compactness of lattice into account, beam position monitor blocks will be directly welded to the vacuum chamber to save space.

Fig.4 is a plan view cross section of typical dipole chamber. It shows the electron beam path, the photon beam exit port, pumping ports for SIP and titanium sublimation pump (TSP), roughing port, and the water-cooled photon absorber. The chamber is tapered outward in such a way that no synchrotron radiation must impinge on a surface other than the photon absorber. Hence there would be no need for continuous cooling of the dipole chamber outer wall. The photon absorber is located outside of the dipole coil directly above a TSP and SIP with high pumping speed. This design allows the major part of the radiation to be intercepted by the photon absorber in such a way that the photons and the desorbed gas are directed down to the pump to prevent them from diffusing into the beam circulation region to interact with circulating electrons. This should result in a longer electron beam lifetime.

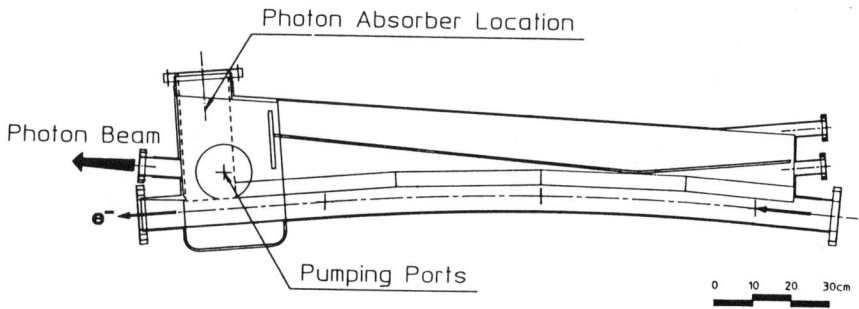

Figure 4. Plan view cross section of typical dipole chamber.

Fig. 5 is a cross section of a typical dipole chamber. It shows antechamber and a wedge shaped slot to allow for photon beam exit. This slot is 10 mm high. Except this slot, the chamber has the same aperture cross section as the straight section chamber has.

As discussed above, the dipole chamber will be a complicated fabrication. This makes the use of extruded aluminium chambers unattractive despite its good thermal properties. Stainless steel however simplifies the fabrication when adding flanges, pumps, etc, and is suitable for constructing the dipole chambers. All other vacuum chambers will also be fabricated of austenitic stainless steel. The chambers must have a relative permeability close to unity even after bending and welding. The dipole and straight chambers will therefore be made of 316 LN stainless steel.

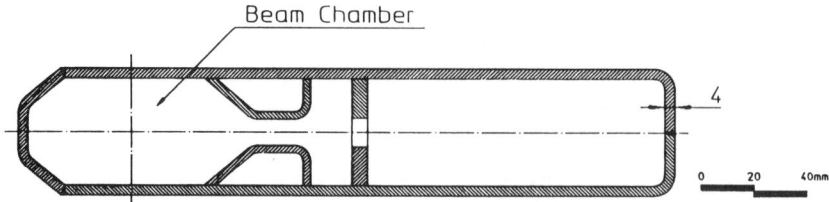

Figure 5. Cross section of a typical dipole chamber.

3. Photon Absorber

The photon absorber plays an important role in the pumping strategies for the PLS. It is hoped that the photon absorber/pump system does a job it is designed to do. The photon absorber should stop the photons and direct as much desorbed gas as possible into the pump underneath. The pump should effectively trap the gas to be a good gas sink.

The water-cooled copper absorber, designed without water/vacuum joints, is of wedge shape with the gas emitting surface offering a large solid angle to the pump[2]. The wedge face has a six degree facet along with steeper facets at the front and rear to accommodate low energy photons and electron orbit error. A slot will be machined in the absorber to define the radiation fan going down the beam lines.

4. Pumping System

In the design and analysis of the pumping system, only one sector is considered for simplicity.

The total surface area for the sector vacuum chamber is about 6×10^4 cm^2. Assuming that, because of appendages and discontinuities, the surface area becomes triple and using 1×10^{-11} torr·l/s/cm^2 as the specific outgassing rate, then a total thermal gas load is 2×10^{-6} torr·l/s.

The photon induced outgassing rate[3] is given by $2 \cdot n^o \cdot \eta \cdot K$, where $K = 3.11 \times 10^{-20}$ torr·l/mol, n^o = number of photons/s, and η is a dosage-dependent molecular yield per photon (mol/photon). Since 10 A·h is considered a reasonable period of beam conditioning, a machine conditioned to 10 A·h will be taken as the point at which photon induced dynamic gas loads are calculated. The calculation is based on the percentage of photons falling upon copper absorbers and on the angle of incidence on these. The angle of incidence dependence of desorption efficiency is assumed to be the same as the one measured at DCI (Orsay) using an aluminum vacuum chamber [4].

We assume that 20% of the radiation becomes stray photons that are continuously scattered at all angles. Gas generated by these stray photons would probably be fairly uniformly distributed and will be treated here like thermal outgassing. Using a value of 4×10^{-6} molecules/photon and calculating the number of photons/s/mA as 1.6×10^{18} photons/s/mA, the gas load at 100 mA due to stray photons is 6.6×10^{-7} torr·l/s.

Similarly, but taking the angle of incidence into account, photon induced gas load from continuous absorbers are calculated as 4.5×10^{-6} torr·l/s. Adding to this the gas load from thermal outgassing and stray photons, the total distributed gas load at 100 mA is approximately 7.2×10^{-6} torr·l/s.

Finally, for any individual discrete photon absorber the gas load can be estimated as 1.9×10^{-6} torr·l/s.

Since the main gases evolved in a clean system due to the photon induced desorption process are found to be mass 2 (hydrogen), mass 28 (mainly CO) and mass 44 (mainly CO_2). Sputter ion pumps will stably pump CH_4 and noble gases which are always present to some degree, as well as active gases. Titanium sublimation pumps will also be used. TSP offers useful additional pumps to deal with gas loads, and their compact size should enable them to be positioned close to the electron beam pipe. The TSPs will be turned on to achieve a better ultimate pressure. Such a pumping system is relatively simple to operate and contamination free.

The distribution of pumps around the ring should ensure that the required average pressure will be achieved under normal conditions. Since the specific chamber conductance is determined to be low (about 25 l·m/s), location of pumping should be such that high conductance is achieved from the high photon induced gas load surfaces to the pumps.

The combination of a large capacity TSP and one 230 l/s triode SIP will be mounted just below each photon absorber, where high local gas loads will be produced. In addition 230 l/s triode SIPs will be installed downstream of the dipoles, while other 120 l/s sputter ion and titanium sublimation pumps are scattered along the ring in such a way that adequate pumping is assured for the entire ring. The pumping speeds of the pumps used per sector are listed in Table 2. In the effective pumping speed calculation, the performance of the pumps in the ultrahigh vacuum region is assumed to be 30% of the nominal values.

Six mobile turbomolecular pumping stations mounted on carts will be used for roughing. The roughing cycle will be completed in two stages; from atmospheric pressure to ~ 10^{-2} torr, and from ~ 10^{-2} torr to below 10^{-6} torr. Conventional cryosorption pumps operating with liquid nitrogen will be

Table 2. Pumping speed per sector.

Pump	Nominal (l/s)	Effective (l/s)	Quantity	Σ * (l/s)
SIP	230	84	3	252
	230	79	3	237
	120	28	13	364
				(853)
TSP		~1400	3	~4200
		~266	3	~798
		~203	10	~2030
		~183	2	~366
		~146	1	~146
				(~7540)

* : Sum of the effective pumping speed

used for pumping the sector from atmosphere to about 10^{-2} torr, at which pressure turbomolecular pumps backed by diaphragm pumps will take over to pump down to below 10^{-6} torr. An interlocked isolation valve will prevent possible sector contamination should failure of the turbo pump occur.

5. Vacuum Monitoring

Conventional Pirani and Penning gauges are required to provide pressure readings and vacuum interlocks from 1 torr down to 10^{-6} torr. Convectron gauges will also be used to monitor continuously system performance from atmosphere down without pressure blind spots. Since the ultimate pressure of the proposed vacuum system should be $< 10^{-10}$ torr, a gauge which will read down to low 10^{-11} torr is desirable. Thus, either hot cathode ionization gauges or inverted-magnetron cold cathode gauges would appear to be suitable for use. Two quadrupole residual gas analyzers are proposed to be installed initially for each sector to quickly diagnose problems *in situ* even with a stored beam. From experience on the other synchrotron radiation laboratory[5], their use has proved invaluable for the storage ring.

6. Pressure Profile

To see design vacuum performance, an analysis has been made of the pressure profile for one sector. The analysis is based on a finite element computer simulations[6] of pressure gradients with the assumptions mentioned above. A magnet cell is divided into about 100 sub-elements. The calculation assumes that the conductances of the sub-elements are adequately calculated by the Knudsen's formulae for short tubes.

Fig.6 illustrates the expected pressure gradient along the electron beam channel through each cell as given by the program. The curves show typical pressures after 10, 60, and 100 A·h of pump down. There are several significant pressure bumps occurring downstream of the dipoles. These are associated with large gas loads but small effective pumping speed due to tight space. The effective pumping

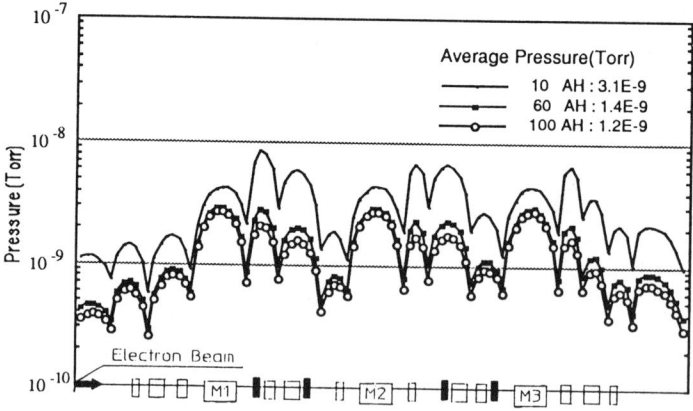

Figure 6. Expected pressure gradient.

capacity in these regions can be augmented to some extent by increasing the pumping capacity in the dipole and the downstream straight chambers to reduce pressure bumps. Apparent slow beam cleaning after 60 A·h appears to be governed by the use of constant thermal outgassing in the calculation. Effectiveness of the photon absorber in preventing gas from entering the circulating beam channel along with the use of lumped ion and titanium sublimation pumps alone, however, makes it possible to achieve average pressure of 3 ntorr after 10 A·h beam conditioning. It is thus expected that with these solutions adopted to vacuum vessels and pumping system design, the goal of obtaining satisfactory beam lifetime (5 hours), after a nominal beam dose of 10 A·h at 100 mA of beam current is a realistic design objective.

7. Chamber Conditioning

To minimize the thermal outgassing rate and photon induced desorption, the vacuum system will be essentially all-metal, and all vacuum chambers will be conditioned prior to installation by using appropriate cleaning techniques including the use of glow discharge and vacuum baking.

Due to the limited gap (2 mm) between the chamber and the magnetic poles, it is very difficult to accommodate baking. However, provision will be made for 150 °C *in situ* bakeout to remove water vapor which is the main residual gas in an unbaked system.

8. References

1. N. B. Mistry, AIP Conference Proceedings No. **171**(1988) 1 .
2. K. Kennedy, AIP Conference Proceedings No. **171**(1988) 52.
3. S. Tazzari, ESRF-IRM-32/84 (1984).
4. O. Grobner, *et al.*, Vacuum **33**(1983) 397.
5. R. J. Reid, J. Vac. Sci. Technol. **20**(1982) 1156 .
6. J. F. J. Van den Brand and A. P. Kaan, ESRF-IRM-61/84 (1984).

LSU ELECTRON STORAGE RING VACUUM SYSTEM DESIGN

Donald E. Geiler

Maxwell Laboratories, Inc., Brobeck Division, Richmond, CA 94804

ABSTRACT

This paper discusses the vacuum system for the electron storage ring system being designed and built for the Center for Advanced Microstructures and Devices (CAMD) at Louisiana State University (LSU). Descriptions are provided for elements of the vacuum system from basic vacuum components to specialized chambers and the pumping system. In addition, vacuum pressure, structural, and thermal analyses conducted in support of the selected configuration are provided.

INTRODUCTION

The LSU electron storage ring system[1] is based on a Chasman-Green double-bend achromat magnet lattice[2] that allows access to 55% of the 2π of synchrotron radiation emitted in the dipole bending magnets. It has three major subsystems: (1) a 1.2 GeV, 400 mA ring, (2) a 200 MeV linac injector, and (3) a beam transport line that elevates the incoming beam from the plane of the linac to the plane of the ring (see Fig. 1).

Table I Storage ring parameters

Energy (GeV)	1.2
Beam current (mA)	400
Number of superperiods	4
Critical wavelength (Å)	9.5
Dipole bending radius (m)	2.928
Dipole bending field (T)	1.367
Circumference (m)	55.2

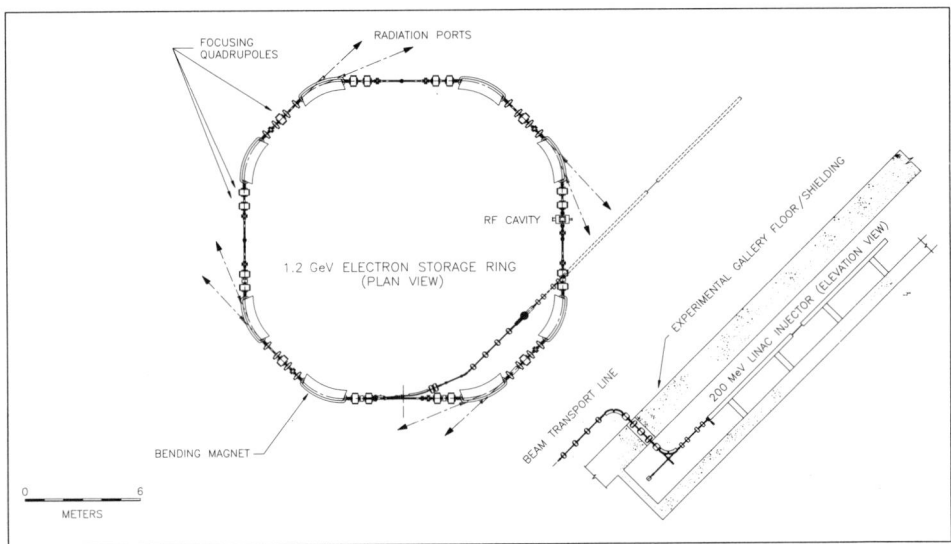

Fig. 1. LSU electron storage ring system layout.

VACUUM SYSTEM REQUIREMENTS

The vacuum system in the beam transport line must maintain a pressure distribution during operation that matches the linac requirement at one end and the storage ring requirement at the other end (see Table II).

The vacuum system in the storage ring must maintain an average pressure of 10^{-9} torr during operation in order to meet the 8-hour electron storage time requirement. During non-operational periods, an average base pressure of less than 5.0×10^{-10} torr must be maintained to prevent recontamination of the interior vessel surfaces by residual gas adsorption.

Table II Vacuum pressure requirements

Beam transport line	
Linac end	10^{-7} torr
Ring end	10^{-9} torr
Storage ring	
Base	5.0×10^{-10} torr

In addition to meeting pressure requirements, the design of the vacuum system addresses geometrical constraints and thermostructural integrity. Components and subassemblies must maintain internal dimensions outside the "Beam Stay Clear" region, while maintaining external dimensions inside "Magnet Stay Clear" boundaries. Within these constraints the vacuum components must be designed to withstand potential collapse from external ambient pressure and thermal stresses with materials degraded due to the effects of heating from synchrotron radiation or as a result of thermal bakeout.

Further design considerations include the selection of materials and processes to provide low thermal desorption and photodesorption values. Finally, vacuum pump sizes and locations must be determined to meet the vacuum pressure specification within the limited space not already allocated to magnets or diagnostic elements.

VACUUM SYSTEM CONFIGURATION: BEAM TRANSPORT LINE

The beam transport line vacuum components are fabricated from three diameters of standard Type 304 stainless steel tubing, with standard conflat flanges and metal seals at the joints. The tube sizes range from 3.81 cm at the linac to 5.08 cm for most of the transport line.

Four vacuum pumps are used in the transport line to obtain the required pressure distribution. Three 45 l/s ion pumps are located in the midsection, and one 230 l/s pump is located on the top of the beam injection dual chamber; its pumping speed is shared with the storage ring. Initial evacuation of the transport line is provided by a 90 l/s turbomolecular pump at a roughing station located approximately one-third the transport line length from the storage ring interface.

The transport line can be isolated from the linac and the storage ring by closing pneumatically operated all-metal gate valves, located at the linac interface and at the beam injection section. These valves are closed for initial pumpdown and are automatically closed when a loss of vacuum is sensed in the linac, transport line, or storage ring.

VACUUM SYSTEM CONFIGURATION: STORAGE RING

Storage ring vacuum vessel

The storage ring vacuum vessel, which is 55.2 m in circumference, consists of three types of chambers: eight 2.3 m 45° sections, curved to a 2.93 m major radius, for the dipole bending magnets; four 3 m straight sections; and four 6.1 m straight sections. One superperiod of the vacuum vessel, together with the magnetic elements, is shown in Fig. 2.

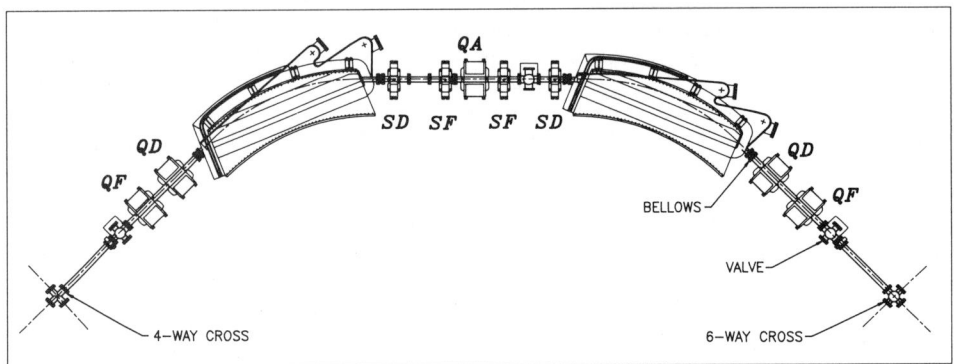

Fig. 2. Storage ring superperiod.

The vessel straight sections are fabricated from standard Type 304 stainless steel vacuum tubing with a 7.62 cm outer diameter and a wall thickness of 0.165 cm. The segments are connected with standard conflat type UHV flanges from 11.43 cm to 15.24 cm outer diameter, depending on location. Tees, four-way crosses, and six-way crosses are provided for pumping and diagnostic purposes. An insert is used within the pumping cross to preserve the vessel cross section and minimize beam impedance effects, while still providing a high pumping conductance. Bellows are used extensively throughout the ring to facilitate joining pipe sections; these bellows, like the pumping crosses, have an inner sleeve to minimize impedance.

Three rf-shorted, in-line sector valves are included: one on each side of the rf cavity, and one diametrically opposite the cavity. In addition, an isolation valve is placed at the end of the transport line vacuum vessel to allow for conditioning and maintenance activities independent of the storage ring vacuum system.

Dipole vacuum vessel

The dipole vacuum vessel (Fig. 3) is a welded assembly consisting of 304L stainless steel top and bottom plates and vertical walls. An inner wall of OFHC copper is used to absorb synchrotron radiation. This inner wall, structurally and thermally isolated from the dipole chamber, includes two copper tubes for cooling water. Short transition pieces with standard conflat UHV flanges are brazed at each end of the dipole chamber. The internal beam chamber is rectangular, with vertical and horizontal dimensions of 5.08 cm and 7.62 cm. The vessel overall height is 5.72 cm. The minimum dipole magnetic gap is 5.90 cm at the pole tips and 6.24 cm at the center. The large outer radius side of the dipole chamber is separated from the beam chamber by an rf absorber which consists of two rectangular bars spaced 1.00 cm apart.

Two 230 l/s ion pumps are mounted on ports located on the bottom of the large outer chamber of each of the eight dipole chambers. Two additional ports are located on top of each dipole chamber for the installation of 1000 l/s titanium sublimation pumps during initial commissioning. Two photon exit ports are located on each of four dipole chambers, one providing an 81 mrad fan of x-ray radiation, and the other an 88 mrad fan. The outer radius copper wall is cooled by the same facility cooling water as the magnet cooling loops.

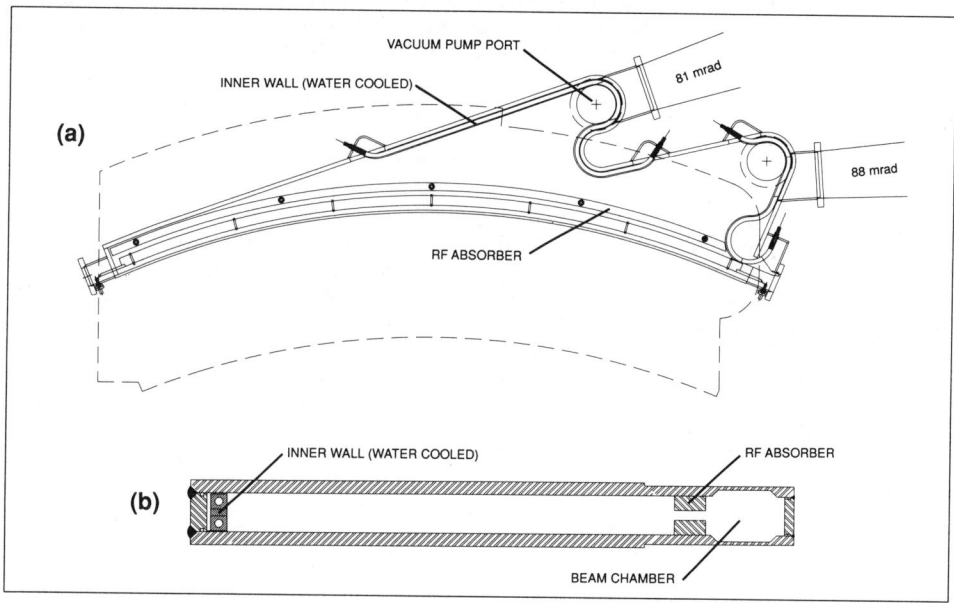

Fig. 3. Dipole chamber: (a) top view, (b) cross section.

Pumping

Ultra high vacuum pumping in the storage ring is provided by twenty-eight ion pumps. In addition to the sixteen 230 l/s ion pumps located on the dipole chambers, there are eleven 140 l/s pumps distributed around the ring on the bottom ports of pumping/diagnostic crosses. Three additional 230 l/s pumps are used. One 230 l/s pump is located on the top of the beam injection dual chamber just prior to the thin septum magnet. This location provides pumping for both the storage ring and the remaining portion of the beam transport line. The other two 230 l/s pumps are located on each side of the rf cavity and serve to pump a section of the storage ring, as well as the rf cavity.

Vacuum pumping for the initial evacuation of the storage ring is provided by three roughing stations. Each station is centrally located on each of the isolated beam sections. An 80 l/s turbomolecular pump backed by a 310 l/min rotary-vane backing pump is mounted on a cart to provide sequential roughing.

Ion clearing

A low profile ion clearing electrode is installed in the bottom of the dipole sections. The electrode consists of a 0.122 cm thick by 3.81 cm wide 304L stainless steel flat plate located just below the beam-stay-clear area. The plate is supported within ceramic insulator blocks that serve to locate the electrode within the beam sections and provide the necessary electrical standoff distance for a minimum of 1 kV bias voltage.

Clearing electrodes in the straight sections are made from 0.476 cm diameter stainless steel rod. The electrode is located in the bottom of the beam tube and is connected to a high voltage vacuum connector. The rod is supported in short cylindrical insulators to provide the necessary electrical standoff for the 1 kV (minimum) bias voltage.

ANALYSIS IN SUPPORT OF DESIGN: TRANSPORT LINE

An analysis was conducted to determine the best arrangement of ion pumps in the transport line. The vacuum pressure distribution was calculated for a variety of arrangements, using a code developed by H. Wiedemann at the Stanford Synchrotron Radiation Laboratory.[3] The outgassing values in the analyses (for both thesport line and the storage ring) assume that all vacuum components will be subjected to a 200°C bakeout and pumpdown prior to installation.

A thermal outgassing value of 1.0×10^{-11} torr-liters/(sec-cm^2) was used.[4] Four pumps distributed along the transport line were found to be sufficient. Although 45 l/s pumps were used in the analysis, the pumping capacity was derated to allow for pumping inefficiency at ultra vacuum levels. One of the four pumps is in the dual chamber located just prior to the thin septum chamber. The 27 l/s pumping speed for this pump represents only 20% of its capacity, since the other 80% is used for pumping the storage ring.

Although the tube varies in diameter, the analysis was based on 2.00 in. OD. The resulting pressure distribution (Fig. 4) meets the design requirement at the linac interface as well as at the storage ring interface.

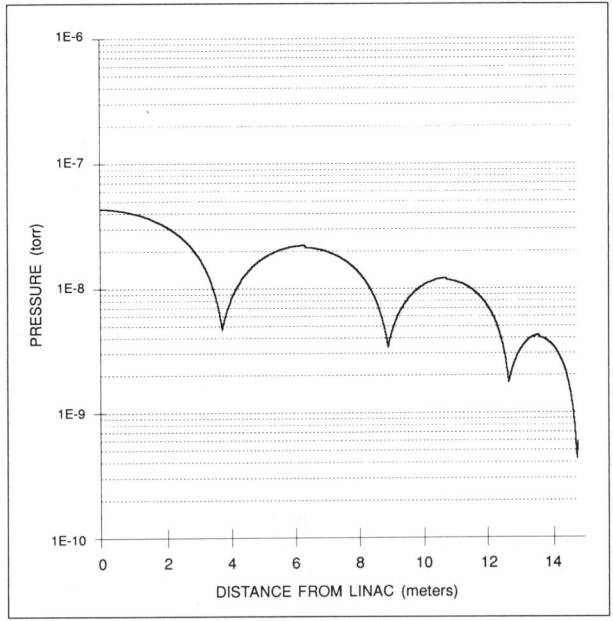

Fig. 4. Transport line pressure distribution.

ANALYSIS IN SUPPORT OF DESIGN: STORAGE RING

Pressure analysis

An initial analysis was conducted to determine the base pressure in the ring due to thermal desorption alone. Because of stringent cleaning and bakeout processes for ring components, an outgassing value of 3.0×10^{-12} torr-liters/(sec-cm^2) was used in the analysis.[5,6] Based on 28 Varian StarCell ion pumps, an average pressure of 2.97×10^{-10} torr was computed for the ring. The pumps consist of sixteen 230 l/s (two per dipole), ten 140 l/s, one 140 l/s shared with the transport line, and one 230 l/s at the rf cavity. These pumps permit easy insertion of NEG cartridges, which can double the pump speed at UHV pressures. All pumps were derated to 60% pumping speed to account for losses in the pumping crosses and their effectivity in pumping gases at ultra high vacuum.

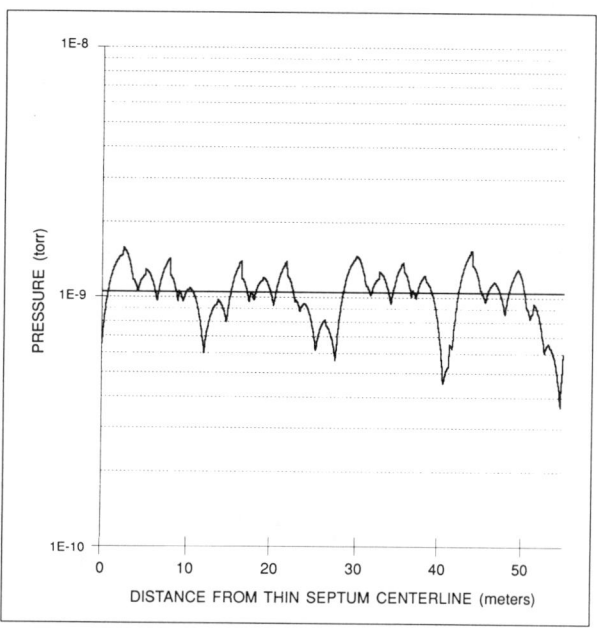

Fig. 5. Storage ring pressure distribution.

Further analyses considered the effect of both thermal desorption and photo-desorption due to synchrotron radiation from the 1.2 GeV circulating electron beam. Assuming beam cleanup of approximately 150 A-hr, a photo-desorption value of 1.0×10^{-6} molecules/photon was used.[7,8,9] The resulting pressure distribution (Fig. 5) averages 1.05×10^{-9} torr.

Dipole chamber structural analysis

Analyses were conducted using the Engineering Mechanics Research Corporation NISA program.[10] The planview geometry was modeled (Fig. 6)

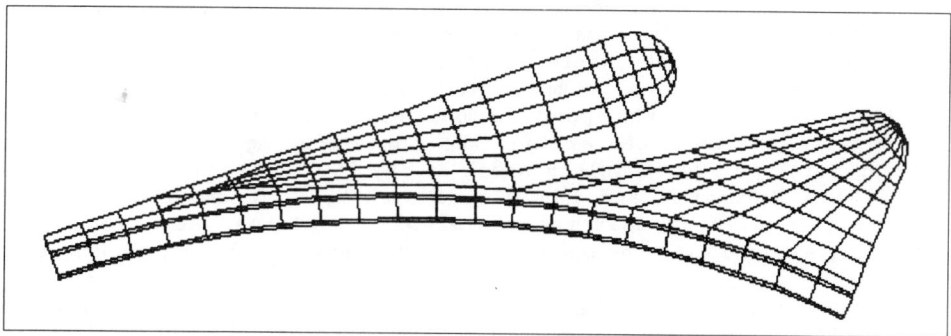

Fig. 6. Dipole chamber FEA model Version 1: planview.

first, with general shell elements that included membrane, bending, and inplane and transverse shear forces. This 276 element, 319 node, 1452 degrees of freedom model represents the stainless steel upper or lower plate (including the cutout in the beam-stay-clear region) with fixed edges under the external atmospheric pressure.

Based on the location of the maximum stresses in the plate, a second structural model ("Version 2") was analyzed for more refinement of the stress distribution through the plate thickness; this cross-sectional model is shown in Fig. 7.

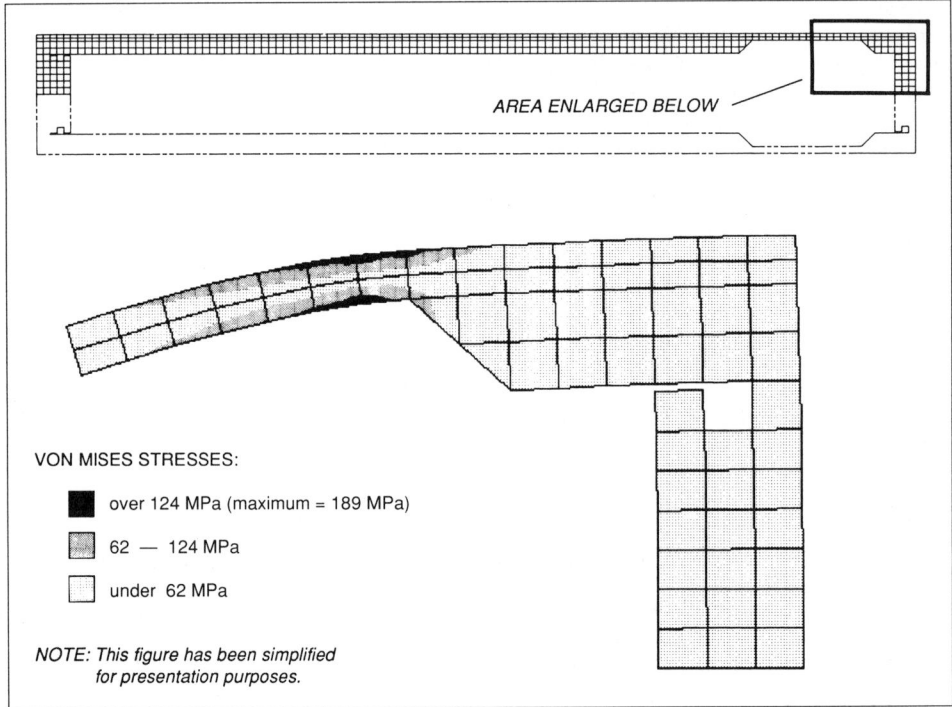

Fig. 7. Dipole chamber FEA model Version 2: stress due to pressure in cross section.

The Version 2 model contains 524 elements and 674 nodes, as shown at the top of Fig. 7. The span of the plate is 38.86 cm, corresponding to the width of the plate where the maximum stresses were found in the Version 1 analysis. The maximum von Mises stress in the stainless steel is 189 MPa and occurs in the upper and lower plate at a location vertically aligned with the beam centerline. The corresponding allowable yield stresses for Type 304 stainless steel at the 200°C bakeout temperature are 207 MPa (minimum) and 248 MPa (average). The computed maximum deflection is 0.10 cm near the center of the plates. At a location vertically aligned with the beam centerline, the deflection is 0.023 cm.

Dipole chamber thermal analysis

A finite element model of the cooling tube and absorber wall was used to compute the temperature distributions arising in the copper wall at the crotch area of the dipole vacuum chamber due to synchrotron radiation. The radiation was assumed as surface heat flux for the purpose of thermal analysis. The beam energy loss is 1.4 kW/m with a vertical distribution of 1.0 mm. The peak heat flux occurs where the angle of beam incidence is 90° to the wall. For the dipole chambers this is the location of the crotch areas of the exit ports. Conservatively, the local heating intensity was assumed to be 8.4 kW/m (six times the beam energy loss of 1.4 kW/m), resulting in a heat flux of 825 W/cm^2. A convective boundary condition was used at the inner wall of the tube with a heat transfer coefficient of 7560 W/m^{2}C°, based on a flow rate of 3.79 l/min in each of the cooling tubes which are brazed to the outer wall of the copper plate. The model was otherwise insulated from the vacuum chamber.

In the crotch area of the vacuum chamber absorber wall, the maximum temperature of 99.8°C occurs at the inner centerline of the copper wall, as expected. At the cooling tube the maximum temperature is 50°C, which provides a safety factor of two against local water boiling at the 3.79 l/min flow rate.

In situ bakeout analysis

After initial assembly and after each atmospheric vent, sections of the vacuum vessel will be baked in situ at approximately 200°C, using electric heating tapes. The vacuum sections correspond to intervals between the sector valves, so that they can be individually heated and pumped to vacuum. Thermal finite element analyses were conducted to optimize the size and location of heat sources for the large dipole magnet vacuum chambers and for the vacuum tubing between dipole chambers.

The dipole vacuum chamber was modeled as a two-dimensional structure with unit thickness. For conservatism in the computation of heating requirements, a section at the widest point of the chamber was used in the analysis. Because of the dipole chamber symmetry about the horizontal axis, only half of the chamber was modeled. Two boundary conditions were considered in the model. In the region of the chamber directly below the dipole magnet pole tips, it was assumed that conduction takes place through the 0.27 cm air gap to the magnet, which is at 25°C. The remaining upper surface of the chamber was assumed to be covered with 1.27 cm fiberglass insulation with free convection of its outer surface to air at 25°C.

The computed heat input meets the 200°C bakeout requirement, and produces a reasonable thermal gradient along the dipole chamber surface. Two heater tapes are required. A 5.08 cm wide heater tape dissipating 24 W/cm^2 is placed along the inner vertical stainless steel wall of the chamber. A 10.16 cm wide heater tape dissipating 24 W/cm^2 is placed on the top stainless steel plate (also on the bottom, by symmetry) just outside the pole tip. Both tapes are approximately the length of the chamber. The resulting minimum temperature of the chamber is 197°C; the maximum temperature, under the tape, is 281°C.

CONCLUSION

Our analyses indicate that the vacuum systems designed for the LSU electron storage ring and beam transport line meet the performance specifications and design considerations. Construction is currently underway, with commissioning scheduled for 1991.

ACKNOWLEDGMENTS

The author would like to acknowledge the contributions of R. Byle, D. Meader, and J. Pearce at Maxwell Laboratories, Inc., Brobeck Division, in the design and supporting analysis for the LSU/CAMD vacuum system. In addition, K. Kennedy of the Lawrence Berkeley Laboratory, N. Dean and H. Wiedemann of the Stanford Synchrotron Radiation Laboratory, and M. Green and W. Winter of the Synchrotron Radiation Center at the University of Wisconsin provided valuable insight.

REFERENCES

1. Maxwell Laboratories, Inc., Brobeck Division, *Electron Storage Ring Conceptual Design Report for Louisiana State University Center for Advanced Microstructures and Devices (LSU-CAMD)*, Brobeck Division Report BD-353-R1 (June 22, 1989).

2. R. Chasman and G.K. Green, Brookhaven National Laboratory, and E.M. Rowe, Physical Science Laboratory, University of Wisconsin, "Preliminary Design of a Dedicated Radiation Facility," *IEEE Trans. Nucl. Sci.*, NS-22:3, 1765 (1975).

3. Undocumented pressure code from H. Wiedemann, Stanford Synchrotron Radiation Laboratory.

4. Y. Strausser, Varian Report VR-51, Varian Associates, Vacuum Products Division (Palo Alto, California: 1968).

5. Ibid.

6. J.C. Schuchman, Brookhaven National Laboratory, "Vacuum System for Room Temperature X-Ray Lithography Source (XLS)," *Advanced Synchrotron Light Source Conference: AIP Conf. Proc.* 171, 130 (1988).

7. Ibid.

8. S. Tazzari, "The European Synchrotron Radiation Facility," *IEEE Trans. Nucl. Sci.*, NS-32:5, 3400 (October 1985).

9. T. Kobari, H.J. Halama, "Photon Stimulated Desorption from a Vacuum Chamber at the National Synchrotron Light Source," *J. Vac. Sci. Technology* A5:4, 2355 (July/August 1987).

10. *Numerically Integrated Elements for System Analysis (NISA II), Version 90.0*, Engineering Mechanics Research Corp. (Troy, Michigan: April 1990).

ALS INSERTION DEVICES*

E. Hoyer, J. Chin, K. Halbach, W. V. Hassenzahl, D. Humphries, B. Kincaid,
H. Lancaster, D. Plate

Lawrence Berkeley Laboratory, University of California
Berkeley, California 94720

* Work supported by the Director, Office of Energy Research, Office of Basic Energy Sciences, Materials Sciences Division, of the U. S. Department of Energy under Contract No. DE-AC03-76SF00098.

ABSTRACT

The Advanced Light Source (ALS), the first US third generation synchrotron radiation source, is currently under construction at the Lawrence Berkeley Laboratory. The low-emittance, 1.5 GeV electron storage ring and the insertion devices are specifically designed to produce high brightness beams in the UV to soft X-Ray range. The planned initial complement of insertion devices includes four 4.6 m long undulators, with period lengths of 3.9 cm, 5.0 cm (2) and 8.0 cm, and a 2.9 m long wiggler of 16 cm period length. Undulator design is well advanced and fabrication has begun on the 5.0 cm and 8.0 cm period length undulators. This paper discusses ALS insertion device requirements; general design philosophy; and design of the magnetic structure, support structure/drive systems, control system and vacuum system.

INTRODUCTION

The Advanced Light Source (ALS), a third generation synchrotron radiation source, is currently under construction at the Lawrence Berkeley Laboratory.[1] This facility consists of a 50 Mev linac, a 1 Hz, 1.5 GeV booster synchrotron and a low-emittance electron storage ring optimized for the use of insertion devices at 1.5 GeV. The use of insertion devices in the low emittance storage ring will produce high brightness beams in the UV to soft X-ray range. Their predicted performance is shown in Fig. 1.[2]

The planned initial complement of insertion devices includes four undulators and a wiggler with the basic parameters given in Table I. To achieve high brightness, the ALS undulator design must meet the stringent requirements in Tables II and III.[3] These requirements are derived from the need for rapid scanning of narrow spectral features and the need to avoid perturbing the electron beam in the storage ring. The wiggler, a high flux device, has reduced spectral requirements but must still meet the storage-ring requirements. The U5.0 Undulator will be used here as an example of a typical ALS insertion device and is shown in Fig. 2 with most major subsystems identified.

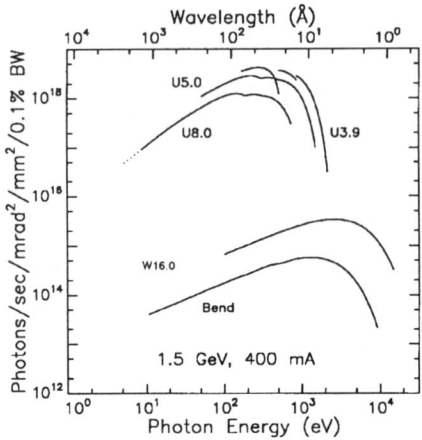

Fig. 1. Spectral brightness as a function of photon energy for the ALS undulators, wiggler and bend magnets. Each undulator curve is the locus of narrow peaks of radiation, tuned by altering the undulator gap, and represents the envelope of the first, third and fifth harmonics.

Table I. Parameters for the planned initial complement of insertion devices for the ALS

Name	Period (cm)	No. of periods	Overall length (m)	Photon energy range (keV)
Undulators				
U8.0	8.0	55	4.6	0.006 - 1.0
U5.0 (2)	5.0	89	4.6	0.052 - 1.5
U3.9	3.9	115	4.6	0.169 - 2.5
Wiggler				
W16.0	16.0	16	2.9	0.5 - 20.0

Table II. ALS undulator specifications based on spectral requirements

Parameter	Value
Useable harmonics	1st, 3rd, & 5th
Brightness requirement	5th harmonic reduction <30%
Spectral broadening requirement	ID broadening ≤ ALS emittance effects
Minimum increment of photon energy	1/10 of 5th bandwidth
Minimum time to go from min. to max. gap (slew)	5 minutes
Maximum photon energy scan rate	1 bandwidth/second

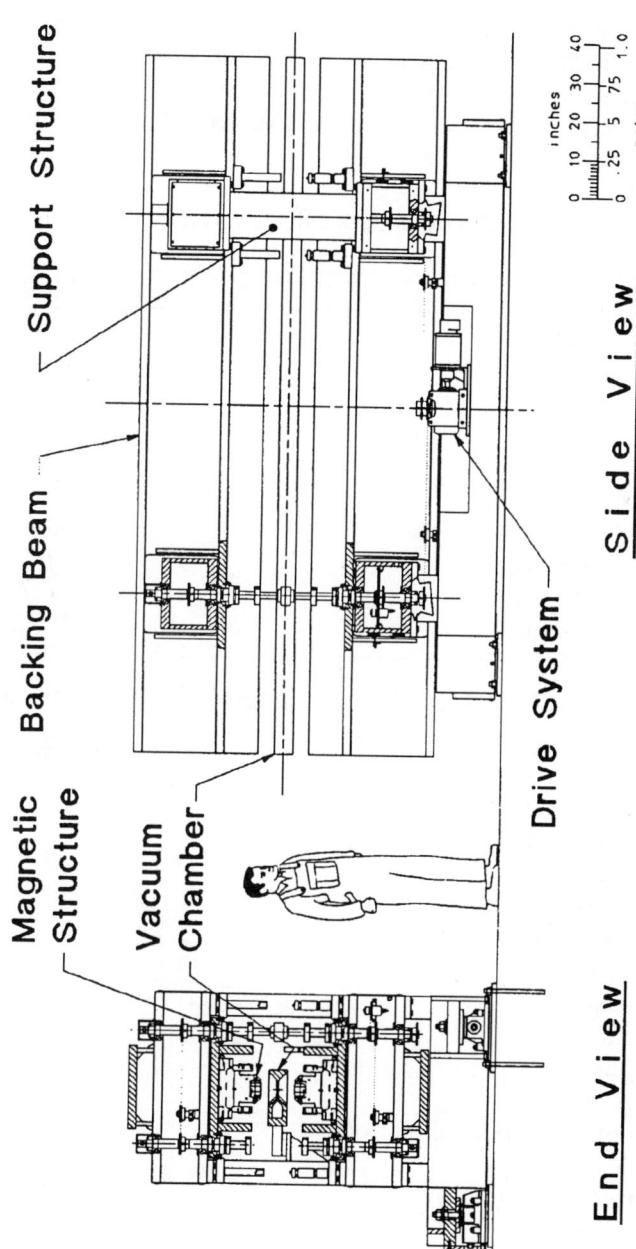

Fig. 2. U5.0 Undulator Design.

The engineering parameters, shown in Table IV for the U5.0 Undulator, are derived from the basic parameter, spectral and storage-ring requirements. The insertion device conceptual designs are nearly complete, and the detail design is well advanced. Component purchase and fabrication has started on the U5.0 and U8.0 Undulators.

Table III. ALS insertion device specifications based on storage ring requirements

Parameter	Limit
$\int B_y \, dl$	100 G cm
$\int\int B_y \, ds \, dl$	100 G cm^2
$\int B_x \, dl$	500 G cm
Integrated quadrupole	50 G
Integrated skew quadrupole	50 G
Integrated sextupole	50 G/cm
Integrated octupole	1 G/cm^2
Required vacuum	10^{-9} Torr

Table IV. U5.0 Undulator engineering design parameters

Parameter	Limit
Maximum peak field (@ 1.4 cm magnetic gap)	0.89 T
Effective peak field (@ 1.4 cm magnetic gap)	0.837
Period length	5 cm
Number of periods	89
Number of full field poles	179
Entrance sequence	0,-1/2,+1,-1
Overall length	455.8 cm
Pole width	8 cm
Pole height	6 cm
Pole thickness	0.8 cm
Number of blocks per half-period (one side of pole)	6
End correction range (B_y)	1,500 G cm
End correction range (B_x)	None
Steering coils (short)	~5 λ long
Dipole trim coils (long)	To 4.5 m
Steering and trim field strength	±5 G
Systematic gap variation	58 μm

DESIGN PHILOSOPHY

The approach taken for the initial complement of ALS insertion devices has been to develop a generic design with the objective of reducing engineering, fabrication, and maintainance costs.

The following commonality exists between the various planned devices:

Magnetic Structure:
- Scaled undulator magnetic configurations.
- Similar backing beams for the undulators.
- Wiggler backing beams two-thirds the length of the undulator backing beams.

Support Structure/Drive Systems:
- Identical support structures for the undulators.
- Support structure shortened for the wiggler.
- Identical drive systems for all devices.

Control System:
- Identical control systems for all devices.

Vacuum System:
- Similar vacuum chamber configurations for the undulators.
- Wiggler vacuum chamber similar to the undulator vacuum chamber.
- Identical pumping systems for the undulators.
- Similar pumping system for the wiggler.

Design, fabrication, testing and installation of the ALS insertion devices takes advantage of the LBL experience with the BL VI and BL X Wigglers that are now operational at SSRL.[4,5]

MAGNETIC STRUCTURE

The magnetic structure provides the required magnetic fields and includes the periodic magnetic structure, end magnetic structures, backing beams and if required auxiliary tuning coils.

The ALS insertion devices incorporate hybrid magnetic configurations consisting of Nd-Fe-B magnetic blocks and vanadium permendur poles. The hybrid design was chosen because there are several advantages over the pure current sheet equivalent material (CSEM) design:
- Fields are dominated by the characteristics of the poles, which can be made very uniform both in size and magnetic performance.
- Errors in magnetic moments of the blocks can be averaged by sorting the blocks for the poles.
- Errors in total magnetic moment of all the blocks of a pole have little effect on the electron beam or the photon spectrum because they contribute equally to adjacent poles and produce no electron steering.
- A higher peak field is achievable. (This is most important for the wiggler.)

For undulators, the objective of the magnetic design is to develop a magnetically well behaved structure which yields a high value of B_{eff} for mid-plane fields. B_{eff} is given by

$$B_{eff}^2 = \sum_{i=0}^{\infty} \left(\frac{B_{2i+1}}{2i+1}\right)^2 \qquad (1)$$

where B_1 is the amplitude of the fundamental, B_3 is the amplitude of the third harmonic, etc.

The magnetic configuration is based on 2-D modeling with the computer code PANDIRA and a 3-D Hybrid theory for hybrid CSEM insertion devices.[6,7] To verify the magnetic design for U5.0, a model was built and tested under a variety of conditions.[8]

The undulator performance criteria is met by tolerances based on the hybrid CSEM insertion device theory. The tolerances established for U5.0 are given as an example in Table V.[9]

Table V. U5.0 Magnetic Structure tolerances

Error Type	Total Tolerance	Error (%)
Spacing CSEM to pole	102 µm	0.08
Pole thickness	50 µm	0.03
Vertical pole motion (gap)	22 µm	0.05
Pole width	100 µm	0.03
Surface easy axis orientation	±2.3 degrees	0.16
	Total:	0.19

Figs. 3 and 4 show the U5.0 magnetic structure which includes:
- Half-period pole assemblies, that consist of an aluminum keeper, a vanadium permendur pole (8 cm wide X 6 cm high X 0.80 cm thick) pinned into the keeper and six Nd-Fe-B blocks (3.5 cm square X 1.7 cm thick in the magnetization direction) bonded into the assembly.[10] This design allows for accurate vertical and longitudinal pole tip placement.
- Assembly sections; that consists of a pole mount fabricated from 5083-H321 aluminum onto which 35 half-period pole assemblies are mounted and accurately positioned.
- Backing beams that are 4.5 m long, stress relieved steel structures, with 81 cm depth and 89 cm width, each beam provides magnetic shielding and holds five assembly sections and two end sections.[11]
- Dipole and steering coils if needed.

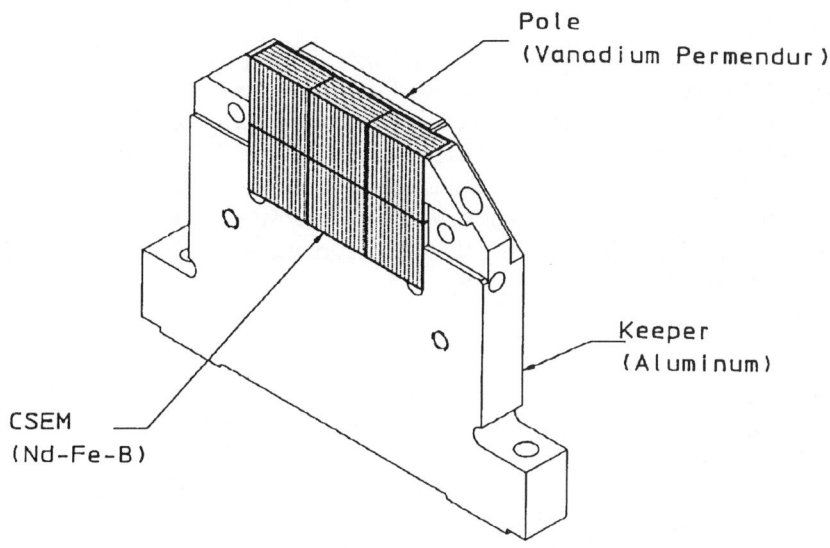

Fig. 3. U5.0 half-period pole assembly

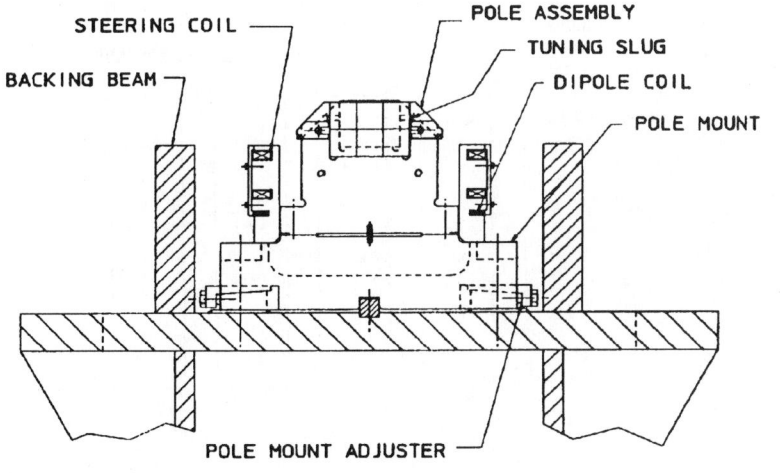

Fig. 4. U5.0 magnetic structure assembly section

The upper and lower backing beams are tied together with low reluctance NiFe hinges to reduce the effect of environmental fields on the electron beam trajectory.[12]

To avoid steering the beam as it travels through the insertion device, it is necessary to control the configuration of the fields at the ends. Fig. 5 shows a schematic of the end magnetic structure that utilizes a system of Nd-Fe-B rotors to fine-tune the fields at the ends of the insertion device. There are four rotors at each end, and a small quantity of Nd-Fe-B at each rotor location. Gap-dependent errors at the ends are small, thus the objective is to determine a single set of orientations for the rotors that minimizes the steering errors introduced by the end magnetic fields over the entire range of gaps.

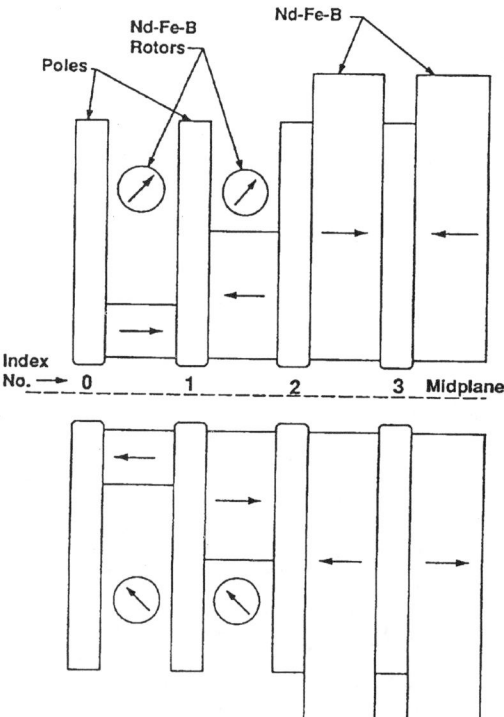

Fig. 5. End rotator configuration.

SUPPORT/DRIVE SYSTEMS

The support/drive systems include the support structure that provides the framework for holding the magnetic structure and the drive system that opens and closes the magnetic gap. Requirements for the support structure shown in Figs. 2 and 6 include the following:
- Support a maximum magnetic load of 84,000 lb. (The loading of a 5 m long insertion device with 10 cm wide poles operating at 1.85 Tesla.)
- Maintain a magnetic gap variation of 46 µm at the smallest gap (14 mm). (U5.0 Undulator requirement is 58 µm.)
- Meet the ALS storage ring, tunnel and adjacent beamlines space requirements.
- Accommodate the vacuum system and its support structure.
- Be capable of being installed, aligned and serviced in the storage ring.

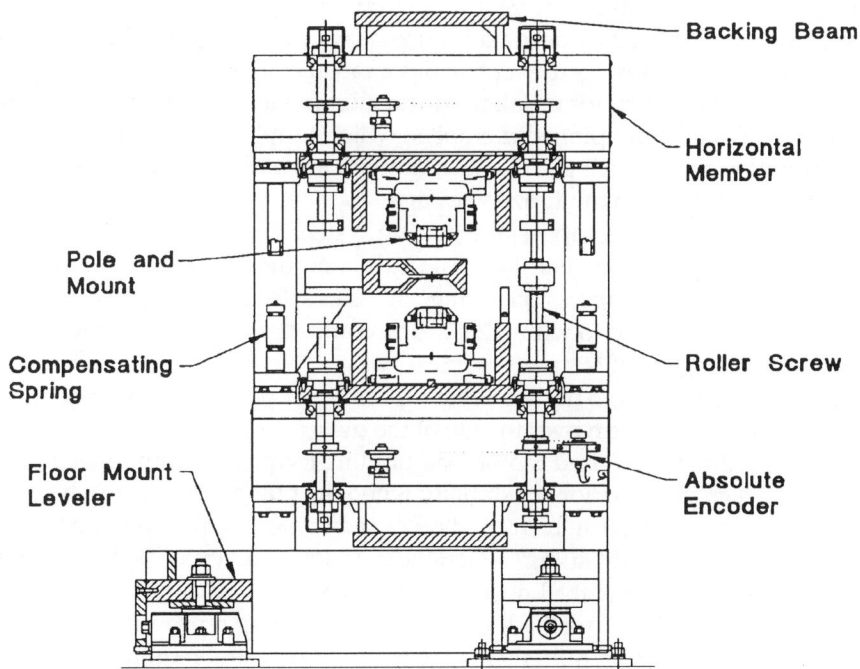

Fig. 6. Undulator end view.

Both 4-Post and C-Frame structures were considered. The 4-Post configuration was selected for the following reasons:
- Greater tunnel aisle clearance.
- Less gap deflection due to a more rigid structure.
- No pole rotation because of symmetrical support loading.
- Better access for assembly and maintenance of components.

The principal advantage of the C-Frame structure is that it would be open on one side allowing magnetic measurements with an external measurement system and the possibility of insertion device installation with the vacuum chamber in place in the storage ring.

As shown in Figs. 2 and 6, the support structure is of rigid construction consisting of a base onto which two lower horizontal members are mounted. Four vertical posts are in turn attached to the lower horizontal members and the two upper horizontal members are attached to the tops of these posts. The horizontal beams pass thru the webs of the backing beams to limit the overall height of the support structure to less than the 8 ft tunnel height. The base is a welded assembly containing a platform for the gear reduction unit and motor. Three y-axis leveling mounts, which include x-axis and z-axis adjustments, provide a kinematic support system. This arrangement provides a satisfactory range of adjustments for all six degrees of freedom for installation and alignment. The support structure is modular with bolted and pinned members, which simplifies fabrication, installation, calibration and servicing. All subassemblies are individually stress-relieved before final machining to minimize warpage.

A magnetic-load compensating spring system is provided to buck the gap-dependent magnetic load.[13] For the U5.0 Undulator, the eight spring assemblies consist of two helical compression springs in series selected to match the gap dependent magnetic load to within 20%. The compensating spring system provides the following benefits:
- Reduced system friction which gives better positional response from the drive system over the life of the device.
- Minimum required motor load holding torque at any magnet gap, which gives stationary stability when the null position is reached. Motor current can be turned off or reduced to minimize motor heating.
- Elimination of "lifting" when the magnetic load exceeds the gravitational weight of the lower backing beam.
- Reduced structure load, which gives better gap reproducibility.

The drive system requirements are set by the spectral requirements and include:
- Capability of opening the magnetic gap with an 84,000 lb magnetic load.
- A step resolution of 1 µm (based on increments of 1/10 of the 5th harmonic for a U3.65 undulator.)
- A maximum scanning speed of 2.3 mm/s (based on a scan rate of 1 bandwidth/s for a 11-cm-period device).

- A magnetic gap range of 1.4 cm to 21.6 cm.
- Opening or closing time must be five minutes or less.
- Gap position determined by an absolute encoder.

Changing the magnet gap in an insertion device requires moving the backing beams. This is accomplished by rotating the 2 mm pitch Transrol roller screws that are mounted to the horizontal beams and support the backing beams. Specifically, the four right-handed roller screws attached to the upper backing beam and four left-handed roller screws attached to the lower backing beam are connected by a shaft coupling and combine to provide equal and opposite vertical motion when rotated. Gap motion begins at the rotation of a stepper motor which is transmitted thru a gear box and a series of sprocket wheels and roller chains to the roller screws. An absolute rotary encoder is coupled to a Transrol roller screw shaft to read the absolute position of the magnet gap.

The drive system has been sized for the maximum possible ALS insertion device magnetic load. Though one revolution of the roller screws changes the gap by 4 mm, the minimum incremental gap motion is 0.1 micron. This is possible because the 200 steps/revolution stepper motor has a 10 micro-step/step capability through the control electronics and the motor rotation is reduced a factor of 30 through the gear reduction unit. At the 2000 step/revolution operation, the motor can easily be driven at a velocity to move the gap from full closed to full open in 1 1/2 minutes. The rotary-encoder selected is a Compumotor AR-23, which has a resolution of 16,384 counts/revolution and is mounted with a step-up ratio of 4.75. This arrangement allows resolution of a gap variation of less than 0.1 µm.

Analysis of the proposed system shows that stick-slip will give a gap uncertainty of less than 0.4 µm.[14] Unidirectional scanning and control of the undulator gap are required because backlash is estimated at 87 µm in gap motion. Scan-to-scan gap reproducibility for unidirectional scanning is estimated to be less than 8 µm.

Drive system protection guidelines include:
- Travel limits set by the closed-loop control system stored in the control program.
- Micro-switches hard-wired to the control system for minimum and maximum gap positions.
- Mechanical stops for minimum and maximum gap positions.
- Full-torque stepper-motor stall capability.
- Mechanical drive components designed to handle full stepper motor torque.
- Full load current sensing (control system will shut down current after a preset time interval).

Insertion device temperature control is important. A vertical temperature gradient of greater than 0.1 degree C in the undulator backing beams produces excessive spectral broadening. Hence, each undulator will have an enclosure, and the temperature in the enclosure will be maintained by circulating the air with muffin fans.

CONTROL SYSTEM

The insertion device control systems are designed to provide sufficient position accuracy, resolution, velocity and range information for the motors and encoders for all anticipated insertion devices. In addition, the control system must control and monitor the dipole and steering correction power supplies, as well as controlling gap dependent rotator positioning, if required. The insertion device control systems are to be integrated into the overall accelerator computor control system.

The insertion device gap must be controlled (via request to the accelerator control system) and monitored by the experimenter using the generated synchrotron radiation. During development, the insertion device is to be capable of being manipulated through the control system by a local computer, so that the necessary control and monitoring algorithms can be determined.

The control system block diagram is shown in Fig. 7. A Compumotor system has been selected for the gap control and is currently undergoing tests. One scheme proposed for the control system is to use five intelligent local controllers (ILCs). One ILC coordinates the activities of the other four and communicates with either the accelerator control system database or an IBM-PC. The ILCs controlling the rotators and power supplies would contain the compensation data tables that give the rotator positions and the coil settings required for each magnet gap.

Provisions for interfacing limit switches are included in the indexer as well as the ability to compensate for backlash and to program acceleration and deceleration curves. The indexer can be programmed to microstep the motor with as many as 25,000 steps per revolution.

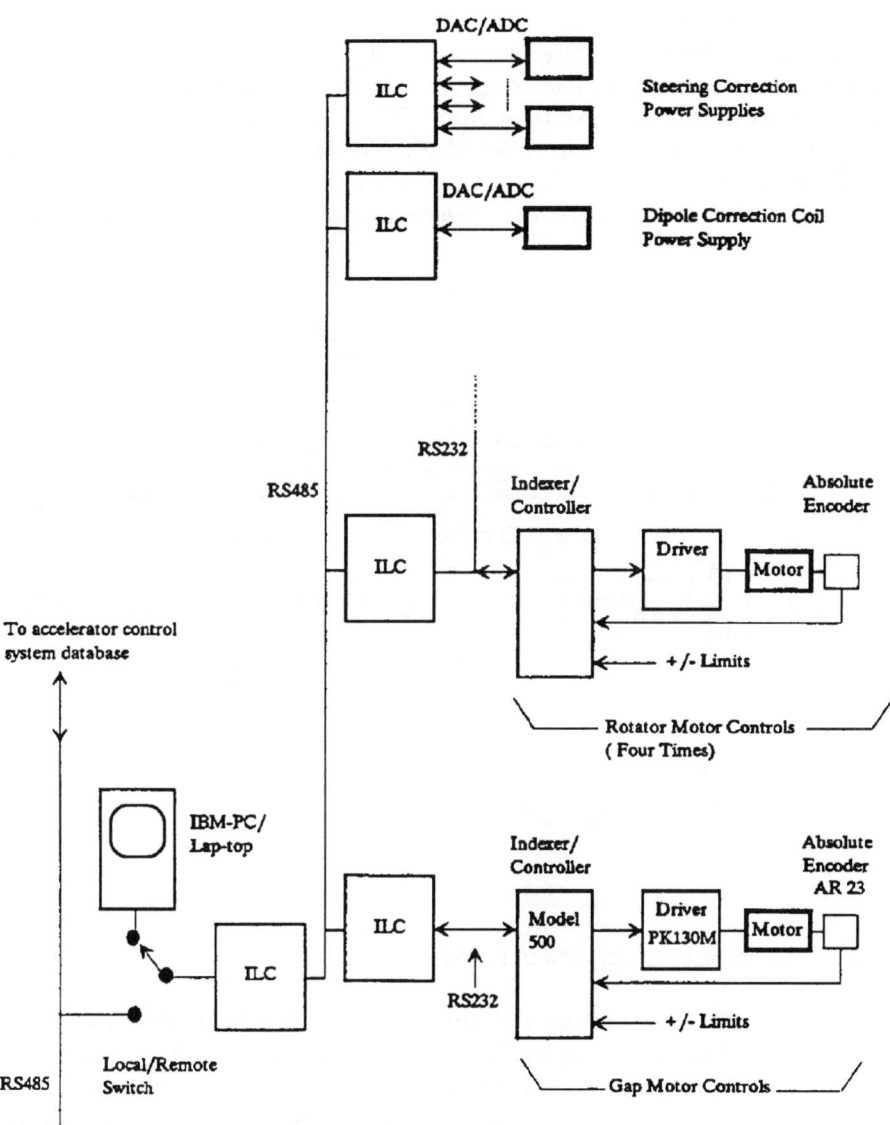

Fig. 7. Insertion device control system.

VACUUM SYSTEM

The objective of the vacuum system is to provide a 10^{-9} Torr vacuum at the insertion device beam aperture. Fig. 8 shows a plan view of an undulator vacuum system. Two vacuum chambers are required for ALS operation, one for commissioning and one for dedicated operation.[15] The commissioning chamber has an elliptical beam aperture of dimensions 1.8 cm vertical x 6.0 cm horizontal. The chamber for dedicated operation, which replaces the commissioning chamber, has a rectangular beam aperture of dimensions 1.0 cm vertical x 6.0 cm horizontal.

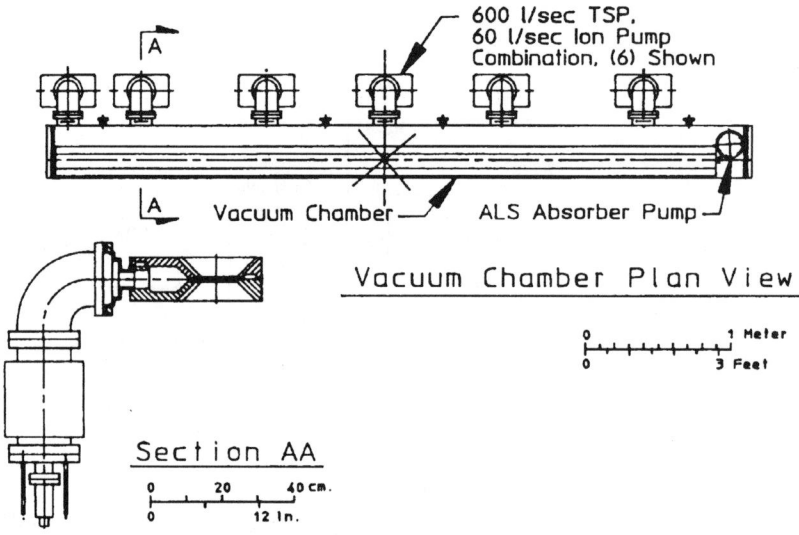

Fig. 8. Undulator vacuum system layout.

The 5.1 m long undulator vacuum chambers will be made of two pieces of machined 5083-H321 aluminum alloy. These two-piece welded chambers are similar to the ALS storage ring sector chambers that are currently under construction. Both the commissioning and dedicated chambers have a total horizontal aperture of 21.8 cm, the inner 6.0 cm provides the circulating beam aperture and the outer aperture allows the bending-magnet synchrotron radiation to pass through the chamber. The radiation is then absorbed by the photon stop located at the exit end of the chamber. Both chambers have an antechamber along the complete length as part of the outer aperture to improve vacuum by an increased conductance. External surfaces of the chambers have pockets machined into them for the magnet poles. The shape allows a minimum magnetic gap of 2.2 cm for commissioning and 1.4 cm for dedicated operation. The undulator vacuum chamber has 6 side ports and one top and bottom port near the exit end of the chamber for vacuum pumps. Several smaller ports are provided for a roughing system, ion gauges and a RGA head. The upstream end of the chamber includes a flange for insertion of NEG pumping strips and for a viewport for remote visual inspection of the aperture.

The vacuum system consisting of six combination 600 l/s titanium sublimation and 60 l/s ion pumps (which give a net pumping speed of 173 l/s each at the antechamber) and an ALS absorber pump of 1450 l/s capacity has a total antechamber pumping speed of 2500 l/s.[16] The pressure distribution at the beam aperture was estimated after 40 Ampere hours of accumulated electron beam operation assuming a thermal outgassing rate of 10^{-11} Torr l/s cm^2, a molecular production rate, due to photon induced desorbtion, of 10^{-5} molecules/photon (for photons of energies greater than 10 eV) and 1.9 GeV - 400 mA storage ring operation.[17] The average pressure distribution is 3×10^{-10} Torr, as shown in Fig. 9. If two NEG strips, 3 cm wide by 450 cm long are inserted into the chamber and activated, the pressure will drop to 1×10^{-10} Torr in the beam aperture.[18]

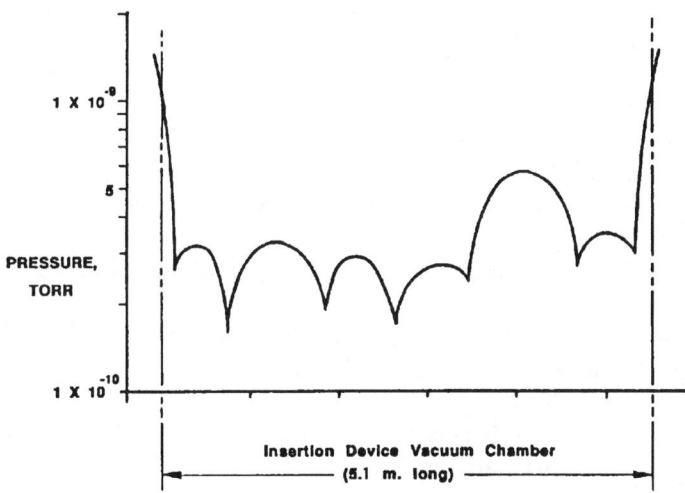

Fig. 9. Insertion device chamber pressure distribution.

The ion pumps will be driven and monitored by ion pump controllers. The titanium sublimation pump filaments will be powered by a single power supply multiplexed to the six pumps. Insertion device vacuum chamber pressure will be monitored in two ways. Ion pump current will be converted to approximate pressure in the local display and accurate pressure measurements will be accomplished at one or two locations with nude ion gauges and an ion gauge controller.

The vacuum chamber and associated pumping system will be supported from the top of the insertion device support structure. Struts will be used for the chamber and spring loaded hangers used for the pumps.

REFERENCES

1. 1–2 GeV Synchrotron Radiation Source, LBL PUB-5172, Rev. (July 1986)
2. An ALS Handbook, Rev. 2, LBL PUB-643 (April 1989).
3. U5.0 Undulator Conceptual Design Report, LBL PUB-5256 (November 1989).
4. E. Hoyer, et al., The Beam Line VI REC-Hybrid Wiggler for SSRL, IEEE Trans. NS-30, p. 3118 (1983).
5. E. Hoyer, et al., The Beam Line X NDFE-Steel Hybrid Wiggler for SSRL, IEEE Catalogue No. 87CH2387-9 p 1508 (1987).
6. K. Halbach, et al., developed PANDIRA, an improved version of POISSON which allows solution of permanent magnet and residual field problems; POISSON is an improved version of TRIM [A. Winslow, J. Computer Phys. 1, 149 (1967)].
7. K. Halbach, Insertion device Design: 16 Lectures Presented from October 1988 to March 1989, LBL Publication V 8811-1.1-16.
8. W.V. Hassenzahl, R. Savoy, U5.0 Model Pole Measurements, (to be published in 1991).
9. R. Savoy, et al., Calculation of Magnetic Error Fields in Hybrid Insertion Devices, LBL-27811.
10. E. Hoyer, Magnetized Neodymium-Iron-Boron Blocks, LBL Specification 734D (April 1989).
11. E. Hoyer, Backing Beam Design Calculations, LBL Engineering Note M6834 (May 1989).
12. E. Hoyer, Flexible Yoke Design, LBL Engineering Note M7039B (July 1990).
13. J. Chin, Magnetic Load Compensating Springs, LBL Engineering Note M6829 (April 1989).
14. J. Chin, Drive System Backlash, LBL Engineering Note M6882 (August 1989).
15. E. Hoyer, Vacuum Chamber Design, LBL Engineering Note M6806 (February 1989).
16. E. Hoyer, Pumping System Design Calculation, LBL Engineering Note M6821 (March 1989).
17. E. Hoyer, CDR Vacuum Chamber Pressure Distribution, LBL Engineering Note M6844 (May 1989).
18. E. Hoyer, Chamber Pressure Improvement with NEG Pumping, LBL Engineering Note M6889 (September 1989).

Carbon and Other Contaminants in Vacuum Systems

Victor Rehn
Physics Division, Research Department, Naval Weapons Center
China Lake, California 93555

ABSTRACT

This paper reviews recent progress toward understanding the physical and chemical mechanisms by which carbonaceous contamination is deposited on surfaces within synchrotron-radiation vacuum systems, especially on optical surfaces. Recent progress made in the prevention or removal of contamination layers, and in the resuscitation of contaminated VUV and soft x-ray mirrors and gratings is reviewed.

HISTORY AND SCOPE OF THE PROBLEM

In this review, an attempt is made to take a somewhat general view of the vacuum contamination problem, its technical origins and physical understanding, its practical consequences and, of course, possible solutions. This is not to imply that all fields of vacuum technology are considered, however. The immediate goal is to help resolve the contamination problem in synchrotron-radiation optical systems, especially for high-brightness insertion-device optical systems. Much, but certainly not all, of the current and recent activity in the field is reviewed, and hopefully synthesized to improve insight. I apologize at the outset for oversights or distortions. Oversights cannot be totally avoided, but I have tried hard to avoid distortions. There has been an effort to reinterpret results in the light of other results, and to clarify the presentation for those who may not be specialists in all of the several fields of science and technology involved.

In many scientific applications of synchrotron radiation (SR), one of the significant advantages cited for SR-based experimentation over other types is the cleanliness of the SR-source, which must, in many cases, be exposed directly to contamination-sensitive experimental samples. Often the cleanliness of SR sources is contrasted with the contaminating effects of electron-beam sources, or x-ray anodes. Yet, practical vacuum environments are not clean environments necessarily: Great effort on the part of experimenters is required to produce clean-vacuum environments. In the early days, the pioneering clean-vacuum experiments of Davisson and Germer[1] demonstrated most of the clean-vacuum techniques that are still practiced today.

On the other hand, in the realm of early high-energy accelerators (pumped by large oil-diffusion pumps), the term "cyclotron varnish" came into the scientific vocabulary. This term was descriptive of the thick, varnish-colored coating that quickly covered the interior surfaces of cyclotrons, synchrotrons and betatrons. With the advent of high-energy electron storage rings in the 1960s and 1970s, the need arose for higher-quality vacuums in order to increase the lifetime of the stored electrons or positrons. Ion pumps replaced oil-diffusion pumps, and the operating pressure was decreased by two or three orders of magnitude. However, contamination films still formed on surfaces, albeit much more slowly, and generally confined to the orbital plane.

With the absence of diffusion-pump oil vapors in the superior vacuum of storage rings, other sources of heavy hydrocarbons were sought out assiduously, and eliminated or minimized in the design and construction of storage rings and beam lines. These improved vacuum practices were sufficient to greatly lengthen the lifetimes of charged particles in orbit, but were not sufficient for the prevention of serious contamination of in-vacuum optical surfaces.[2,3,4,5,6,7,8]

The problem of contamination of optical surfaces in vacuum systems was recognized in the area of space applications, and attacked in the early 1970s. Methods for cleaning optical surfaces contaminated in vacuum systems were developed and tested.[9,10,11] The cleaning methods developed for the space program were applied to optical surfaces for synchrotron-radiation beam lines in the 1980s,[12,13,14,15,16,17,18] and ultimately to *in situ* cleaning of optical surfaces.[19,20,21,22] An alternative cleaning process involving only SR in an oxygen atmosphere was suggested much earlier,[5] but only one report testing this concept has appeared to date.[23] Methods for retarding or preventing contamination of SR optical surfaces have been discussed,[5] but no reports of tests of such concepts have yet appeared. Progress in the development and implementation of methods, hardware, and procedures for removal of contamination, either outside or inside the vacuum, is reviewed in the final section of this presentation.

Progress toward understanding the physical and chemical mechanisms involved in the deposition of contamination layers on internal surfaces of vacuum system has progressed somewhat slowly. Cyclotron varnish was commonly attributed to hydrocarbon pump-oil vapors that were polymerized on surfaces, possibly by stray charged particles. The concept of hydrocarbon cracking and polymerization was assumed to carry over to storage rings, as well. Finally, Auger-electron spectroscopic measurements made at the Synchrotron-Radiation Center, Stoughton, WI, showed that contamination films on SR optical surfaces are primarily carbon with small amounts of oxygen, nitrogen, and, in cases where a titanium-getter pump was located nearby, titanium.[8] Further, comparison of the transmission spectrum of contaminated SR beam lines with EXAFS spectra of amorphous carbon and graphite showed the EXAFS spectra are consistent with a primarily graphitic carbon film.[4,24] More recently, the graphitic nature of the film has been questioned on the basis of measurements of the VUV reflectance, which appears to be higher than that of graphite crystals.[14]

The deleterious effects of contamination films on the spectral transmission function of beam lines was clearly apparent. In the soft x-ray range, research in the region of the K-absorption edge of carbon was largely stymied by the K-edge features of the carbon contamination, making interpretation of spectral data in that spectral vicinity difficult or impossible.[2,3,4,5] At much lower photon energies, the strong absorption of the carbonaceous contamination in the VUV range also inhibited research, especially on gas-phase beamlines where the carbon contamination formed more quickly.[14]

A major step in understanding the physical and chemical mechanisms involved in SR-based carbonaceous contamination of optical surfaces occurred in 1983 with the development of a simple, rate-equation-type model for the deposition.[25] In spite of the simplifications implicit in the linear rate-equation model, recent measurements of both etching[23] and deposition[26] of carbon-contamination films have supported the validity of the HASYLAB model. Hence, at present there appears to be an adequate physical and chemical model of the contamination process to serve as a guide for those interested in either prevention of contamination, or resuscitation of contaminated surfaces. In the next section of this presentation, the model will be reviewed in detail for the purpose of emphasizing its predictions for various types of SR beam lines, and of examining relevant experimental data.

The purpose of the present review is to highlight areas of recent progress, and also areas of critical need for progress in understanding and controlling the formation of contamination films in vacuum systems, particularly on SR optical surfaces. It is hoped that such a review might help in the direction and focussing of effort toward the more productive concepts, methods, or techniques that are applicable to the practical problem of vacuum contamination.

PHYSICAL MECHANISM OF VACUUM CONTAMINATION

In his 1978 workshop paper, "Beamline Chemistry", D. A. Shirley[4] outlined the basic surface chemistry of carbon-contamination formation in terms of a surface-catalysis process similar to the well-known Fischer-Tropsch reaction. In the Fischer-Tropsch reaction, $CO + H_2$ is converted to hydrocarbon molecules in the presence of a hot, active catalytic surface such as Ni or Pt. However, an unwanted side reaction produces carbon, which poisons the catalytic surface: $2CO$ is converted to $CO_2 + C$. The carbon film produced is described as tenacious, and irreversibly bonded to the catalytic surface. This mechanism could be involved in the formation of the first monolayer of carbon in cases where the surface temperature rises above 200 C. Subsequent carbon layers, however, form on the carbon surface, and do not have the benefit of the active catalytic metal surface. As pointed out by Shirley, the first step in the reaction is cleavage of a bond in the adsorbed molecule, and that step may be accomplished by VUV or soft-x-ray photons, or by fast electrons. Hence the reaction responsible for deposition of carbon on SR optical surfaces may be quite different from the classical Fischer-Tropsch reaction.

The synchrotron-vacuum community was alerted to the importance of residual CO, CH_4, other hydrocarbons and other carbon-containing gases. As pointed out in 1983 by Lichtman,[8] CO is generated within vacuum systems by reaction of residual oxygen with carbon impurity in hot W filaments (ion gage filaments, for example), or in stainless steel. Evolution of CO from hot filaments may be avoided by the use of LaB_6-coated filaments, which operate at a much lower temperature, and evidently do not evolve CO.

Ample evidence has been presented to show that the carbonaceous contamination film produced in SR vacuum environments forms only on surfaces exposed to SR,[2,3,14,25,26] although one report also illustrates back-streaming of a turbomolecular pump onto an optical surface that was located opposite the pumping orifice.[14] The predominance of carbon in the contamination film was shown in early Auger-electron spectra of contamination layers.[8] It has also been shown convincingly that the formation reaction is localized on the surface, as opposed to the nearby vacuum region.[14,26] Hence the process that dominates the contamination of SR vacuum surfaces is one that requires the presence of carbon-containing molecules and SR photons together on a surface.

The publication in 1983 of the HASYLAB study of the carbon contamination of mirror surfaces exposed to synchrotron radiation is currently the scientific basis of our understanding of the contamination process. To my knowledge, nothing published since that time invalidates the basic principles of the HASYLAB model, and supporting evidence has been reported. On the basis of measurements on Au-coated Si mirrors exposed to SR in two

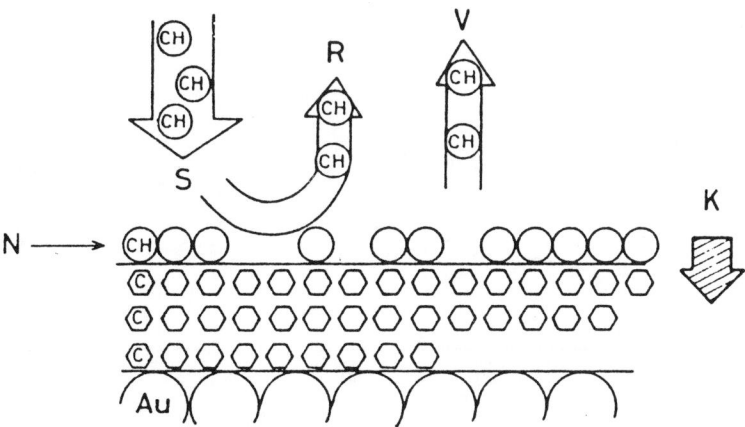

Fig. 1. Contamination Growth from Cracked Hydrocarbons. S = Incoming molecular flux; R = Reflected molecular flux; V = Evaporated molecular flux; K = rate of molecular cracking; N = surface density of adsorbed, uncracked molecules. Ref. 25.

different beam lines, the HASYLAB group formulated a model for the carbon-film deposition process, as illustrated in Fig. 1.[25] Residual gas molecules adsorb onto the mirror surface, residing there for a time that depends on the gas, the surface, and the surface temperature. During this residency time, a fraction of them will be cracked by SR photons or SR-excited photoelectrons from the underlying surface, producing atomic carbon and volatile fractions. Uncracked molecules may evaporate from the surface, leaving no contamination.

A further feature of the HASYLAB model is the association of the cracking with photoelectrons from the surface, rather than with SR photons directly. The measured yield of photoelectrons as a function of time during exposure was correlated with the observed thickness of the contamination layer, showing a gradual reduction of the high photoelectron yield from the Au-mirror surface to the lower yield of the C-contamination surface. In the history of a contamination layer, the HASYLAB model predicts a period of rapid growth to a thickness of about 5 nm, during which the primary source of cracking photoelectrons is the underlying metal. Subsequently, the growth rate diminishes to one characteristic of the lower photoyield of a carbonaceous contamination. Examples of such a history are shown in Fig. 2.[25]

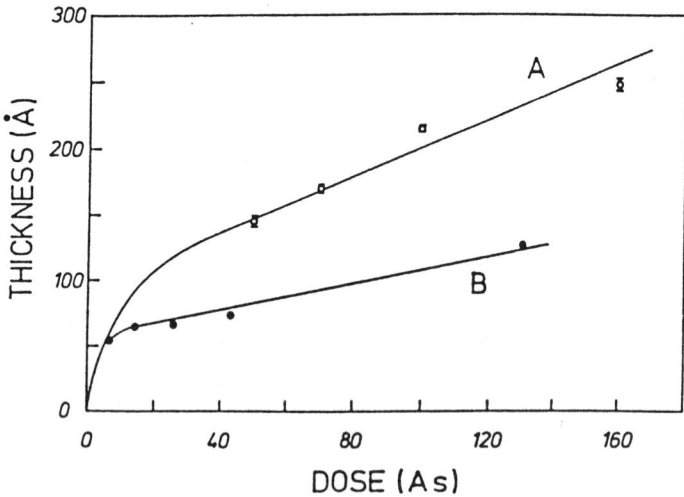

Fig. 2. Thickness of contamination layer as a function of dose. Curve A represents growth at $1-2 \times 10^{-5}$ Torr, 29° C. Curve B represents growth at $1-2 \times 10^{-6}$ Torr, 25° C. Ref. 25.

As illustrated schematically in Fig. 1, the overall rate of deposition of C atoms is modelled by a linear rate equation that includes the arrival rate, S, the reflection rate, R, the evaporation rate, V, and the cracking (or contamination) rate, K. Only single cracking processes are included in the model, and no account is taken of the possibility that the adsorbed molecular layer may be more than a single monolayer thick. The fraction of surface sites occupied by uncracked, adsorbed-gas molecules in the surface monolayer is n. Both R and S are assumed to be proportional to the gas pressure, p. The contamination rate, K, is assumed to be proportional to the flux of SR-excited photoelectrons, I. They make no attempt to include the energy dependence of the (primary or secondary) photoelectrons, nor the energy dependence of the cracking cross section of the adsorbed molecules. The evaporation rate, V, is assumed to be controlled by a simple latent heat of evaporation, E_L: $V \approx n\,T\,\exp[-E_L/kT]$. In steady state, the contamination rate derived from this simple model is:

$$K = C_1\,p\,I\,[C_2\,p + C_3\,T\,\exp(-E_L/kT) + I]^{-1} \qquad (1)$$

in which the C_i are constants. Eq. 1 is the central equation of the HASYLAB model.[25] For a certain choice of constants C_i, the general behavior of K is illustrated in Fig. 3, where the K, is plotted versus cracking flux, I, for several pressures of the crackable gas.

Two limiting cases of Eq. (1) are worthy of more detailed examination.

(a) *The High-Pressure Limit*, in which the occupation of surface sites by crackable molecules is high in steady state, and the contamination rate is controlled by the intensity of cracking (photoelectron) flux. In this limit, we assume that:

$$C_2\,p \gg C_3\,T\,\exp(-E_L/kT) + I. \qquad (2)$$

Both pressure dependence and temperature dependence cancel out, producing a "pressure-temperature saturation" regime. For pressure above some value, or temperature below some value, further pressure increases or temperature decreases will not significantly increase the contamination rate, which becomes proportional to the photoelectron flux only:

$$K \simeq 1. \tag{3}$$

This limit obtains either at high partial pressures of contaminating gas, or at low temperatures and low flux. The latter conditions lead to a long residency time of an adsorbed molecule on the surface, and both conditions ensure a high occupancy of surface sites by adsorbed, uncracked molecules This condition permits the carbon film to form as fast as the cracking flux allows, independent of temperature or pressure. Behavior consistent with this limit is shown in Fig. 4, where the thickness of the contamination film, corrected for equal doses of SR, is shown to be approximately independent of pressure.[25]

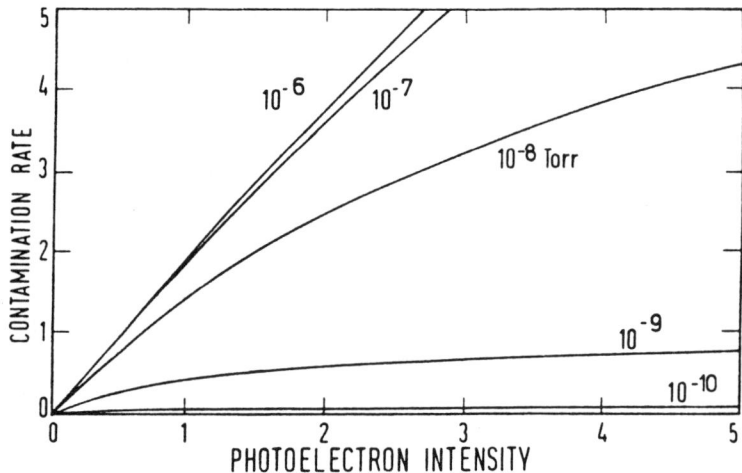

Fig. 3. Plot of Eq. (1) vs. I for several pressures. The constants (C_i) were chosen to show pressure saturation above 10^{-7} Torr. Ref. 25.

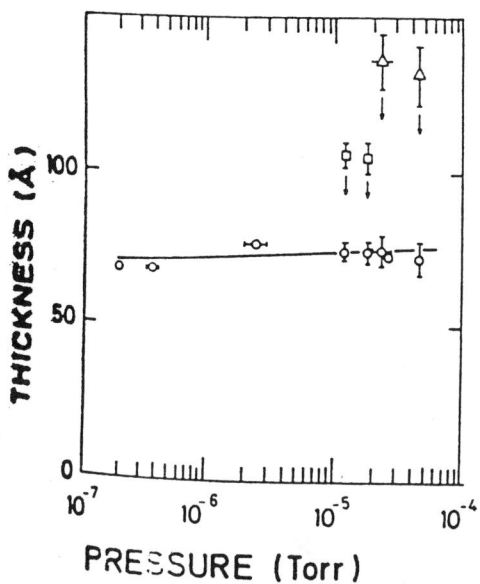

Fig. 4. Thickness of contamination layer vs. pressure for equal dose. The lack of a strong pressure dependence indicates parameters appropriate to the pressure-saturation regime. Ref. 15.

(b) <u>The High-Flux Limit</u>, in which the contamination rate is controlled by the pressure of crackable gas molecules. In this limit, we assume that:

$$I \gg C_2 p + C_3 T \exp(-E_L/kT). \qquad (4)$$

Here, the cracking flux is so intense as to maintain a low occupancy of surface sites by uncracked molecules. Both temperature dependence and intensity dependence cancel out, and K is proportional to the pressure only:

$$K \approx p. \qquad (5)$$

This constitutes an "intensity-temperature saturation" regime. For cracking flux (proportional to SR intensity) above some value, further increases in intensity will not increase the contamination rate significantly, and K becomes proportional to pressure alone. Behavior consistent with this limit is shown in Fig. 5.[26] This limit is the hope of the future for extremely high intensity insertion-device beam lines. It predicts that the contamination rate of optical surfaces may be kept within manageable limits if adequately low partial pressures of contaminating gases is maintained, along with exclusion of certain high-volatility gases such as CO and CH_4.

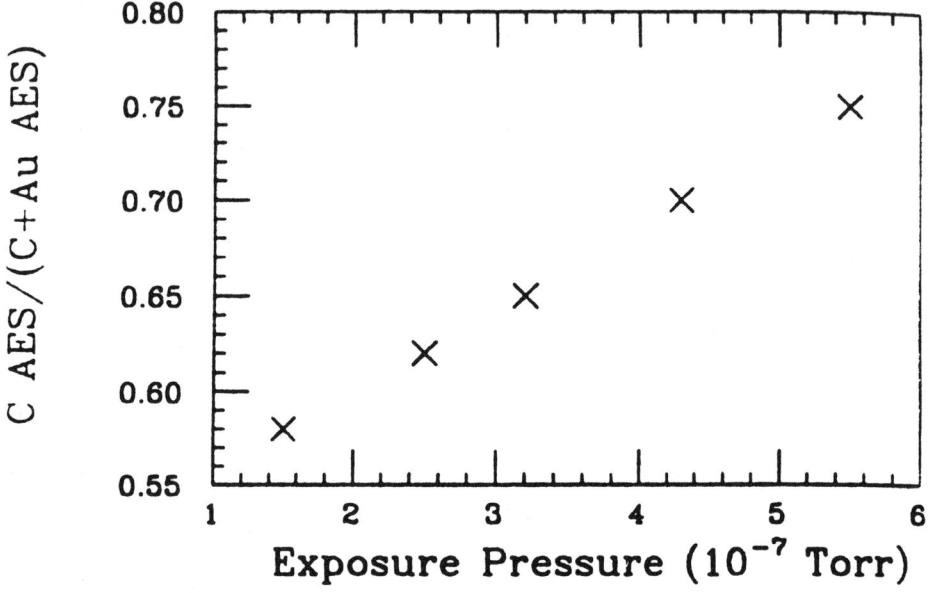

Fig. 5. Dependence of the carbon Auger signal strength for a given exposure on the pressure of methyl methacrylate. Ref. 26.

Complications within the model: The flux of cracking (primary or secondary) photoelectrons is proportional to the incident photon flux, and is dependent upon the photoelectron yield of a surface region of thickness equal to a few multiples of the photoelectron escape depth. The photoelectron yield, in turn, is determined by the photon energy, and by the composition and density of the escape region. For example, transition metals typically have a high density of states near the Fermi level, which produces a high photoyield. Adsorbed simple molecules, on the other hand, generally have a relatively low photoyield.

The escape-region composition is determined by the history of the contamination layer, by the rate of adsorption of gas molecules, by and in the early stage, by the underlying metal. Hence, initially the escape region will be metal, but will change later to a mixture of carbon and adsorbed gas. The adsorbed-gas concentration may or may not be negligible, depending on the kinetics of adsorption, evaporation, and cracking. The escape depth is a function of the energy of the photoelectrons, also. This function is generally represented by the well known "universal curve", a U-shaped curve with a broad minimum in the range of 50-100 eV. The *effective* depth of the escape region will be controlled by the cracking cross section of the adsorbed gas as a function of electron energy, which depends on the adsorbed gas. Thus, if higher electron energies are required for cracking adsorbed-gas molecules, then low-energy secondary electrons, for which the escape depth is long, will be ineffective, and the effective depth of the escape region will be reduced. This leads to a *double reduction* in the cracking rate: less photoelectrons because of the ineffectiveness of the low-energy secondaries, and a less photoelectrons emanating from a thinner effective escape region. None of these considerations are included in Eq. 1, however.

Oversimplifications: In addition to complications that lie within the framework of the HASYLAB model, there are other possible complications with processes that are ignored by the simple linear-rate-equation assumptions of the HASYLAB model. As indicated above, multiple-step cracking processes may be important for heavier molecules such as heavy hydrocarbons. Also, at high pressures, low fluxes, or low temperatures, the adsorbed layer may become thicker than a single monolayer. In sum, quantitative agreement between contamination rates and predictions of the HASYLAB model are unlikely. However, the importance of the HASYLAB model in predicting functional dependence on pressure, temperature, and flux should not be underestimated.

Experimental support: Using SR Auger-electron spectroscopy, Rosenberg and Mancini have shown the real-time buildup of the C layer during exposure to each of seven organic gases or vapors.[26] Among seven organic gases tested, a strong dependence of contamination rate on gas was reported, as shown in Fig. 6.[26] The contamination rate was well correlated with the propensity for decomposition of the gas molecule to produce CO. In agreement with the HASYLAB study, it was found by Rosenberg and Mancini also that a temperature increase of 100° C or so produces about a five-fold reduction in the contamination rate. This is an important confirmation for those interested in retarding the vacuum-contamination rate.

246 Carbon and Other Contaminants in Vacuum Systems

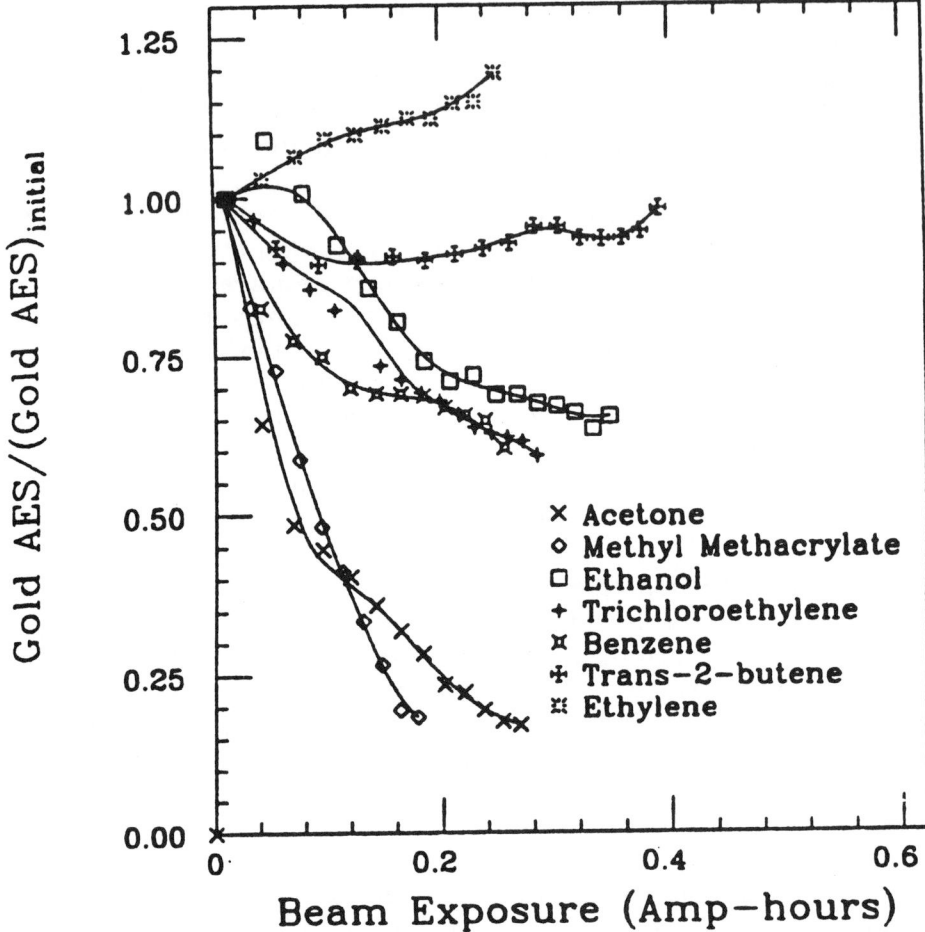

Fig. 6. Variation of the gold Auger signal strength with exposure during photolysis of various gases at p = 4 × 10⁻⁷ Torr. Ref. 26.

Optical Effects of Contamination on Mirrors. It is important to understand the deleterious optical effects of contaminated mirrors.[27,28] The most obvious change comes from the change in reflectance between the clean mirror and the contaminated one, as represented by the reflectances of the metal (e. g., Pt or Au) and carbon, respectively.[27] The degradation in the low-energy reflectance is illustrated in Fig. 7, where the deep dip in the reflectance near 7 eV (177 nm) is characteristic of carbon.[14,15]

Fig. 7 (a). Photograph of a contaminated undulator mirror. Area A was illuminated by the undulator radiation. Area B was opposite the inlet to a turbomolecular pump.

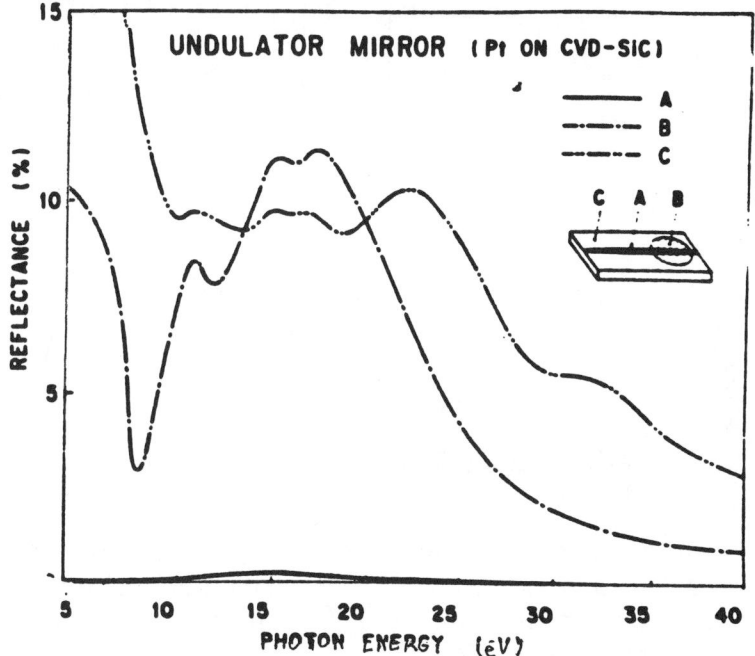

Fig. 7 (b). VUV reflectance spectra of three contaminated areas. The mirror is Pt-coated CVD-SiC. Ref. 15.

Near the K-absorption edge of carbon (284 eV, or 4.4 nm) is another deep reflectance dip with spectral features on the high-energy side.[24,29,30] Depending on the incidence angle and the thickness of the C layer, significant reflectance changes associated with K-edge of carbon may extend from as low as 220 eV to more than 1000 eV, as illustrated in Fig. 8.[17] Note that contamination of the Pt surface actually increases the reflectance in the approximate range 100 eV - 250 eV due to the high reflectance of carbon in this range. In the range 250 eV - 1000 eV, the reflectance degradation produced by contamination is very severe, as Fig. 8 illustrates.

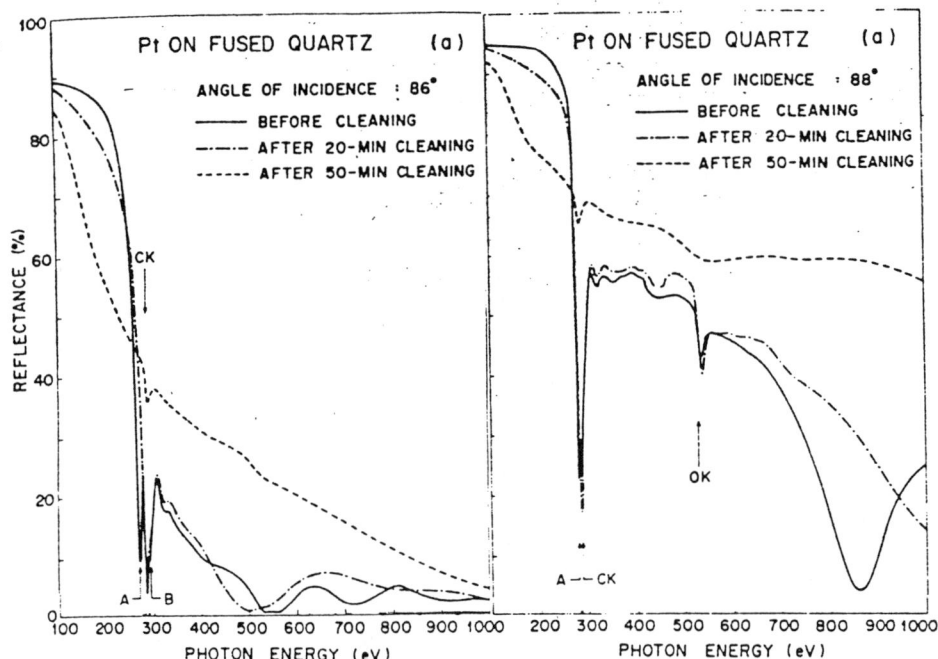

Fig. 8. Soft x-ray reflectance spectra of a contaminated Pt-coated quartz mirror before, during and after removal of contamination. (a) 86° incidence. (b) 88° incidence. Ref. 17.

Note that Fresnel-type interference is observed between energy reflected from the underlying metal interface and energy reflected from the contamination-film surface, as illustrated in Fig. 9.[17] The position and angle dependence of the interference minima were used for determining the thickness of the contamination layer, giving a value of 39 ± 1 nm in this case.[17] After partial removal of the contamination layer, a fit to the interference minima in the reflectance spectrum gave a thickness of 21 ± 1 nm.[17] Hence it is found possible to obtain quite accurate values of the thickness of the contamination layer from the spectrum of Fresnel-type interference above the K-absorption edge of carbon.

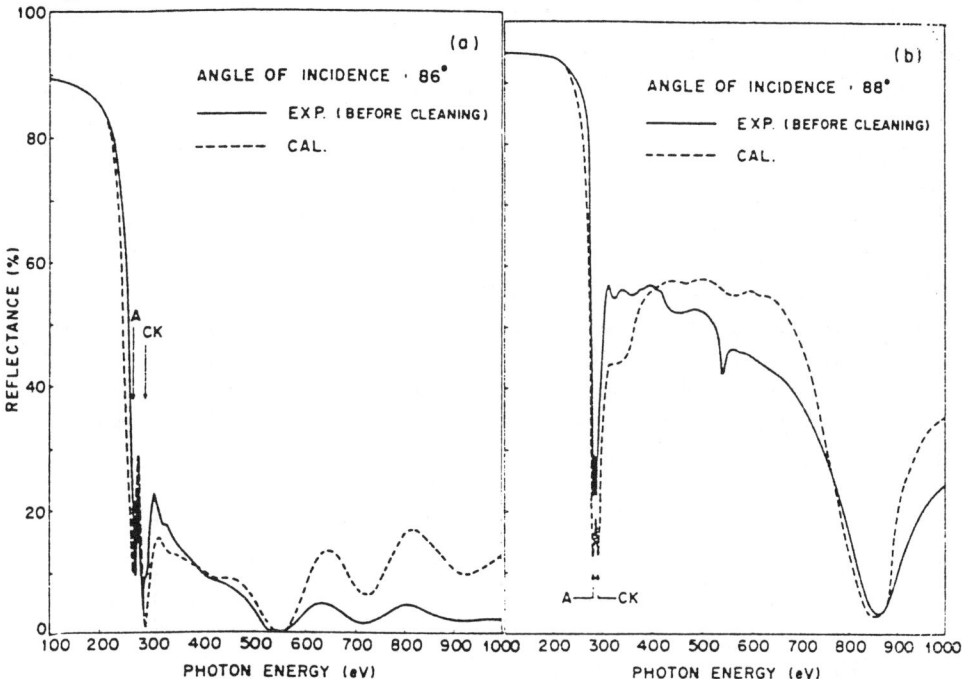

Fig. 9. Interference effects in the contamination layer, theory and experiment. (a) 86° incidence. (b) 88° incidence. Ref. 19.

At still higher energies, the reflectance of carbon at incidence angles normally used for metal-surfaced mirrors lies typically about a factor of two lower than for Pt or Au due to its smaller critical angle for total external reflection. For very energetic x-rays, carbon becomes quite transparent, and the reflection from the underlying metal may dominate that reflected from the contamination surface.

A second concern in the optical performance of contaminated mirrors is the diffuse reflection (optical scattering) induced by the contamination layer. Our understanding of diffuse reflection was advanced by recent studies of the angle-resolved scattering (ARS) by soft x-rays.[31,32,33] The general angular characteristics of radiation reflected from mirrors are illustrated in Fig. 10.[31] For a given incidence angle, the intensity reflected specularly is reduced from that incident on the mirror by two factors: absorption and diffuse reflection (scattering). The dashed line indicates the predicted reduction due to scattering if the optical surface has a random roughness of rms height σ. If the mirrors were perfectly smooth ($\sigma = 0$) the dashed line would coincide with the abscissa, and the narrow specular peak about the incidence angle would be broadened only by the diffraction limitation for the area of mirror illuminated and the wavelength of illumination. The peaks of the experimental ARS curves all lie below the dashed line by an amount determined by the Fresnel reflectance factor, which accounts for the absorbed energy. The energy scattered by surface roughness is not absorbed, but is redistributed in the wings of the ARS curve beyond the diffraction limitation. The distribution of diffuse reflection of soft x-rays by mirrors is now quite well understood on the basis of the Rayleigh-Rice vector perturbation theory.[32,33] Analysis of ARS data with the Rayleigh-Rice theory is becoming a most powerful technique for

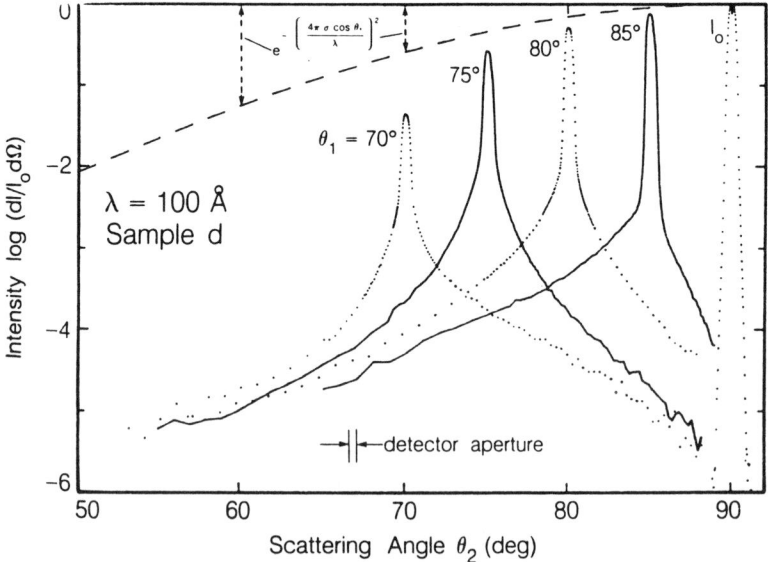

Fig. 10. Angle-Resolved Scattering of 124 eV radiation by a slightly rough mirror at several angles of incidence. Ref. 31.

characterization of optical surfaces.

It is clear that diffuse reflection creates two problems: reduction of specular intensity, and increase of stray-light intensity. While the reflection loss has been discussed above, the increase in stray light due to the contamination may be even more serious for applications requiring a high signal-to-noise value. Many soft x-ray optical systems suffer from very high stray light, although SR beam lines, with their high-brightness source, are potentially considerably superior in this respect. Note, for example, the I_0 curve in Fig. 10, in which the stray light eminating from the source lies more than six orders of magnitude below the I_0 peak for 124 eV radiation. However, reflection from available SR mirrors degrades that considerably in the range up to 1000 eV.[31,32,34,35]

Degradation in the performance of optical multilayers as a result of vacuum contamination has been reported recently.[36] Multilayers designed for use as soft x-ray filters and dispersing

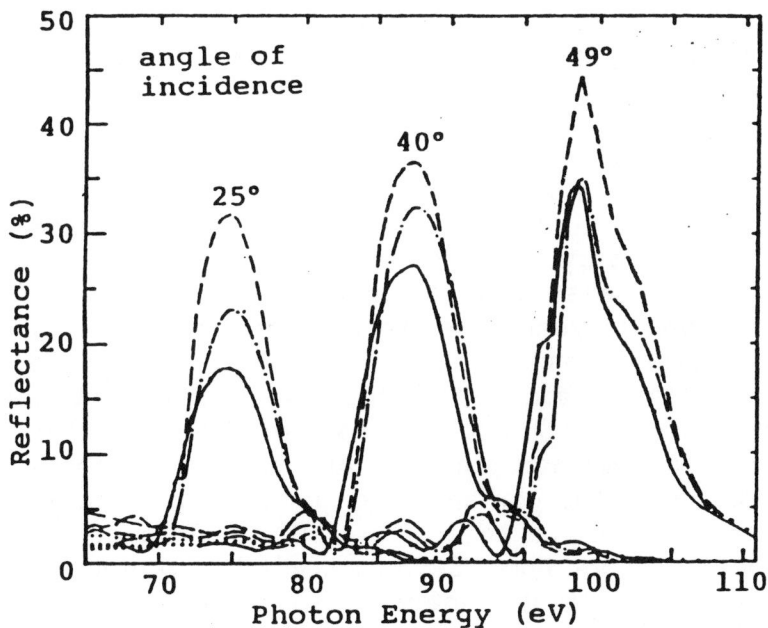

Fig. 11. Reflectance spectra of an optical multilayer reflector at incidence angles of 25°, 40°, and 49° before (solid line) and after (dashed line) irradiation in undispersed SR for 26 Hr. at a beam current of 175 mA. Ref. 36.

elements are sensitive to any additional surface layer. In Fig. 11, the effects of exposure to undispersed SR for 26 hr are shown for one Mo-Si multilayer deposited on a SiC substrate (exposure = 4.55 A-hr, totalling about 4×10^{20} photons / cm^2).[36] The three pairs of reflectance curves were taken with incidence angles of 25°, 40°, and 49°, respectively. The solid curve of each pair was taken before exposure, and the dashed curve afterword. The temperature rise during the exposure was about 100° C. No increase in surface roughness was associated with the contamination, within the detection limits of the Wyco-profilometer measurements. It is clear that the effectiveness of soft x-ray multilayer optical elements can be jeopardized by relatively short exposures to undispersed SR in beamline 12C of the Photon Factory. No detailed analysis of the contaminated multilayer has been reported to date: The physical and chemical mechanisms of multilayer degradation are not yet known.

PROGRESS IN PREVENTION OF CONTAMINATION.

Logically, the first priority following the identification of a problem would be the search for methods of prevention. In early discussions of the contamination problem, prevention was discussed, and at least one concept for prevention or retardation of contamination was suggested.[5] That concept was the local reduction of the partial pressure of contaminating gases with, for example, a nearby cryogenic surface. This simple concept has been used successfully in the relatively poor vacuum of electron microscopes in years past to lengthen the useful lifetime of contamination-sensitive surfaces. In spite of general agreement that this concept had merit, I know of no implementation or test of this concept for SR applications in the 12 years intervening.

Seven years ago, the HASYLAB study showed that a five-fold retardation of the contamination rate is obtainable simply by warming the mirror to 100° - 130° C. This gentle warming probably will not harm SR mirrors if the temperature is not permitted to fluctuate widely. This suggestion was discussed at a conference on SR vacuum in 1983,[8] but has not yet been implemented or tested, either, I believe. Implementation would be quite simple in some cases. For example, the Pt-coated Cu mirrors used in the 4° and 8° beam lines at SSRL were originally thermostated to operate slightly above room temperature (about 35° C). That set point could be increased to 100° C or so.

It would seem that greater attention to prevention of contamination is warranted. The cost of cleaning or replacement of optical elements in SR beam lines is very high in both dollar cost and lost beam time.

PROGRESS IN REMOVAL OF CONTAMINATION.

As mentioned above, in the early 1970s vacuum contamination was removed from contaminated optics by use of an oxygen glow discharge.[9,10,11] Atomic oxygen generated in the plasma was identified as the active agent. Similar glow-discharge techniques were applied to SR vacuum contamination on replica gratings in 1982.[12,13] Since that time a great deal has been accomplished in developing a practical and efficient method for removal of

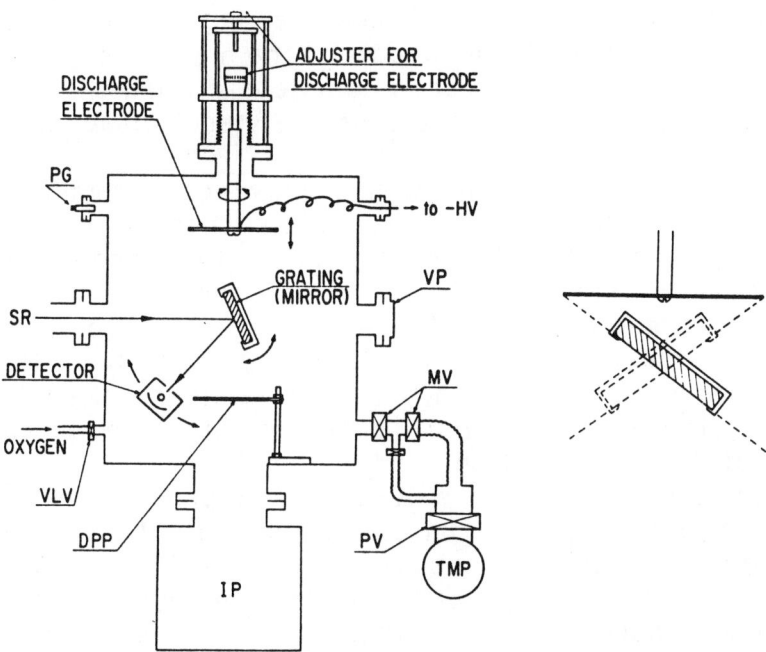

Fig. 12. (a) Schematic diagram of the DC oxygen-discharge chamber for removal of contamination from optical elements with in situ optical evaluation. VP = view port; DPP = detector protection plate; VLV = variable leak valve; MV = manual valve; PV = pneumatic valve; PG = Pirani gage; TMP = turbomolecular pump; IP = ion pump. (b) Orientation of optical surface and discharge electrode to avoid sputter damage to the optical surface. Ref. 15.

contamination without damaging high-quality optical surfaces.

The use of a DC glow discharge has been reported by Koide and coworkers in Japan.[14,15,16,17] The DC technique is illustrated in Fig. 12.[15] With the chamber walls and the object to be cleaned at ground potential, the discharge electrode is biased to about -400 V, and the chamber is filled with oxygen to a pressure in the range 0.2 to 0.4 Torr. The distance between the discharge electrode and object to be cleaned may be varied in accordance with the size of the object, and the discharge power desired. It was found to be critically important that the sensitive surface of the object to be cleaned face away from the discharge electrode, as illustrated in Fig. 12(b), or else the reflectance was degraded, presumably by material from the electrode sputtered onto the sensitive surface. Both SR-induced contamination and contamination resulting from backstreaming of a turbomolecular pump were successfully removed, and the high reflectance of the optical surface was recovered completely when the discharge was continued a few minutes past the visual disappearance of contamination.

It has been noted that partial removal of contamination may degrade the VUV reflectance below that of the original contaminated surface, while changing its reflectance spectrum to more closely resemble that of glassy carbon.[9,10,11,12,13,14,15] (This effect also occurs at the C K edge, but over only a narrow spectral range. See Fig. 8.) This observation implies that either the contamination layer is non-uniform, or more likely that the action of the oxygen plasma changes the structure or chemical nature of the contamination.

Example results of the DC-discharge process are shown in Fig. 13 for a Pt-coated SiC undulator mirror.[15] The mirror is the same mirror illustrated above in Fig. 7. Fig 13 shows the reflectance of the stripe of contamination associated with the undulator radiation, marked "A" in Fig. 7. The reflectance prior to cleaning is barely visible above the abscissa, while after an 80 min cleaning using 25 - 50 W, about 80% of the reflectance of clean Pt was recovered. Other examples showed full recovery. Note, however, that overcleaning is possible. The curve in Fig. 13 taken after 180 minutes of exposure to the oxygen glow discharge shows a lower reflectance than the 80-min reflectance curve. It was suggested that oxidation of the cleaned surface may be the cause of this "overcleaning" effect. ARS measurements of 21 eV radiation show that discharge cleaning does not increase the VUV

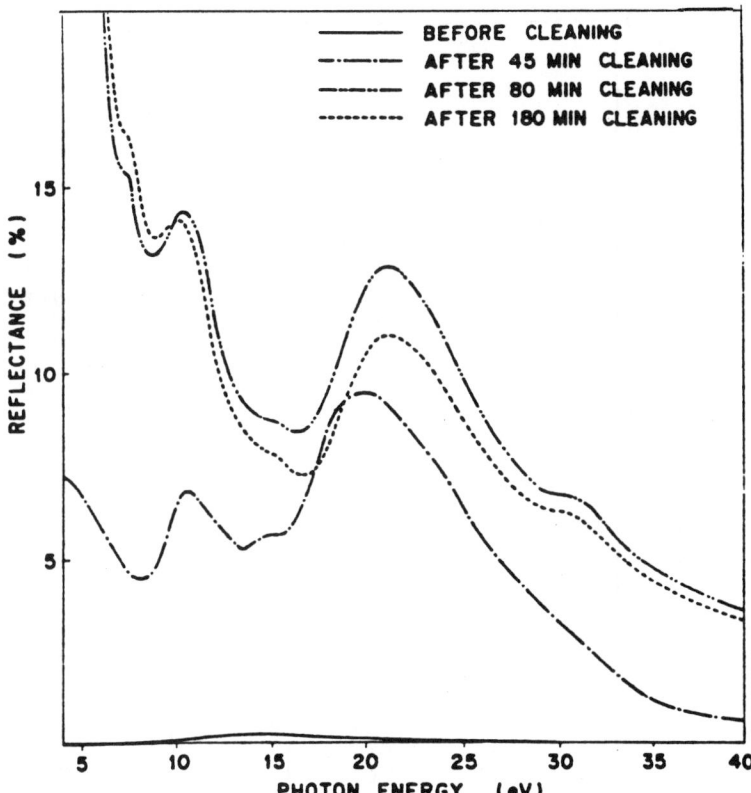

Fig. 13. UVU reflectance spectra before, during and after removal of the contamination layer from Part A of the mirror shown in Fig. 7(a). Ref. 15.

diffuse reflectance.

The resuscitation of contaminated diffraction gratings is illustrated in Fig. 14. The grating had been used in a Seya-Namioka monochromator for poor-vacuum, gas-phase experiments. It was highly contaminated.[15] Cleaning with about 28 W for 30 min increased in the efficiency of the m = +1 diffraction order by a factor of more than 2.5, while decreasing the intensity of the unwanted m = 0 order by a factor of two.

From a consideration of the contamination process as it occurs on gratings, it is apparent that contamination can gradually change the blaze efficiency, shift the blaze wavelength, and lower the reflectance efficiency of the grating. Fig. 15 illustrates the process schematically as a contamination layer is

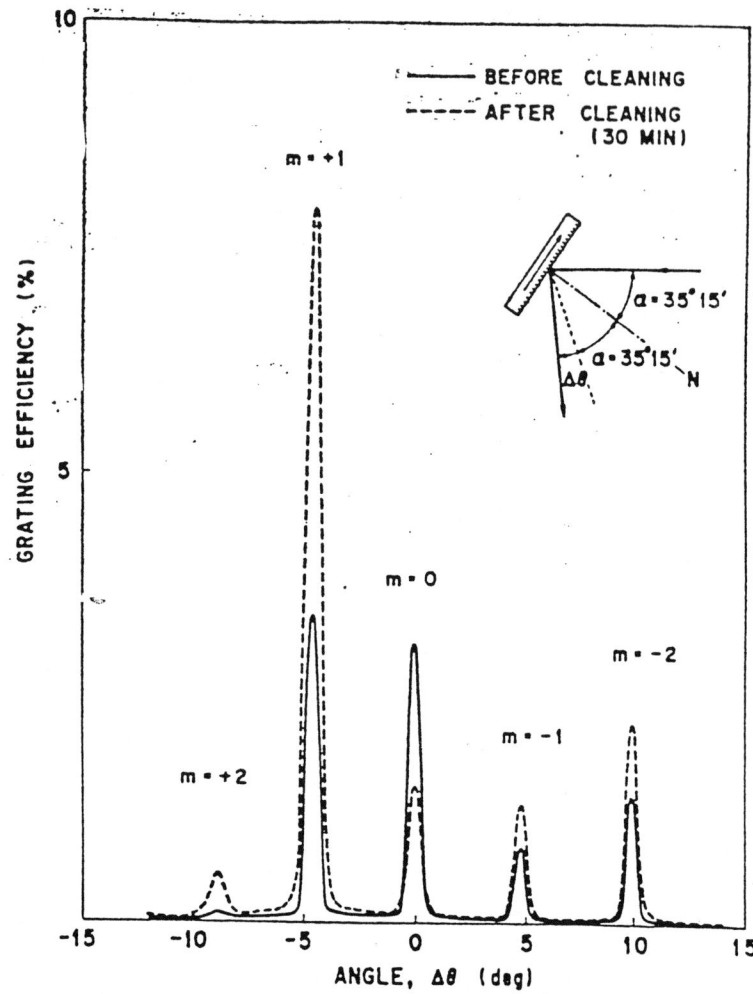

Fig. 14. Resuscitation of the efficiency of a VUV grating by removal of contamination. Ref. 15. hv = 22.6 eV.

built up on a blazed grating by use in SR vacuum. Fig. 15(a) is appropriate for blazed gratings using of an "inside" or positive diffraction order, as is customary with VUV monochromators such as the Seya-Namioka monochromator. In this orientation, part of each reflecting facet of the groove is shadowed from the incident radiation, and consequently should receive very little contamination. On the facet areas that are fully exposed to incident SR, the contamination layer is expected to grow thicker. Thus, the grating-groove shape is changed, jeopardizing the blaze wavelength and the blaze efficiency, along with the reflection

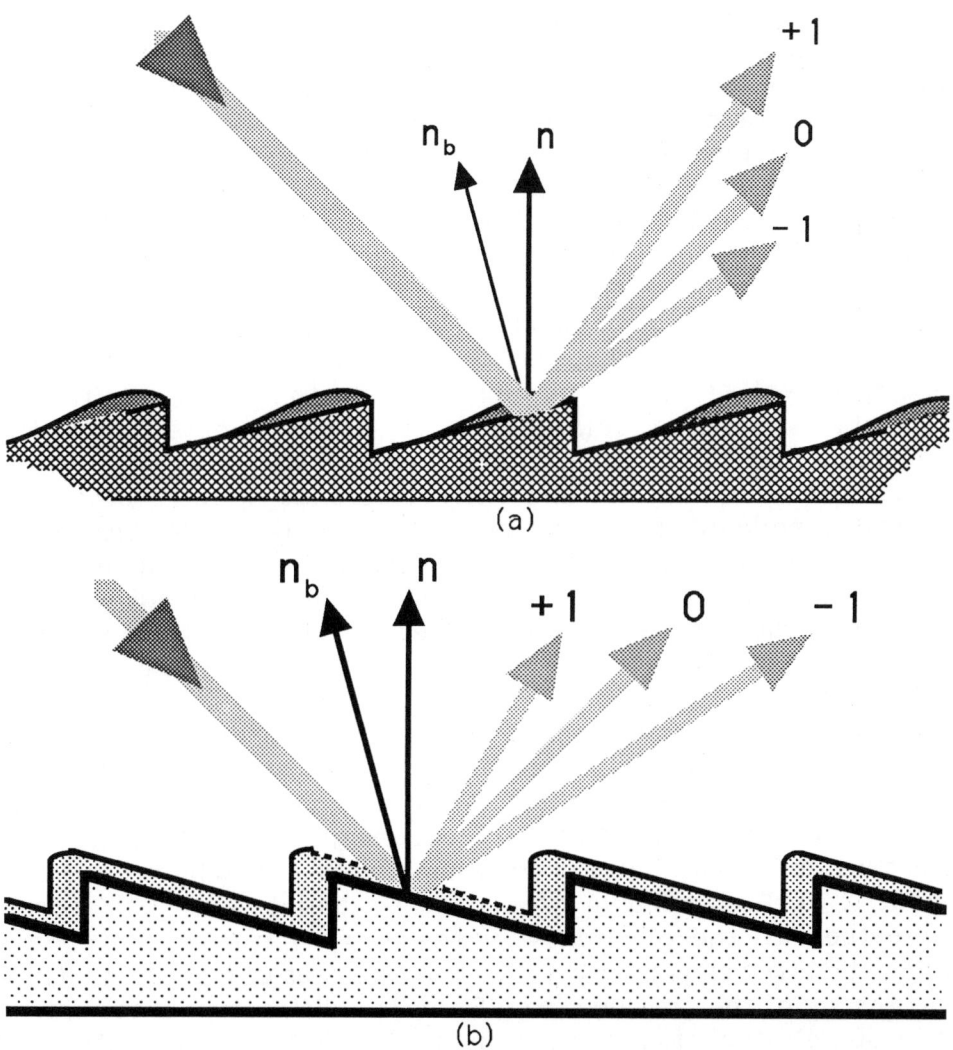

Fig. 15. Contamination on blazed gratings. (a) Grating used in a positive order. (b) Grating used in a negative order.

efficiency.

For blazed gratings used in an "outside" or negative order, the geometry is illustrated in Fig. 15(b). Here no part of the groove is shadowed from the incident radiation, but the flux density will be greater on the "riser" part than on the facet part of the groove, especially for grazing-incidence applications. If the contamination variables are not in the intensity-temperature-saturation regime described above, the contamination layer will grow thicker on the riser, but the shape of the grating groove would be preserved approximately. (A rounding of the sharp corners may be expected, but would produce only a small effect on blaze efficiency.) Thus we are lead to expect that gratings used in outside-order geometry will suffer less degradation of the blaze efficiency and less shift of the blaze wavelength, but about the same degradation of the reflection efficiency as gratings used in inside orders.

Soft x-ray range. Evaluation of contamination removal via DC oxygen glow discharge has been reported in the soft x-ray range, as well.[17] Using similar cleaning techniques to those described above, excellent results were reported in both reflectance and ARS evaluations. In Fig. 8 above is shown the resuscitation of the 86° and 88° reflectance spectra of a Pt-coated fused-silica mirror by DC glow-discharge cleaning. Partial cleaning, represented by the 20-min curve, changed the spectrum, including shifting the interference minima, but was far from adequate resuscitation. After 50 min of cleaning, the reflectance spectrum was essentially identical to that of a clean Pt mirror.

Grating resuscitation by DC glow-discharge cleaning has been shown to be effective in the soft x-ray range, as well. Fig. 16 shows the grating-efficiency improvement for a 1200 groove-per-mm grating used with an incidence angle of 88°. The grating was considered not severely contaminated, as it had been used for about 2 yr in a vacuum of pressure about 3×10^{-9} Torr. Nevertheless, the efficiency of the $m = +1$ diffraction order improved after cleaning by as much as an order of magnitude near the energy of the grating blaze, E_b.[17] The diffraction pattern for $h\nu = 362$ eV radiation is shown in Fig. 17 for the same grating in the same geometry.[17] The efficiency improvement for the $m = +1$ order of about an order of magnitude is evident, along with a small decrease in the unwanted $m = 0$ peak.

In-situ removal of contamination. Contamination removal *in situ* without removal of the optical element from its normal place in the SR beam line was first reported in 1987.[19] For application to *in situ* cleaning, two significant advantages of the rf discharge over the DC discharge are: (1) The discharge electrodes are external to the vacuum system so that contamination from that source is obviated. (2) The rf-excited plasma may be passed through a vacuum system to optical

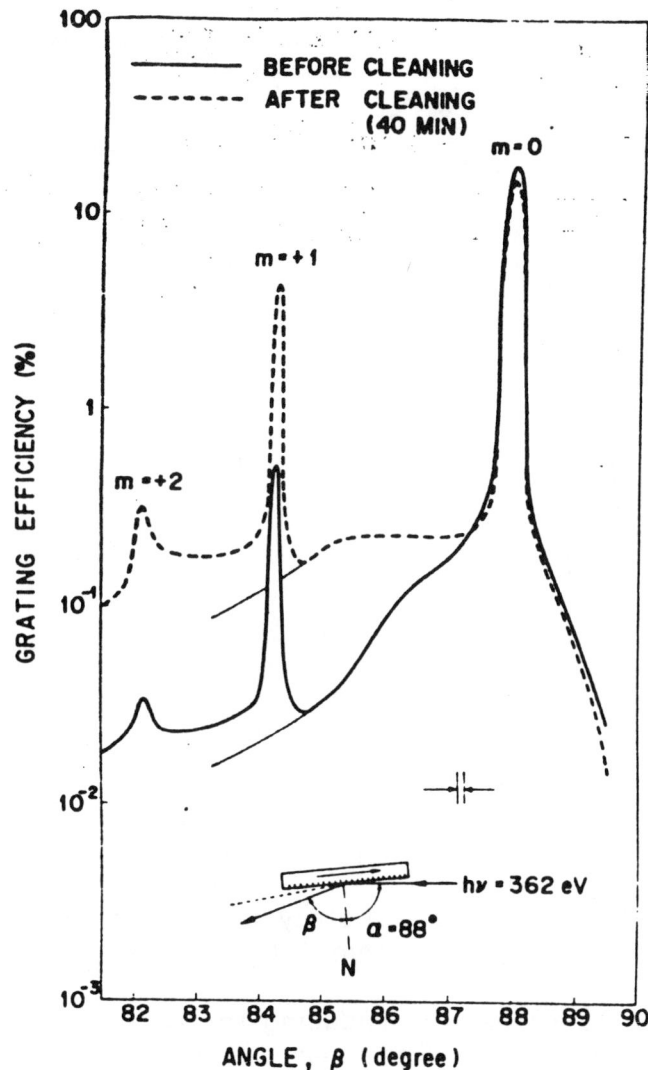

Fig. 16. Resuscitation of the efficiency of a soft x-ray grating by removal of contamination. Ref. 17.

elements that are not in line-of-sight with the source. Against these advantages lies the difficulty of determining the degree of success, and hence the end point of the cleaning effort.[21]

An rf glow discharge was used in moist oxygen (2% H_2O vapor) with total pressure between 0.1 and 0.5 Torr.[19] The small amount of water vapor was shown to increase the cleaning rate

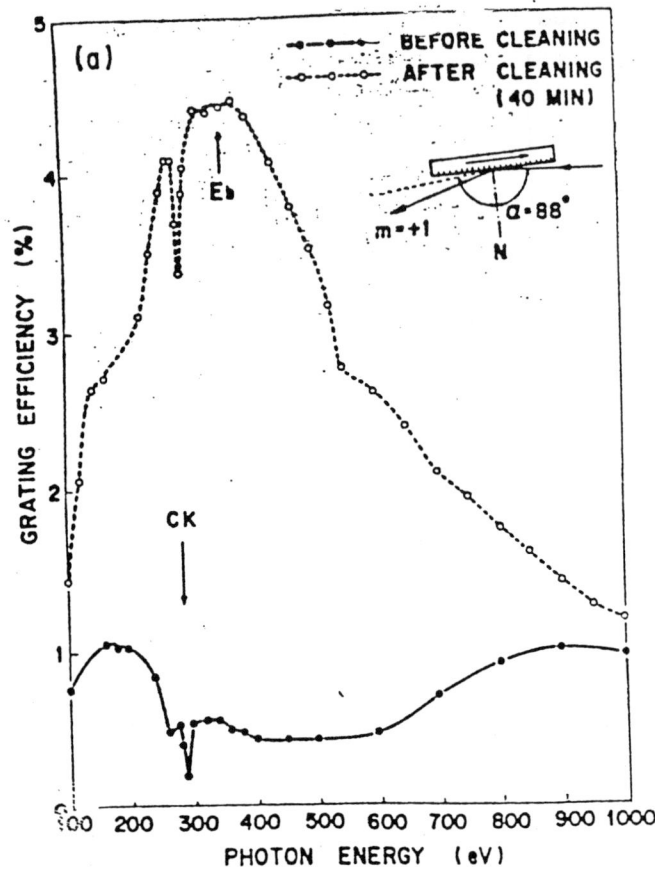

Fig. 17. Improvement of stray-light performance of a soft-x-ray grating by removal of contamination. Ref. 17.

by as much as 40-fold by reducing the recombination rate of atomic oxygen. Recently, Rosenberg and Crossley have reported that use of pure water vapor in an rf-excited glow discharge provides effective contamination removal, and in fact shows the highest removal rate for their test material, PMMA.[21] It should be noted that the H_2O-vapor exposure should be carefully controlled to avoid the necessity of a prolonged bakeout of the vacuum chamber after cleaning.

The discharge unit was built on a 2.75" Conflat™ vacuum flange so that it could be attached simply to any available port in a mirror chamber or monochromator.[19,20] The unit is illustrated schematically in Fig. 18. The rf frequency used was 13.5 MHz, with discharge power up to 100 W. The discharge was typically run at various pressures for up to 24 hr. The gas flow was maintained at about 20 sccm, and the pressure was maintained at a chosen set point by control of pumping speed. The operating pressure was changed about once each hour in typically 0.1 T increments. Results have been so satisfactory that rf-discharge cleaning modules have been fitted to 12 of the 16 VUV beamlines, and at least 4 x-ray beamlines at the National Synchrotron Light Source (NSLS), Brookhaven, NY.[37]

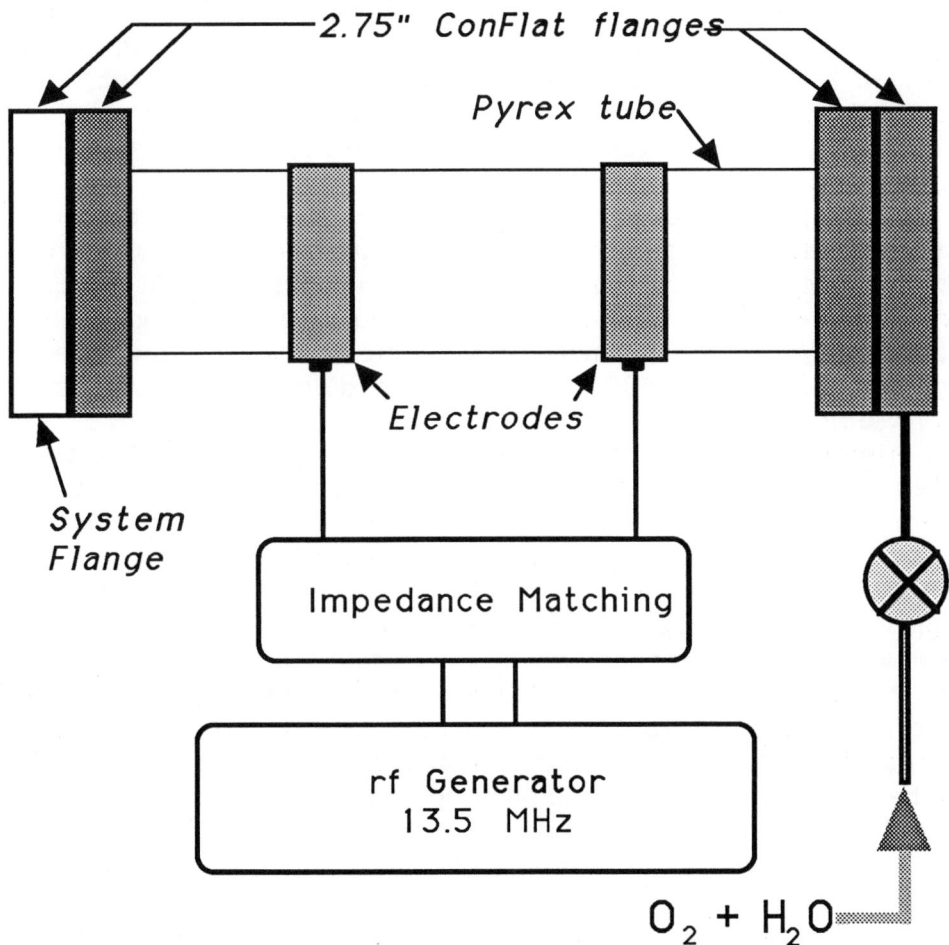

Fig. 18. Schematic diagram of an rf oxygen glow-discharge generator built on a 2.75" Conflat flange for attachment to an available port near a contaminated optical element. Ref. 19.

Characterization of the glow-discharge plasma via analysis of the residual gases and via fluorescence spectroscopy have been reported as possible indicators of progress of the *in situ* cleaning process.[20] The residual-gas analysis requires careful sampling of the high-pressure plasma, and then suffers from difficulty in separating the CO and N_2 peaks: Trends in the amplitude of the peaks representing masses 16 and 32, compared with 28 and 44, indicated progress, but appeared to offer no clear indication of an end point. Fluorescence spectra for the wavelength range 290 – 600 nm offered a rich spectrum that changed with time during cleaning. Again, however, no clear end point of the cleaning process was identified. On the positive side, overcleaning effects have not been reported for the rf-discharge technique.

Initial efforts at Stoughton to implemented *in situ* cleaning on the Mark V Grasshopper monochromator achieved with only partial success.[20] Continuing efforts on the part of the Stoughton group implemented DC discharge cleaning on an aluminum Seya-Namioka-type monochromator have achieved greater success.[38] Implementation of the DC discharge method on other beamlines is planned for the near future.

Implementation of *in situ* DC discharge cleaning has also been reported by Koide and coworkers.[22] All five optical elements in the Seya-Namioka beamline 12A at the Photon Factory, and all except the recently renewed M_0 and M_3 mirrors in the grasshopper beam line 11A were cleaned. A 50-fold flux enhancement was reported for the Seya-Namioka beam line, and nearly complete flux recovery at the carbon K edge was reported for the grasshopper line. Because of the use of dry oxygen, the original ultrahigh vacuum pressures were restored without bakeout. However, due to a non-ideal geometry for the electrode in the grasshopper line, some sputtering of Cr onto Codling mirror was observed.

Finally, SR-induced etching of carbon films in low-pressure oxygen gas has been reported,[23] and could be considered for an alternative method of removing carbonaceous deposits from SR vacuum interiors. Using oxygen pressures between 0.01 and 0.2 Torr and undispersed SR with photon energies up to about 1240 eV, etching rates between 10 and 30 nm / min were reported. The etching rate was linear in SR intensity, with a pressure dependence proportional to $[p(O_2)]^{0.5}$. It was ascertained that no etching occurred in areas shaded from the SR source, or by passing the SR beam parallel to the surface. Hence a surface photochemical reaction is indicated, possibly including dissociation of O_2 on the surface, followed by creation of CO or CO_2 molecules. Using a stylus-type profilometer, the etch depth as a function of position across the edge of a shadowed area was studied and compared with a theoretical model. The edge sharpness observed was actually slightly sharper than predicted for a surface photochemical reaction, and much sharper than predicted for a gas-phase reaction near the surface. The Fresnel interference fringes that should accompany edge diffraction were not observed, however.

In summary, the surface photochemical model suggested by these experiments is not in disagreement with the HASYLAB model for deposition of carbonaceous contamination, but further experiments are needed before satisfactory agreement between etching and deposition processes can be considered to be in consonance.

As practical method for resuscitation of contaminated SR optical components, the requirement that both undispersed SR and at least 0.01 Torr of oxygen be present in the beam line at once will be difficult to achieve. Extrapolating, a lower oxygen pressure, e. g. 10^{-6} Torr, reduces the etch rate to 0.1 Å / min or 6 Å / hr, but also reduces the load on the differential pumping system that must protect the SR source ring from $p > 10^{-9}$ Torr. Continuous cleaning of carbon from differentially pumped beam lines might be possible unless the presence of the oxygen absorption interferes with the intended application of the SR.

Note added in proof. During the discussion following presentation of this paper, it was remarked by C. Pruett that R. A. Rosenberg and coworkers have not been able to reproduce the high rates of carbon removal reported by Koide, et al,[14-17] for the DC-discharge process. The Rosenberg group has been able to show that a small amount of fluorine gas in the discharge, such as might be generated from teflon-coated wires in the discharge region, greatly enhanced the rate of carbon removal. This remark is repeated here in the workshop spirit as an unproved, tentative conclusion. Further experiments are being conducted by both groups in order to clarify the matter.

REFERENCES

[1] C. J. Davisson & L. H. Germer, Phys. Rev. **30**, 705 (1927), and many papers following over the next decade.

[2] R. Z. Bachrach, S. A. Floodstrom, R. S. Bauer, V. Rehn & V. O. Jones, Nucl. Instru. Meth. **152**, 135 (1978).

[3] V. Rehn & V. O. Jones, Optical Engineering **17**, 504 (1978).

[4] D. A. Shirley, "Beam line chemistry", in the *Proc. of the Workshop on X-Ray Instrumentation forSynchrotron Radiation Research*, SSRL Report #78/04, p. VII-80 (May, 1978), (unpublished).

[5] V. Rehn, et al, "Report of the Working Group on Mirrors" in the *Proc. of the Workshop on X-Ray Instrumentation forSynchrotron Radiation Research*, SSRL Report #78/04, p. VII-1 (May, 1978), (unpublished).

[6] D. E. Aspnes & V. Rehn, "In-situ monitoring of mirror surfaces in synchrotron radiation applications", in the *Proc. of the Workshop on X-Ray Instrumentation forSynchrotron Radiation Research*, SSRL Report #78/04, p. VII-67 (May, 1978), (unpublished).

[7] J. Stöhr, L. Johansson, I. Lindau & P. Pianetta, Phys. Rev. **B20**, 664 (1979).

[8] R. Avery, Editor, *Proc. of the Workshop on Synchrotron-Radiation Vacuum*, Lawrence Berkeley Laboratory, 25-26 July 1983 (unpublished).

[9] R. B. Gillette & B. A. Kenyon, Appl. Optics **10**, 545 (1971).

[10] W. R. Hunter, G. N. Steele & R. B. Gillette, Appl Optics **12**, 2800 (1973).

[11] W. D. Beverly, R. B. Gillette & G. A. Cruz, "Removal of a hydrocarbon contaminant film from spacecraft optical surfaces using a radio-frequency-excited oxygen plasma" in *Thermal Control and Radiation, AIAA Progress Series*, C. Tien, Ed. (MIT, Cambridge, MA, 1973), Vol. 31, p. 159.

[12] W. R. McKinney & P. Z. Takacs, Nucl. Instru. Meth. **195**, 371 (1982).

[13] A. Mikuni, S. Asaoka, K. Soda & H. Kanzaki, "DC discharge cleaning of replica gratings" in *Activity Report of Synchrotron Radiation Laboratory*, Institute for Solid State Physics, University of Tokyo, p 71 (1983), (unpublished).

[14] T. Koide, S. Sato, T. Shidara, M. Niwano, M. Yanagihara, A. Yamada, A. Fujimori, H. Fukutani, A. Mikuni, H. Kato & T. Miyahara, Nucl. Instru. Methods **A 246**, 215 (1986).

[15] T. Koide, M. Yanagihara, Y. Aiura, S. Sato, T. Shidara, A. Fujimori, H. Fukutani, M. Niwano, H. Kato, Appl. Optics **26**, 3884 (1987).

[16] T. Koide, M. Yanagihara, Y. Aiura, S. Sato, H. Kato, H. Fukutani, Physica Scripta **35**, 313 (1987).

[17] T. Koide, T. Shidara, M. Yanagihara, S. Sato, Appl. Optics **27**, 4305 (1988).

[18] I. Davoli, Zhu Zhi Ji, P. Chiaradia, S. Priori & M Fanfoni, "A glow discharge process for optical elements cleaning", Laboratori Nazionali di Frascati report # LNF-88/30(R), (1 June 1988, unpublished).

[19] E. D. Johnson, S. L. Hulbert, R. F. Garrett, G. P. Williams & M. L. Knotek, Rev. Sci. Instru. **58**, 1042 (1987).

[20] E. D. Johnson & R. F. Garrett, Nucl. Instru. Meth. **A266**, 381 (1988).

[21] R. A. Rosenberg & D. B. Crossley, Nucl. Instru. Meth. **A266**, 386 (1988).

[22] T. Koide, T. Shidara, K. Tanaka, A. Yagishita & S. Sato, Rev. Sci. Instru. **60**, 2034 (1989).

[23] H. Kyuragi & T. Urisu, Appl. Phys. Letters **50**, 1254 (1987).
[24] D. Denley, P. Perfetti, R. S. Williams, D. A. Shirley, & J. Stöhr, Phys. Rev. **B21**, 2267 (1980).
[25] K. Boller, R.-P. Haelbich, H. Hogrefe, W. Jark & C. Kunz, Ncul. Instr. Methods **208**, 273 (1983).
[26] R. A. Rosenberg & D. C. Mancini, Nucl. Instru. Meth. **A291**, 101 (1990).
[27] V. Rehn, "X-ray mirrors", in *Proc of the Workshop on X-Ray Instrumentation forSynchrotron Radiation Research*, SSRL Report #78/04, p. VII-13 (May, 1978), (unpublished).
[28] V. Rehn, "Focusing, filtering & scattering of soft x-rays by mirrors" in *Low Energy X-Ray Diagnostics*, D. T. Attwood & B. I. Henke, Eds, Am. Inst. of Phys. Conference Proceeding #75 (1981), p 162.
[29] J. F. Morar, F. J. Himpsel, G. Hollinger, G. Hughes & J. L. Jordan, Phys. Rev. Letters **54**, 1960 (1985); J. F. Morar, F. J. Himpsel, G. Hollinger, J. L. Jordan, G. Hughes & F. R. McFeely, Phys. Rev. **B33**, 1346 (1986); R. A. Rosenberg, P. J. Love & V. Rehn, Phys. Rev. **B33**, 4034 (1986).
[30] F. Sette, Y. Ma, G. Meigs & C. T. Chen, "Multiple exciton structure at K(C) in diamond" in the *Proc. 9th Int. Conf. on VUV Rad. Phys.*, Hononlulu, 17-21 July 1989
[31] H. Hogrefe, R.-P. Haelbich & C. Kunz, Nulc. Instru. Meth. **A246**, 198 (1986).
[32] H.-G. Birken, C. Kunz & R. Wolf, in the *Proc. 9th Int. Conf. on VUV Rad. Phys.*, Hononlulu, 17-21 July 1989
[33] P. Z. Takacs & E. L. Church, Nucl. Instru. Meth. **A291**, 253 (1990).
[34] V. Rehn, V. O. Jones, J. M. Elson & J. M. Bennett, Nucl. Instru. Meth. **172**, 307 (1980).
[35] J. M. Elson, V. Rehn. J. M. Bennett & V. O. Jones, Proc. SPIE **315**, 193 (1982).
[36] M. Yamamoto, M. Yanagihara, A. Arai, J. Cao, T. Mizuide, H. Kimura, T. Namioka, "Reflectance degradation of soft x-ray multilayer filters upon exposure to synchrotron radiation", in the *Proc. 9th Int. Conf. on VUV Rad. Phys.*, Hononlulu, 17-21 July 1989 (Phys. Scripta, 1990).
[37] E. D. Johnson in*NSLS Users' Newsletters*, July, 1990. The author is indebted to Dr. Johnson for sending him this information.
[38] M. Bissen, J. Welnak, M. Marsi, C. Pruett & R. A. Rosenberg, "In-situ oxygen DC discharge cleaning of the Al Seya monochromator grating", abstract of presentation to the SRC Users' Meeting, Stoughton, WI, (Oct, 1990) (unpublished). The author is indebted to Dr. Rosenberg for sending me these results.

THE DESIGN OF ESRF FRONT ENDS

Trevor Mairs and Jean Claude Biasci
ESRF, BP220, 38043 Grenoble, France

ABSTRACT

As part of phase 1 the ESRF will be installing 15 Front Ends in 1992, which will be split into 7 insertion device Front Ends and 8 bending magnet Front Ends. The design of the Front Ends is limited by the high power loads (500W/mm^2 at 13m from the source and a total of 10KW per Front End) and the time restrictions imposed by a short time schedule.

The ESRF has adopted a standardised approach where all insertion device Front Ends are identical and capable of handling the high power densities of undulators and/or the high total power of wigglers. This simplifies the manufacture and installation of such a large number of Front Ends in the short time available. This approach has necessitated the use of some novel compromises

The design choices made for high power absorbers, beryllium windows, vacuum protection systems and other vacuum equipment within the Front End are described. Future possible developments on phase 2 beamlines at the ESRF are discussed.

INTRODUCTION

As one of the new generation synchrotron sources the ESRF has encountered many demanding design constraints due to the high power and small size of its photon beams. These are particularly demanding in the Front Ends where power densities reach approximately 5 times those reached at other synchrotron sources. This has led to the necessity to develop further ideas used at other sources in order to cope with these high power loads.

In addition the ESRF plan to install 15 Front Ends prior to commissioning of the storage ring and up to a further 12 Front Ends 18 months later. These large numbers have highlighted the need for a large amount of standardisation which in turn has led to a design that is not tailored for each beamline, but one which can cope with all types of insertion devices. There are 3 types of Front Ends at ESRF, being:- insertion device (ID) Front Ends, ultra high vacuum (UHV) Front Ends and bending magnet (BM) Front Ends, which will be installed in quantities of 6,

1 and 8 respectively during 1991/92.. The design of the Front Ends was only started in January 1989, and therefore, it is clear that this short timescale imposed some compromise solutions.

ESRF have recently ordered the major components (items 3-14 in figure 1) for the first 15 Front Ends and have almost completed the drawings for the remaining components which are essentially only cylindrical vacuum tubes and associated supports.

DESIGN DEFINITION

The ESRF defines the Front End as the part of the beamline between the tangent point of the storage ring where the photon beam exits and the exterior of the shielding wall. The Front End consists of equipment common to all beamlines and initially does not include any experimental equipment.

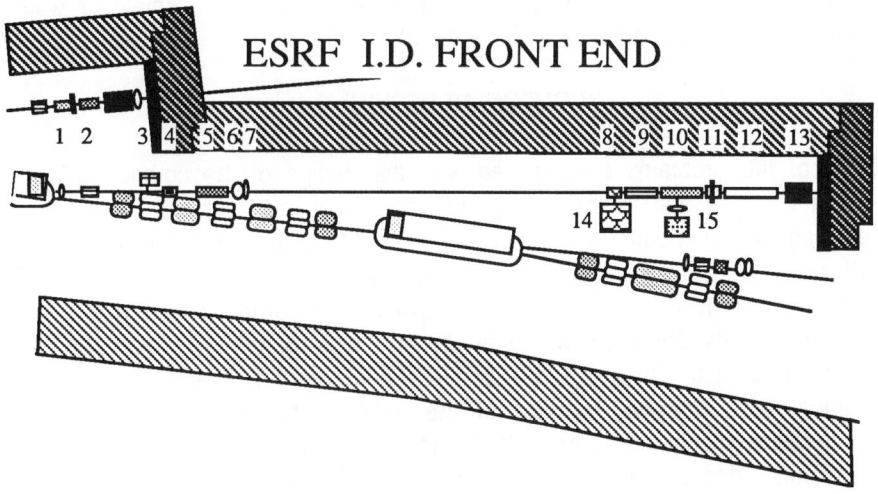

Figure 1 General layout of ESRF insertion device Front End.
1   manual all metal valve
2   photon beam position monitor 1
3   ion pump 400l/s
4   fixed absorber
5   movable absorber
6   pneumatic all metal valve
7   fast shutter
8   fixed absorber
9   photon beam position monitor 2
10 slit/filter assembly
11 beryllium window
12 experimental space
13 beamshutter
14 ion pump 400l/s
15 turbo pump (not permanent)

The insertion devices presently under development at ESRF are capable of producing very high power densities on Front End components which using present technology could not be absorbed or transmitted to the experiments. Therefore, ESRF has imposed on itself a power limit for the Front Ends defined by a maximum total power for each beamline of 10KW and a maximum power density at 13 metres from the source of 500W/mm^2.

Figure 1 gives a schematic layout of the ESRF ID Front End. It is essentially the same layout as other sources except that there is significantly more distance between the tangent point and the shielding wall. This allows a large distance between the two position monitors which helps to ensure beam stability as these position monitors are used for feedback purposes.

The ID Front End is designed for use with all insertion devices presently envisaged ranging from undulators of field 0.42T length 5m and period 40mm to wigglers of field 1.5T length 2.4m and period 200mm. A beam of ±2.3mrad can be extracted. This means that at the shielding wall the vacuum chambers are relatively large at CF200.

COMPONENT DESIGN

Most of the problems encountered with the design of the components in the Front End are due to the high power loads involved. Table 1 shows the power densities incident on components at 11 metres (position monitor 1), 13 metres (movable absorber) and 23 metres (slit/filter assembly and beryllium window) for various different types of insertion devices.

All components are designed to be able to absorb total power in normal operation and also in event of a miss-steer of the beam. There are mechanical restraints on the size of the miss-steer, which were used to define the size of chambers and absorbers within the Front Ends.

The main design difficulties involved the following items which are described more fully:

MOVABLE ABSORBER

The movable absorber is used to stop the photon beam when desired and is, thus, an absorber plate capable of absorbing full power, but also capable of being moved up or down. This item is situated at 13metres from the source and, thus, it can be seen from table 1 that it must be capable of absorbing power densities up to 500W/mm^2. These power densities require that the absorber plate is positioned at a grazing incidence to the photon beam. ESRF knows from experiments on the crotch absorbers that it is possible to absorb approximately 50W/mm^2 at normal incidence[12]. Hence, the grazing angle required to absorb 500W/mm^2 is approximately 5 degrees. Unfortunately a simple flat plate at this angle would be about 1 metre long which would be difficult to manipulate. Therefore, a solution as shown in figure 2 was adopted, which reduces the length of the absorber plates to 520mm. This solution takes advantage of the fact that the beams with larger widths also have lower power densities.

ID period mm	field T	ID length m	total power KW	beam width at 11m (mm)	power density at 11m (W/mm2)	beam width at 13m (W/mm2)	power density at 13m (W//mm2)	beam width at 23m (W/mm2)	power density at 23m (W//mm2)
40	0.42	5	2.0	3.1	599	3.7	429	6.5	137
50	0.58	5	3.8	5.2	667	6.1	478	10.8	153
70	0.83	5	7.9	10.2	688	12.1	492	21.4	157
80	0.93	5	9.9	13.1	674	15.4	483	27.3	154
100	1.09	3.8	10.3	19.1	480	22.5	344	39.9	110
150	1.34	2.5	10.3	35.2	260	41.6	186	73.7	59
200	1.49	2	10.1	52.2	173	61.6	124	109.1	40

Table 1: horizontal spot size and powers for various insertion devices at ESRF. Energy 6GeV, Current 100mA, gap 20mm

The cooling is by water flowing through continuous lengths of tube, thus avoiding water to vacuum joints. Figure 3 shows the profile for the absorber plates. Actuation is by means of a standard pneumatic cylinder with one exhaust fitted with a quick exhaust valve. This actuation system means that it is possible to intercept the beam within 55ms. The ESRF are presently investigating methods of closing a beamline in emergency due to a vacuum failure without dumping the electron beam, hence the requirement for fast actuation of the movable absorber.

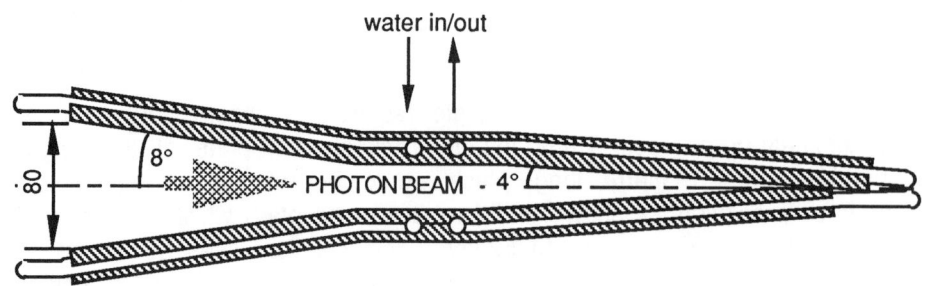

Figure 2 plan view of insertion device movable absorber plates

A finite element analysis of this profile was carried out using ANSYS[1] which showed that the maximum temperature reached in the worst case is 285°C and this in turn gives a maximum stress of 380MPa. This stress level is 5 times the yield strength of OFHC copper. Therefore, it is necessary to use a dispersion strengthened copper alloy. Glidcop AL15[7] with yield strength 380MPa at 20°C and thermal conductivity of 340W/m°C was originally chosen as the most suitable material.

A prototype movable absorber was manufactured using this material and is presently undergoing functional tests and thermal tests under an electron beam welder.

There are a number of materials which can be used for this type of absorber and the ESRF have evaluated three such materials, Glidcop AL15, Outokompu DS-Cu [8] and a Trefimetaux[9] alloy. All of these options need careful control to avoid possible problems:- vacuum outgassing, brazing ability and material availability.

Series production of 8 of these units will be finished in November 1991. Glidcop AL15 low oxygen grade declad copper will be used with the following precautions being taken. Each piece of material used will have an outgassing test performed on it to ensure that it is suitable for vacuum use. Each piece will have a structural analysis carried out to ensure that the dispersion strengthening is homogeneous. It will then be imperative to maintain material traceability during the manufacturing process.

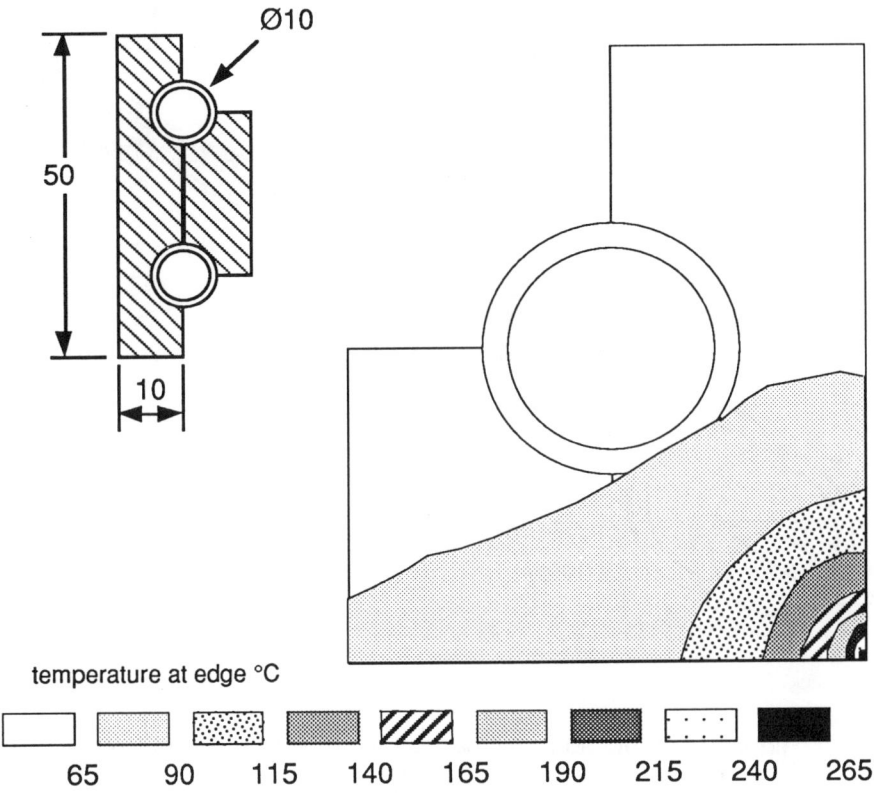

Figure 3 Profile of movable absorber plate and temperature profile at 500W/mm2

BERYLLIUM WINDOW AND SLIT/FILTER ASSEMBLY

The ESRF decided as a policy to have a beryllium window separating the experimental vacuum and the machine vacuum on all beamlines. This is feasible on most beamlines because the experiments are based on hard X-rays. Of the first 15 beamlines only one or two will be UHV beamlines without a beryllium window.

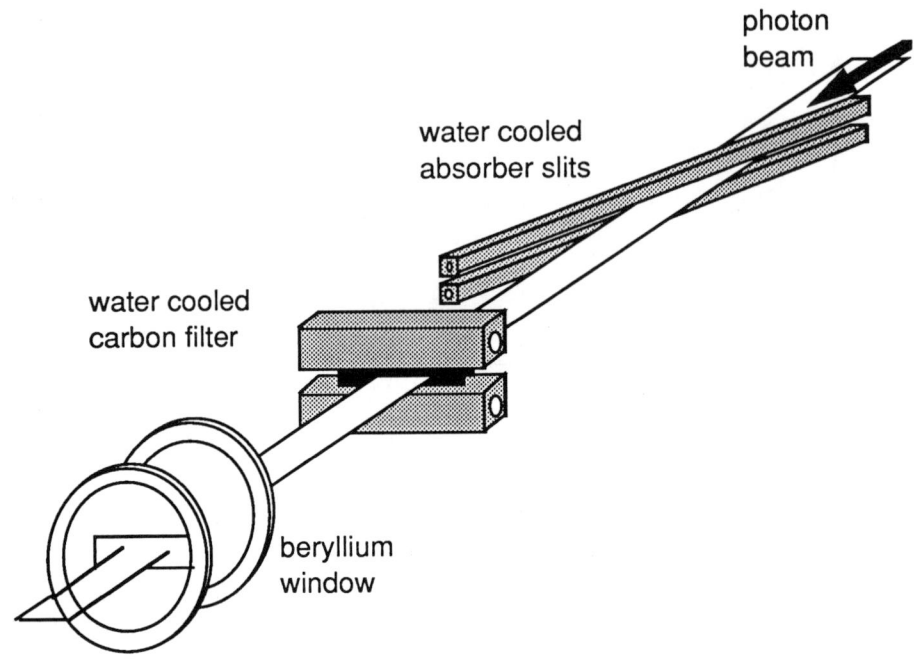

figure 4 ; principle of slit/carbon filter/beryllium window assembly

ESRF faces two problems here:

(i) In 1988 there was no European manufacturer of beryllium windows. However, ESRF persuaded two European companies to manufacture prototype windows to test their capabilities. Their prototypes were successful and were tested for thermal problems under laser beam. (see below)

(ii) Even though most X-rays pass through beryllium a certain amount of energy is absorbed by the beryllium and in the case of ESRF this would be greater than 500W in the worst case which was clearly unacceptable for a foil of 0.5mm thick.

This problem has been encountered at other synchrotron sources[2,3,4] but ESRF power densities are too high to use a direct copy of other solutions, and therefore, a synthesis was made. However there are some characteristics of the ESRF beam that can be used to its advantage. The beam is small in height, only 2mm at the beryllium window, and the beam stability is predicted to be good, ±0.2mm.

Even though the nominal height of the beam is only 2mm there are always some low energy photons that diverge at a greater angle. These photons also happen to be those harmful to the beryllium window, but can be absorbed by a vertical slit, see figure 4. A vertical slit of 4mm is designed for ESRF which is also used to protect downstream components (Be window, cooling for filters and beamshutter) in case of beam miss-steer, thus an efficient water cooling can be designed. After the slit a number of carbon filters are used to reduce the low energy photons that are incident on the beryllium window. The number of filters is variable up to a maximum of 5 filters. This is the only difference between the ID Front Ends and will be variable according to insertion device used for that beamline and the experiments to be carried out. In this way it is possible to reduce the absorbed power to levels of less than 100watts.

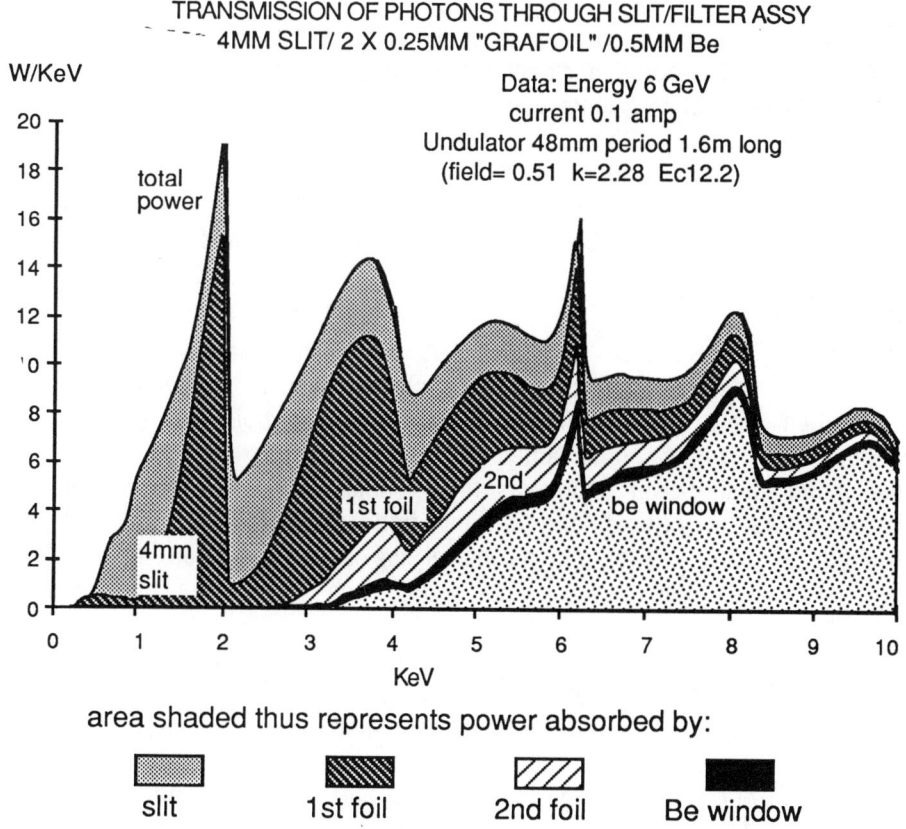

Figure 5 Transmission of X-rays through beryllium window

The effect on transmitted photons is shown in figure 5. The spectral distribution was calculated using RADIA[10] and the absorbtion of the carbon filters and Be window by SectEff[11]. The effect is clearly significant below 8KeV, but above this level only a 10-20% reduction is seen. It can be seen from figure 5 that the 4mm slit absorbs a part of the first two harmonics, but this is not worrying because the Be window would eliminate these first two harmonics anyway even if the slit was not there. If the user requires these two harmonics a UHV Front End would be installed without a Be window. Clearly each beamline needs to be considered separately to decide the number and thickness of filters that are acceptable to the users. The slit height can be varied between 4mm and 7mm and the number of filters can be varied between zero and five.

Similar graphs can be drawn for wigglers and higher power undulators, which show that the heat load on the window can be reduced to manageable levels. The problem that occurs for the high power insertion devices is that it is not possible to cool the carbon adequately. As can be seen the 1st foil absorbs a significant amount of energy. Previously it has been suggested that the filters can be considered to be cooled by radiation. Unfortunately even with the thinnest pyrolitic carbon available the temperature would reach 2000°C and the carbon would evaporate completely within 10 hours. The conductivity of pyrolitic carbon is particularly good and can be used to cool the carbon, and so, ESRF has chosen to water cool the filters.. For the first 7 ESRF insertion devices the power densities absorbed on the carbon filters are low compared with those already experienced at KEK[3]. Two dimensional iterative calculations have been made for these cases which show maximum temperature in the first carbon foil of 1600°C. The material is not isotropic and the thermal conductivity changes significantly with temperature., and therefore, for small period high power undulators (40 - 60mm period) the cooling of the carbon filters is critical and further finite element analysis will be required to choose exact dimensions of the slit and carbon filters.. The ESRF will investigate this problem when such a powerful undulator is proposed.

ESRF also performed some tests on the prototype beryllium windows received. The windows were tested under laser beam with a thermal camera showing the temperature distribution on the beryllium. The results for this work have not been fully analysed, but the windows withstood power densities of at least 1.5 times the maximum expected for the first 15 Front Ends. Initial calculations show that for the worst case presently envisaged the window will absorb 90W at a power density of 2.5 W/mm^2. Out of the four windows tested one window failed. The type of failure was a leak in the beryllium of 10^{-5} mbar l s^{-1} directly under the laser beam. If this failure mechanism is repeated under X-ray heat loads there would not be a catastrophic vacuum failure into the storage ring. Although any leak is totally unacceptable the Front End fast shutter and associated all metal gate valve would be able to isolate the Front End safely without any effect on the storage ring vacuum if a failure of this nature occurs.

FAST SHUTTER

Fast shutters have been available on the market for some time and the technology is relatively well known. ESRF made tests on two shutters from European suppliers and made its choice accordingly. However, ESRF are in the lucky position of having 10 metres between the most likely failure point, the beryllium window, and the fast shutter, and therefore, the time for the propogation of a wave will be relatively large, around 20ms. This allows us to consider the following sequence of events:

	elapsed time (ms)
1 Catastrophic vacuum leak at beryllium window	0
2 Leak detected at window	2
3 Movable absorber closes	55
4 Wave travels through acoustic delay line (ADL)	40
5 Fast shutter triggered	42
6 Fast shutter closes	55
7 Wave reaches fast shutter	60

The aim is to close the movable absorber before the fast shutter thus protecting the shutter plate of the fast shutter. This means that the storage ring does not have to be completely stopped due to a vacuum failure in the Front End.

Tests have shown that the time taken for a wave to travel through a vacuum chamber varies according to entry pressure[5,6], but also depends heavily on the type of failure. For example tests have shown that the time for an atmospheric wave admitted to an acoustic delay line by a butterfly valve to propagate through the ADL is 80ms, but if admitted by an electromagnetic valve it is 40ms.

Presently it is believed that it is possible to shut down a Front End due to a vacuum leak without dumping the electron beam in about 90% of failures. For the remaining 10% of cases the electron beam will be dumped. What is left is a control problem to decide under what circumstances the electron beam should be dumped.

MANUFACTURE OF ESRF FRONT ENDS

It was clear that the ESRF would be unable to assemble 15 Front Ends within the short time schedule demanded in house. Therefore, as adopted for other parts of the machine, ESRF is using industry to manufacture and assemble the major parts of the Front Ends. The Front End components are grouped into two modules, one near the storage ring including components 3-7 in figure 1 and the other near the shielding wall which includes items 8-14.

The manufacture of these modules includes assembly and test of the vacuum system, installation and test of the bakeout system and all wiring and piping. In principle it should be possible to install these modules in the storage ring tunnel quickly and efficiently. This will become more critical during the installation of phase 2 when ESRF installs a further 12 Front Ends, but this has to be done during machine shutdowns.

FUTURE DEVELOPMENTS

Presently the Front End technology is in front of experimental technology as far as coping with high heat loads and high power densities is concerned. The present design is capable of coping with 10KW with a power density of 500W/mm^2 @ 13 metres with a reserve on the protection of the beryllium window which remains to be tested.

When experimental mirrors are capable of withstanding these power loads an option may be to remove the beryllium window.

REFERENCES

1. G. Marot, ESRF internal report, Beamline Movable Absorber Thermomechanical Study, 1989
2. Swortti & Thomlinson, informal report, X-Ray Filters for Synchrotron Radiation, 1984
3. Sato, Asaoka, Nagakura & Kanaya, A Beryllium window assembly for the 53 pole wiggler beamline at the Photon Factory, 1988
4, Shen, Bedzyk, Keeffe & Schildkamp, CHESS internal report, A heat transfer study for beamline components in high power wiggler and undulator beamlines. 1988
5. Sato, Kakizaki, Miya, Morioka, Yamakawa & Ishii, Nuc Inst A240(1985) 194-198, Transit Times of Pressure Waves in an Acoustic Delay Line
6. Jean & Rauss, Le Vide N°111 1964, Protection Contre Les Rentrees d'Air
7 Gldcop AL15 low oxygen grade declad, SCM Metals Products inc, USA

8. DS-CU, OUTOKOMPU OY, Finland
9. TREFIMETAUX, France
10. RADIA, P. Elleaume and J. Chavanne, ESRF, France
11. SectEff, P. Elleaume, ESRF, France
12. G. Marot, ESRF internal report 89 09, Design of ESRF Absorbers, 1989

FREE-ELECTRON LASER SOURCES OF EXTREME-ULTRAVIOLET RADIATION
AND THEIR VACUUM REQUIREMENTS*

Brian E. Newnam
Chemical and Laser Sciences Division
Los Alamos National Laboratory
Los Alamos, New Mexico 87545

ABSTRACT

Recent development of free-electron laser (FEL) component technologies should enable these devices to operate in the extreme-ultraviolet, well below 100 nm. When fully developed, FELs represent the next generation of coherent-radiation sources with peak- and average-power outputs surpassing those of any existing, continuously tunable photon source by many orders of magnitude. An rf-linac-based, multiple-FEL facility, spanning the spectral range from 1 nm to 100 µm, is proposed. To enable such a facility to operate without significant degradation over long periods, contamination of certain of the FEL components must be prevented. Requirements for ultra-high vacuum and restricted contamination from outgassing from chamber walls are discussed.

INTRODUCTION

Free-electron lasers appear to be the natural finale in the progression of light sources based on radiation from relativistic electrons passing through magnetic undulators. Since 1977, more than one-dozen FEL oscillators and amplifiers have produced coherent radiation over a broad spectral range extending from 240 nm in the ultraviolet[1,2] to millimeter wavelengths as shown in Fig. 1. Even shorter wavelengths (down to 106 nm) have been produced by coherent spontaneous emission from FEL amplifiers, albeit at very low power.[3,4] These successes have encouraged FEL researchers to examine the parameters required for FEL operation in the extreme ultraviolet (XUV) below 100 nm where no powerful, tunable, coherent-radiation source presently exists.

Extending FELs to ever shorter wavelengths, however, is inherently difficult since the gain decreases monotonically with the square-root of the wavelength,[5,6] and below 100 nm the available mirrors for resonators have comparatively low reflectance, generally <50%. With such mirrors, the small-signal power gain for a single pass through the magnetic undulator must exceed 400% to reach the threshold for oscillation. Such high gain can be attained with large peak electron-pulse current (>100 A) and long undulators comprising several hundred periods. However, long undulators have small homogeneous gain bandwidth proportional to 1/N, where N is the number of periods. Since the gain is limited by both the transverse electron beam emittance and longitudinal energy spread, it is essential that the accelerators deliver very bright (ratio of peak current and norm-

---

*Work supported by Los Alamos National Laboratory Program Development Funds and performed under the auspices of the U. S. Dept. of Energy.

alized emittance-squared) electron beams.

Storage rings and rf linear accelerators (rf linacs) with high-brightness injectors are presently the only sources of electron beams with sufficiently high peak current and low emittance to drive XUV FELs. Storage rings (SR) with long straight sections for XUV FEL oscillators are being constructed by Madey's[7-11] group at Duke University and designed by the DELTA[12,13] group at the University of Dortmund. These systems should provide good spectral stability for spectroscopic measurements and have variable pulsewidth capability. System design and component development for rf-linac-based XUV FELs have been pursued by Newnam[14-25] and co-workers at Los Alamos National Laboratory since 1984. More recently, Ben-Zvi et al.[26] at Brookhaven National Laboratory have proposed a VUV-FEL amplifier (100-200 nm) to be driven by a superconducting rf linac. Rf-linac FELs should generate high average power ($\sim 10^2$ W) in the XUV depending on the duty factor ($\sim$100 times more power than SR FELs), and a series of FELs can be driven simultaneously by one linac.

The essential elements of an rf-linac-FEL oscillator, as illustrated in Fig. 2, are the electron beam produced by an accelerator, a magnetic undulator in which the electrons oscillate transversely and emit (or absorb) light, and the resonator mirrors. By varying either the electron energy, the magnet separation, or the magnetic field on axis, the wavelength can be tuned continuously over a broad range. The laser radiation is coherent, polarized, and nearly diffraction limited. The typical temporal format is a 10- to 1000-µs macropulse train of 1- to 10-ps micropulses with ~1- to 100-ns separation and with an arbitrary macropulse repetition rate. By slight detuning of the resonator length, the spectral bandwidth of radiation from FEL oscillators can be varied from the narrow Fourier-transform limit of the ps pulses to several per cent with sidebands.

Los Alamos National Laboratory,[27-30] Stanford University,[31,32] and Boeing Aerospace Company[33,34] have operated rf-linac-driven FEL oscillators in the visible and near infrared for a cumulative total of several thousand hours. Experience with these systems has provided invaluable insight and data with which to design rf-linac-driven FEL oscillators (and possibly high-gain, single-pass amplifiers)[35,36] for operation in the XUV spectral range. An important distinction of the Los Alamos IR FEL is its very high 500-A peak current capability and correspondingly large single-pass optical gain from a short, 1-m undulator (250% small-signal gain/pass at 10 µm with 300- to 400-A peak current).[28,29] Such are the magnitudes of the parameters that will be required for operation of FEL oscillators in the XUV. With recent significant improvements to the primary components, it only remains to construct a high-brightness linac with energy ≥200 MeV to verify the feasibility of operating below 100 nm.

The accelerator and optical components and numerical simulation codes that will be needed to design and construct the next-generation XUV-FEL light sources have been under active development at Los Alamos for the last five years. These components include the laser photoinjector, multifacet XUV resonator mirrors, and high-precision undulators. For each of these technologies, operation in high vacuum is imperative. Photoemissive surfaces with high quantum efficiency (≤10%) and reflector films susceptible to oxidation and/or

280 Free-Electron Laser Sources of Extreme-Ultraviolet Radiation

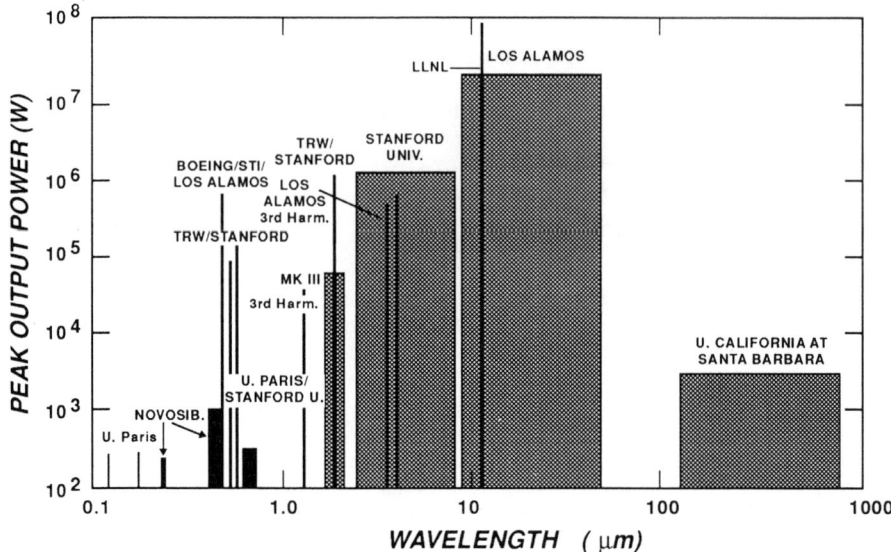

Fig.1. Succesful free-electron laser operation in the near-ultraviolet to the far-infrared provides the experience base for extension below 0.1 µm. (The weak output at the two VUV wavelengths was coherent spontaneous emission from electrons bunched within an undulator by an external 532-nm laser.3)

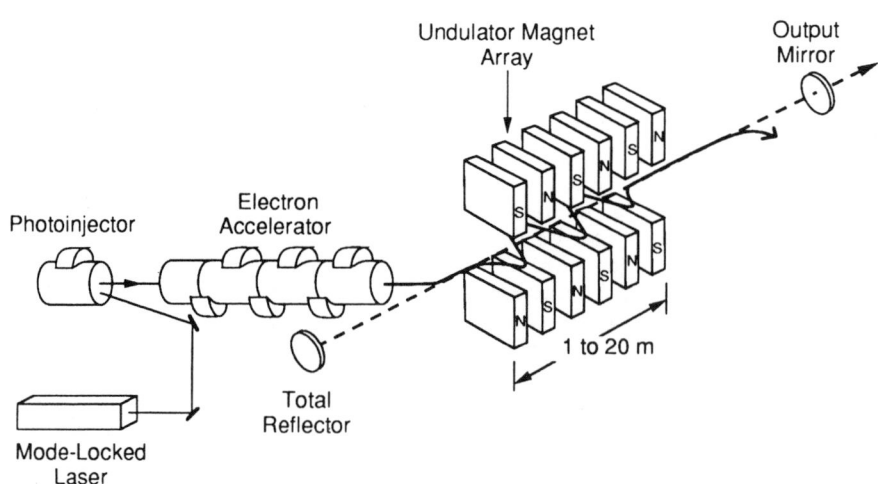

Fig. 2. Basic schematic of an rf-linac-driven free-electron laser oscillator showing the laser photoinjector, rf linac, magnetic undulator, and optical resonator.

carbon film buildup, e.g., Al, Si, Rh, and Ag, will require ultra-high vacuums to maintain satisfactory performance over a conveniently long period. Less stringent vacuum levels will suffice if photocathodes with moderate quantum efficiencies, e.g. 10^{-3}, can be used. For the reflectors, carbon contamination will have to be controlled locally with ion pumps, and coatings may have to be cleaned and refreshed periodically. Finally, an adequate vacuum must exist in the small-diameter beam tubes within the magnetic undulators to prevent electron-beam emittance growth as a result of ion trapping. For single-pass machines, moderate vacuum levels should be adequate, but the usual 10^{-9} to 10^{-10} Torr UHV level will be necessary in storage-ring FELs to maintain long beam lifetime.

In the following two sections, we will describe the design and operating characteristics of an XUV-FEL user facility as proposed by Los Alamos National Laboratory, and then discuss the vacuum environments necessary for the three FEL components introduced above.

PROPOSED XUV-IR FEL USER FACILITY

<u>FEL Configuration</u>: Based on considerable technical progress to date (primarily supported by the U.S. Strategic Defense Initiative Ground-Based Laser Program plus Department of Energy investments in future XUV-FEL extensions), Los Alamos National Laboratory has designed a series of FELs with output ranging into the soft x-ray spectral region for integration into a proposed XUV-IR FEL national user facility for scientific experimentation and industrial applications. This design uses one high-energy rf linac (≤ 1 GeV) to drive a series of seven FEL oscillators that will produce trains of coherent, picosecond pulses spanning the spectral range from 1 to 500 nm. A second series of three FEL oscillators, driven by a separate 60-MeV rf linac synchronized with the first, will produce wavelengths from the visible to the far infrared from 0.5 μm to 100 μm. Over the range of high single-pass FEL gain, the predicted peak- and average-output powers, especially below 300 nm, should surpass the capabilities of any existing, continuously tunable photon source by four to seven orders of magnitude. In addition, the spectral bandwidth may be sufficiently narrow, e.g. $\leq 10^{-4}$ at 100 eV, that use of a monochromator may be optional for many applications.

The conceptual design of the proposed Los Alamos XUV FEL facility is shown in Fig. 3, and design specifics are given in Table I. The shortest-wavelength oscillators are ordered first in the sequence since they require the highest-quality electron beam; the gain at longer wavelengths is less affected by beam degradation. Even so, all of the oscillators are designed to perturb the electron beam energy only very slightly, with the energy-extraction efficiency being 0.1% or less. Further beam degradation by wakefield effects in the beamline and magnetic undulator will be avoided by minimizing discontinuities. The number of oscillators may be increased arbitrarily, consistent with the amount of accumulated energy spread and/or emittance degradation in the electron beam that can be tolerated. The operating wavelengths of each of the FELs will either be tuned as a group by varying the electron energy or tuned independently over a smaller range by adjusting the undulator gaps.

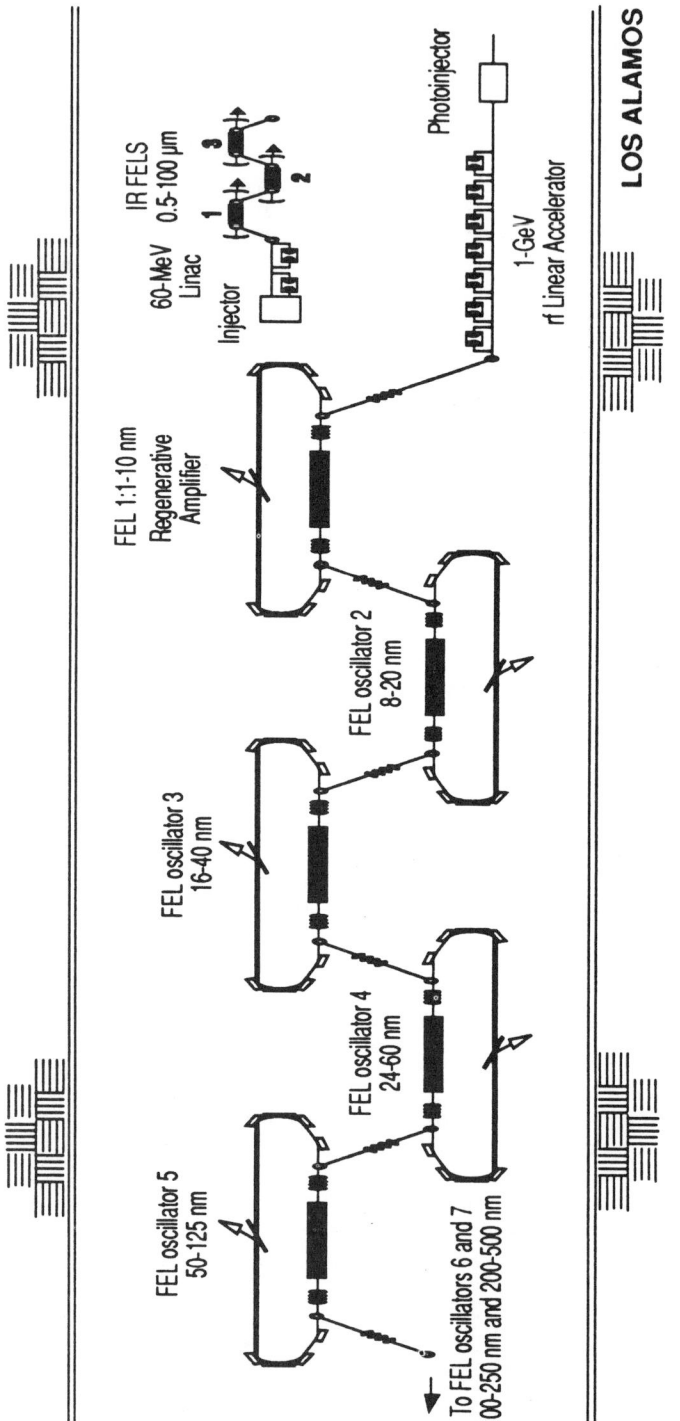

Fig.3. Source configuration of the proposed Los Alamos XUV Free-Electron Laser National User Facility producing coherent radiation from 1 nm to 100 μm. A series of seven XUV/UV FEL oscillators spanning 1 to 500 nm is driven by one 1-GeV rf-linac. A second 60-MeV rf linac, synchronized with the first, drives a series of three vis/IR FEL oscillators to cover the 0.5- to 100-μm range. The FELs are sited below ground to provide radiation shielding, and a series of user-laboratory suites are in an above-ground facility adjacent to the FELs.

Table I. Parameters for the proposed Los Alamos rf-linac-driven xuv-ir FEL user facility

Electron beam	
Energy:	≤1 GeV, adjustable
Peak current:	300 to 400 A in ~10-ps micropulses
Normalized emittance: (90% of electrons)	10π mm-mr for <10 nm 30π mm-mr for >50 nm
Energy spread:	0.1% to 0.2%, FWHM
Undulators	
Length:	5 m for 50-nm, 12 m for 4-nm
Period:	1.6 cm (and 4.5 mm)
Peak axial field:	7.5 kG, 2-plane sextupole focusing
Undulator parameter, K:	1.1, peak
Resonator mirrors	
End mirrors:	R ≥40%, multifaceted reflectors metal coatings on Si for <100 nm CVD SiC for ≥60 nm, Al for ≥120 nm
Intracavity hyperboloids:	Au coating on SiC or Si substrates

A mirror gallery has been devised to deliver radiation from any of the ten FELs to each of six experimental rooms devoted to physical-, chemical-, materials-, biological-, and medical sciences as well as industrial applications.[37] According to the participants at the 1988 OSA Topical Meeting on FEL Applications in the Ultraviolet,[38] the unique capabilities of such a facility will open numerous, new research opportunities in these disciplines.

At this time, the rf linac is designed for room-temperature operation. All of the past Los Alamos infrared FEL experiments have used an L-band (1.3 GHz), side-coupled, standing-wave rf linac operated at slightly above ambient temperature. Design calculations for advanced, standing-wave linac structures have been performed from 100 MeV to 1 GeV for an XUV FEL. Cryogenic and superconducting (at 4K) cavity options also are being examined, because they offer potential advantages of cw macropulse operation, improved pulse-to-pulse stability, and reduced electrical cost due to lower power dissipation in the linac structure. However, the capability of either of these alternate configurations to generate several-hundred amperes of peak current has not yet been demonstrated.

<u>Predicted XUV-IR FEL Output</u>: We have performed 3-D numerical simulations using the FEL code FELEX[39] and its derivatives to predict

the single-pass and multiple-pass gain in XUV-FEL resonators, and the spectral bandwidth, output power, and spectral brightness versus wavelength. Tables II-IV provide an abbreviated summary. Electrical power cost considerations will limit the year-long average duty factor to 1% for ambient-temperature linac structures, but occasional operation at ≥10% duty is feasible with proper cooling of the accelerator cavities, and 100% duty may be possible if either a cryogenic or a superconducting linac design is adopted. Practically, the duty factor will be limited by laser-induced thermal distortion of the FEL resonator mirrors.[40] The latter limitation can be avoided if enough (~10^4) single-pass gain can be realized in FEL amplifiers based on self-amplification of spontaneous emission (SASE).

Table II. Radiation properties of the proposed Los Alamos rf-linac-driven xuv-ir FEL user facility

Micropulse duration:	10-20 ps (FWHM) compressible to <2 ps
Micropulse repetition rate:	3×10^7 Hz
Macropulse duration:	100-µs, Rep. @ 100 Hz (1% duty)
Facility wavelength span:	1 nm to 500 nm, 7 FEL oscillators (0.5-100 µm, 3 IR FELs w/2nd 60-MeV linac)
Spectral bandwidth:	1 cm^{-1} Fourier-transform limit for single, 10-ps pulses 10^{-4} w/25 klystrons, possibly 10^{-5} up to ~1% if sidebands are allowed
Polarization:	Linear, w/circular/elliptical options
Temporal coherence:	Limited by micropulse Fourier trans.
Spatial coherence:	Near diffraction-limited focusability

<u>Comparison with Synchrotron Light Sources</u>: It is of interest to compare the predicted stimulated-emission output from FELs with the spontaneous-emission output generated from existing and planned synchrotron light sources that have magnetic insertion devices (wigglers and undulators). Figure 4 presents a comparison of time-averaged spectral brightness (photons/s/(mm-mr)^2/0.1% bandwidth) delivered on target as a function of wavelength, and Table IV lists the flux and power output at 100 nm predicted for an rf-linac-FEL along with that of two storage-ring devices. At 10 eV (124 nm) the FELs will produce 10^4 to 10^6 higher average spectral flux on target; at 100 eV (12 nm), the FEL advantage will decrease to a factor of 10^2 to 10^4, depending on the FEL duty factor. Moreover, the corresponding peak flux output of the rf-linac FEL (@ 1% duty) will exceed that of synchrotron undulators by an additional factor of 10^3.

Table III. Photon power and brightness predictions for the proposed Los Alamos rf-linac-driven xuv-ir FEL user facility

Peak power at target: bw ~1 cm^{-1}, single pulse and ~10^{-4}, macropulse average	1 to $\geq$10 MW, 12 to 100 nm >20 MW, for >100 nm
bw $\geq$1%, w/sidebands	10 to $\geq$50 MW, 12 to 100 nm, 5 MW at 4 nm
Average power at target: 1% duty	1 to >10 W, w/o sidebands 30 to 200 W, w/sidebands
Photon flux at target: from 1 - 500 nm, resp.	$10^8 - 10^{15}$ photons/10-ps pulse $10^{15} - 10^{20}$ photons/s, aver/1% duty
Spectral brightness (Peak):	10^{28} photons/s/(mm-mr)^2/bw
Spectral brightness (Aver.):	10^{22} photons/s/(mm-mr)^2/bw/1% duty bw ~10^{-4}, macropulse average

Table IV. Output comparison of rf-linac-driven FEL and synchrotron radiation sources at 100 nm

	SSRL WIGGLER [a]	ALS UNDULATOR [b]	RF-LINAC XUV FEL [c]
Spectral flux (Photons/sec) at target	10^{13}	4×10^{14}	10^{19}
Average power (W) at target	10^{-5}	10^{-3}	20
Peak power (W) at target	10^{-2}	10^{-1}	10^7
Average & peak spectral brightness at target (photons/sec/(mm-mr)^2/bw)	$10^{12}, 10^{15}$	$10^{16}, 10^{19}$	$10^{22}, 10^{28}$

[a] Stanford Synchrotron Research Laboratory wiggler.[41] 0.1% spectral bandwidth after monochromator with 10% efficiency.
[b] Predicted performance of undulator U8.0 designed for the Advanced Light Source storage ring at Lawrence Berkeley Laboratory.[42] 0.1% spectral bandwidth after monochromator with 10% efficiency.
[c] Single-pass, rf-linac FEL operated at 180 MeV with 1% duty factor Multiply listed FEL powers by 10X if driven by a 500-MeV beam. FEL spectral bandwidth = 0.01% limited by rf-power noise.

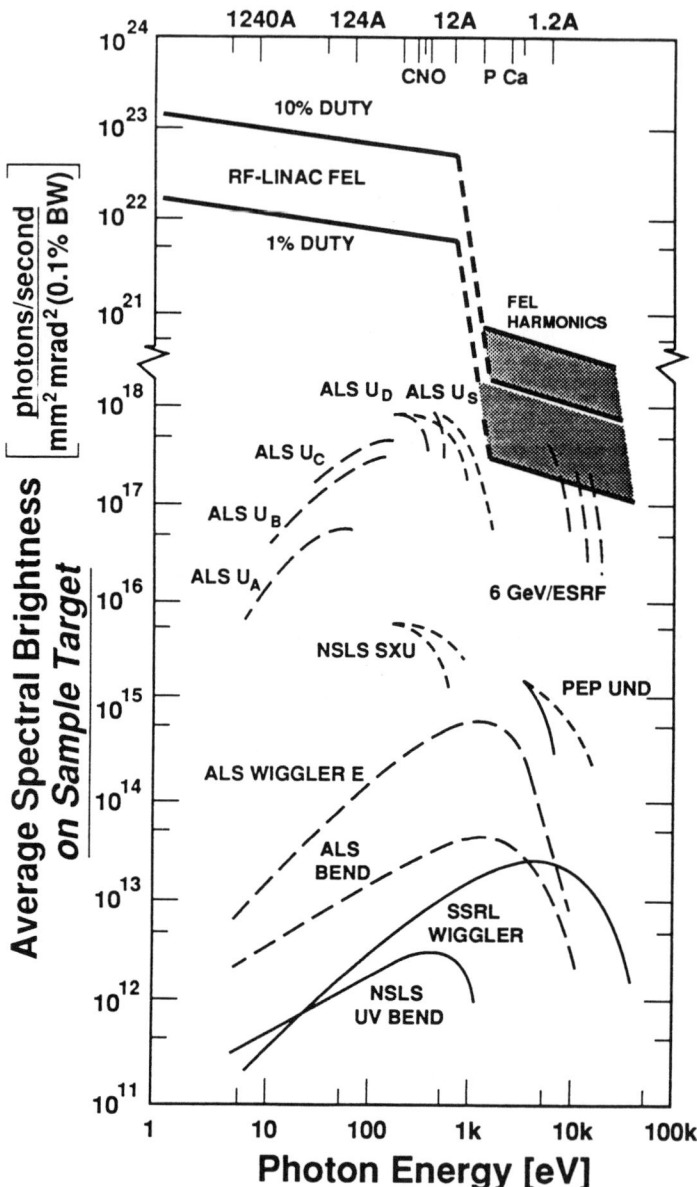

Fig.4. Time-average spectral brightness (delivered on target) of FELs will far exceed that of the most powerful storage rings designed with insertion devices (undulators and wigglers). The FEL curves were calculated for the Los Alamos rf-linac FEL design, and a monochromator efficiency of 10% was applied to the published undulator output curves. To convert the time-average curves to peak values, the appropriate conversion factors are 3×10^5 for the FEL output @ 1% duty and $\sim 3 \times 10^2$ for the storage-ring insertion devices.

XUV-FEL SYSTEM VACUUM REQUIREMENTS

<u>Overview</u>: For future FELs built to operate in the XUV, the vacuum environment will have a major impact on three primary components. These are the electron injector, resonator mirrors, and the undulator beam tube; their respective vacuum requirements are summarized in Table V. Furthermore, the vacuum requirements will depend to a large extent on what type of electron source is used. Storage rings are inherently not concerned with photoinjectors, but instead they must have high enough vacuum to assure a sufficiently long lifetime for the recirculating electron (or positron) beam. On the other hand, rf linacs, as single-pass systems (or at most several-passes in a recyclotron mode), do not pose electron-beam lifetime concerns, but a contaminated photoinjector cathode will reduce the available beam current and lower the FEL gain.

Table V. Vacuum requirements for XUV-FEL oscillators for two bright electron sources

FEL Component	RF-Linac	Storage Ring
Photo-Injector	10^{-10} T	N.A.
Resonator Mirrors	$\sim 10^{-10}$ T	$\sim 10^{-10}$ T
Undulator Tube	$< 10^{-3}$ T	$\leq 10^{-9}$ T

Contamination of the resonator mirrors must be prevented and/or controlled in FEL oscillators driven by both SR and rf-linac electron sources. Mirror contamination will reduce the reflectance, raise the gain threshold for lasing, and decrease the steady-state output power. With sufficient single-pass gain ($>10^4$) to obtain exponential rise of the laser power through a long amplifier undulator, mirror problems would not exist. However, such high gain will very difficult to attain. Although a resonator is more complex than a simple amplifier, it allows for reduced requirements on the electron beam and undulator. The single-pass gain ($<10^3$) required for XUV-FEL oscillators can be achieved with 10-100X lower electron beam brightness and undulators with half as many periods as needed for an amplifier starting from noise. Additionally, contaminated external beam-steering mirrors will reduce the useful power on target for either oscillator or amplifier configuration, and excess absorption of the incident energy will distort the mirror surface figure, thereby degrading the beam brightness.

In a storage ring, the vacuum quality within the narrow undulator beam tube must again be good enough to prevent significant degradation of the beam lifetime via scatter losses. In an rf-linac FEL, it must be only good enough to preclude plasma effects that could increase the emittance. The following paragraphs provide additional details on the vacuum requirements for each of these three elements.

Photoinjector Cathode: The paramount importance of a high-brightness electron beam (defined as 2 X ratio of peak current and normalized emittance-squared) for short-wavelength FELs has been emphasized in the Introduction. As an example, Goldstein et al.[24] determined from 3-D numerical simulations that a 12-nm FEL oscillator will require a beam brightness of 1.2×10^{11} A/(m-r)^2, corresponding to a peak current of 300 A and normalized transverse emittance of 23π mm-mr (90% of particles), with energy spread of 0.1% (FWHM) at 535 MeV. Until the development and experimental demonstration of the laser photoinjector by Fraser and Sheffield,[43-45] electron beams with such high brightness had not been generated in an rf linac. [The beam brightness capabilities of thermionic, dispenser, and semiconductor photocathode injectors are 0.08, 2, and 30×10^{11} A/(m-rad)^2, respectively.[45]]

As shown in Fig. 5, a photocathode positioned on the end wall of the first accelerating cavity is irradiated by a train of ps-duration, visible to ultraviolet laser pulses. By applying a very high accelerating gradient, e.g., 30 MeV/m, the photoelectrons attain a relativistic energy of 1 to >3 MeV in the first cavity, thereby minimizing their susceptibility to perturbations that cause emittance growth. The following photocathodes with varying levels of quantum efficiencies (QE) are being used in FEL injectors: CsK_2Sb (≤10% QE) at Los Alamos National Laboratory,[45,46] yttrium (0.1% QE) at Brookhaven National Laboratory,[47] and LaB_6 (0.01% QE) at Stanford University.[48]

The CsK_2Sb cathode is most susceptible to poisoning by contamination and must be operated in high vacuum of 10^{-10} Torr in order to have a practical lifetime with QE >1%. Recent lifetime measurements of this cathode (see Fig. 6), while mounted within the photoinjector cavity at 2×10^{-10} Torr and without laser irradiation, indicated that the QE declined only from 10% to 8% in a 36-hour period followed by no change until the test ended at 68 hours.[49] However, in the presence of full rf-drive power, Los Alamos experimenters found that CsK_2Sb photocathodes have lifetimes limited to one to two days. The degradation occurs only when the accelerating rf power is applied. It is suspected that water vapor and carbonaceous gas compounds, desorbed from the cavity walls of the photoinjector, poison the cathode. Evidence of substantial outgassing, with both rf power applied and drive-laser illumination of the cathode, was the large pressure rise measured in the photoinjector cavity.[49] Also, addition of H_2 and N_2 gases at pressures >10^{-8} Torr had no deleterious effect. These observations are in accord with the experience at NSLS[50] and CERN[51] where the synchrotron radiation striking the metal chamber walls causes photostimulated desorption of H_2O, CO, CO_2, and CH_4. As the photodesorption yield of these gases declines with accumulated radiation dose, the beam lifetime is observed to increase.

With the objective of producing 5-10 nC of charge per each ~10-ps micropulse, Los Alamos scientists have determined that the lifetime of semiconductor photocathodes can be increased by use of more stable cathodes and by exposing the photoinjector accelerating cavities to high rf fields for long periods. Their first Cs_3Sb photo-

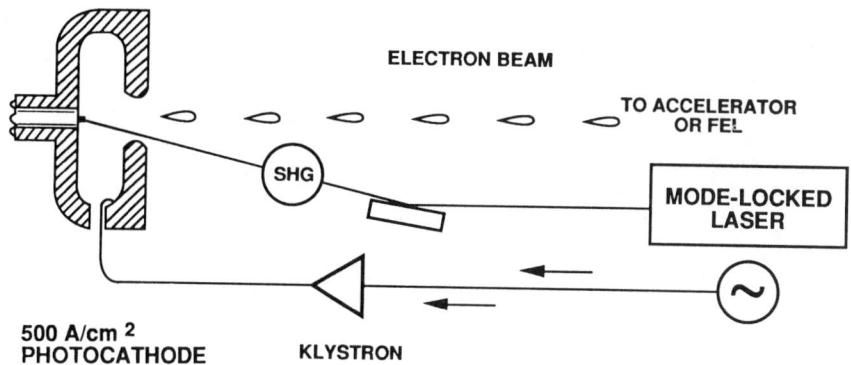

Fig.5. Basic concept of the laser-irradiated photoinjector. The photocathode is mounted in the first accelerator cavity where emitted picosecond electron pulses are rapidly accelerated by a high-gradient electric field, thereby minimizing emittance growth.

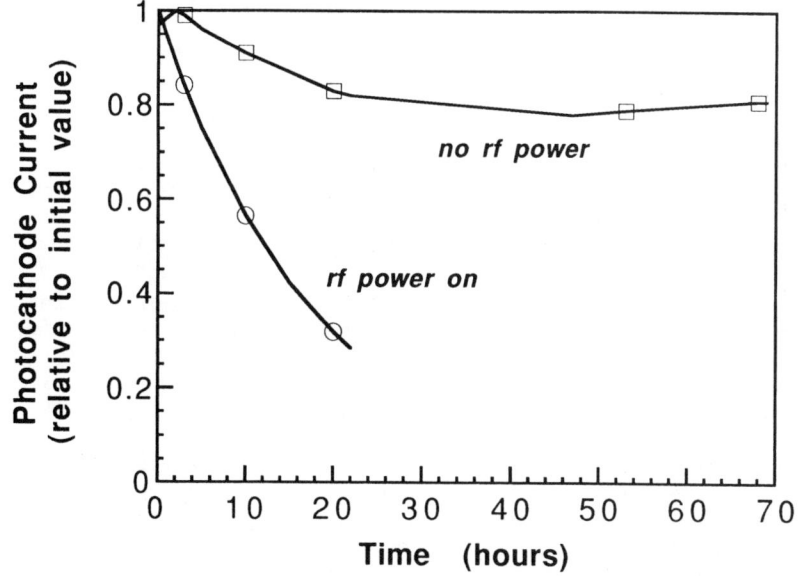

Fig.6. Initial tests of $CsK_2Sb:O$ photocathode current emission when mounted on the wall of the first injector linac cell: with and without rf accelerating power. Glow-discharge treatment and ongoing electron-induced desorption of contaminants is expected to greatly lengthen the cathode lifetime. After Volz.49

cathodes produced 1-A average current, but due to loss of Cs, they only had a short, 0.5-hr lifetime (1/e). Addition of K in the cathode fabrication process resulted in $CsK_2Sb:O$ surfaces with greatly reduced Cs loss rates which immediately extended the lifetime to two hours. With continued operation of this cathode formulation within the first accelerator cavity, an 18-hour lifetime with full rf-drive power was obtained at a given laser intensity.[46] The effective lifetime has been extended even more by raising the drive laser power level enough to compensate for the declining quantum efficiency. Further improvement is expected when glow-discharge cleaning of the cavity walls is implemented to augment electron-beam desorption of contaminant gases. This procedure has proven effective in conditioning the Brookhaven NSLS storage rings.[52]

It appears that semiconductor photocathodes with high quantum efficiency are more susceptible to degradation by contamination than metal photocathodes, e.g. Y. However, to generate electron pulses with large charges, e.g. 5-10 nC/pulse, as just implied, it is necessary to have the product of the drive-laser pulse energy and QE be sufficiently large. The experience of the Los Alamos FEL team is that a photoinjector with a CsK_2Sb-cathode irradiated with 532-nm laser pulses produces a micropulse charge of $Q(nC/pulse) = 4.5 QE(\%) \times E(\mu J/pulse)$.[53] Although it becomes increasingly difficult and expensive to generate higher-energy, cw-modelocked laser pulses in the green or ultraviolet, it is possible to trade-off low-QE cathodes with longer lifetimes for a more powerful and expensive drive laser.

The concentrated efforts of several research groups[46-48] to implement photoinjectors on rf linacs designed for FELs should eventually make rf linacs reliable sources of high-current, low-emittance electron beams. If the high brightness of such beams can be maintained without degradation during acceleration to high energy, then numerical simulations predict that FEL gain should be high enough for oscillator operation at wavelengths as short as 4 nm.[6,24] The first FEL experiments with a photoinjector, conducted by a Stanford University-Rocketdyne team, demonstrated lasing at 3.1 μm with laser-enhanced electron emission from a LaB_6 rf gun.[48] Meanwhile, a high-current photoinjector has been integrated into the Los Alamos infrared FEL system. As reported by Feldman et al.,[54] initial measurements of the beam quality at 17 MeV were in agreement with predicted values for the normalized design emittance, ~30π and ≤50 π mm-mr for 5-nC and 10-nC bunches, respectively.

<u>Resonator Mirrors for XUV FELs</u>: FEL oscillator operation in the XUV will require resonator mirrors with sufficiently high retro-reflectance to provide a substantial advantage over a single-pass amplifier. By setting the reflectance requirement at an arbitrary value of ≥40% for each of the two end mirrors, the mirror losses will have to be offset by single-pass gain >600%, which is attainable.

Below 100 nm, few materials have reflectance exceeding 40% for normal incidence. For wavelengths from 60-100 nm, polished chemically vapor-deposited (CVD) silicon carbide (SiC), deposited on hot SiC substrates, exhibits a reflectance between 40% to 50% at normal incidence.[55] In the same spectral region, the reflectance of polished type-I diamond reportedly exceeds this by a few per cent.[56] From 80 nm to 100 nm, the reflectance of unoxidized Al films goes from 40% to >90%.[57] Finally, in the range from 10 to 20 nm, multilayer reflectors of Mo and Si are available commercially with reflectance typically between 40 to 50%.[58]

An additional requirement for FEL resonators is that the total beam-induced thermal distortion of the mirror surfaces must be restricted to a small fraction (~1/4) of the operating wavelength. Thus, high mirror reflectance ≥95%, i.e., absorptance ≤5%, is essential.[24,40] The multifacet XUV metal mirror design proposed by Newnam[59,60] should satisfy both requirements in certain spectral regions (retroreflectance ≥40% and ≤5% absorptance per facet). This mirror system exploits the phenomenum of total external reflectance (TER) at large angles of incidence. As pointed out by Vinogradov et al.,[61] materials with refractive index (both real and imaginary parts) sufficiently less than unity can provide surprisingly high retroreflectance by using a sequence of reflections beyond the critical angle, typically larger than 60°. Unoxidized Al and crystalline Si, for example, have the necessary optical constants to attain ≥40% retroreflectance from 35 nm to 100 nm.

An FEL ring resonator incorporating multifaceted retroreflectors is shown in Fig. 7. The appropriate reflective coatings will depend on the spectral operating range: ~35-100 nm using Al films or crystalline silicon, and 9-14 nm with Ag, Ru, and Rh films.[59-61] Other single-layer films that could exhibit sufficiently high TER in the wavelength range intermediate between 14 and 35 nm include Y and Se.[62] An important advantage of these single-layer reflectors is that they can be deposited on the multifaceted Si or SiC mirror substrates without removing them from the ring resonator.

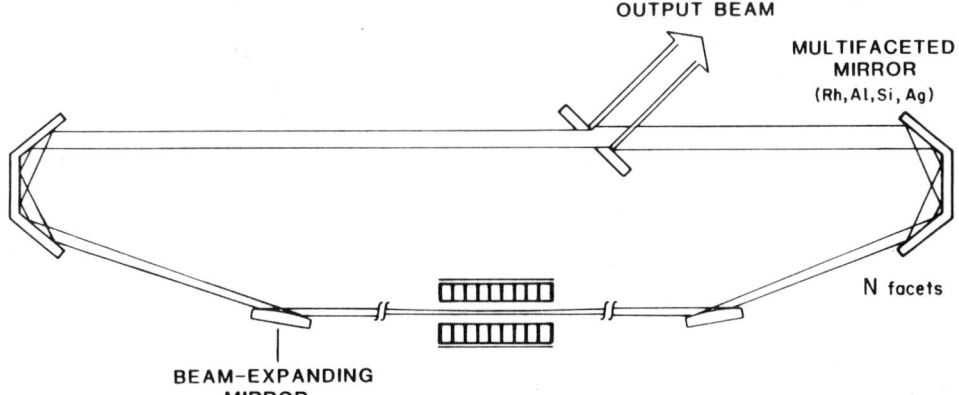

Fig.7. Multifacet, all-metal mirrors based on total external reflectance at large angles of incidence (~80°) will provide the necessary ≥40% retroreflection for FEL ring oscillators over broad spectral ranges in the XUV. Intracavity, grazing incidence, hyperboloidal mirrors diverge the beam to reduce beam-induced thermal distortion on the multifacet mirrors and allow shorter resonator lengths. The off-axis paraboloidal figure of the upper facet of the multifacet mirrors collimates the reflected beam. After Newnam.[59]

Los Alamos researchers have experimented primarily with aluminum thin films deposited with an electron gun on polished silicon substrates in an ultra-high vacuum (UHV) of ~10^{-10} Torr to minimize surface oxidation. At 58.4 nm using a He-discharge source and 80° incidence, Scott[63] measured the *in situ* reflectance of single, Al-coated mirrors to be 98.7 ± 2%. Under similar conditions, the net retroreflectance of a nine-facet Al-coated retroreflector, was 89 ± 3%, in agreement with the single-facet measurement. This value of retroreflectance is more than a factor-of-two higher than reported for any other XUV mirror in the vicinity of 60 nm. At 30.4 nm, additional UHV single-mirror measurements indicated that this same nine-mirror array should have a retroreflectance value of ~33%.

Contamination by oxide and carbon epifilms can severely increase the absorption of metal films and thereby prevent TER. Using an ellipsometer, Scott et al.[64-66] measured the growth rates of oxide epifilms *in situ* on Al and Si films in different vacuum environments. Figure 8 shows that a monolayer of oxide grows on Al in about one hour in a constant 10^{-8} Torr partial-pressure environment of oxygen (or water vapor, not shown). At five times higher pressure (2 x 10^{-7} Torr), an oxide monolayer on Si takes more than an hour to form (Fig. 9). Fortunately, at sufficient vacuum levels with low oxygen and water partial pressures, the oxidation of aluminum films proceeds at a much slower rate. For example, a two-week exposure of a fresh aluminum film in a 2 x 10^{-9} Torr vacuum, primarily with residual He, resulted in formation of only 1/4 of an oxide monolayer (see Fig. 10).[64] The net result on the retroreflectance is shown in Fig. 11. Measurements after four weeks indicated negligible additional growth of the oxide layer.[67]

The growth of carbon contaminant films on optical surfaces irradiated by intense x-rays and effective cleaning techniques for non-oxidizing materials are well documented.[68-70] However, carbon-film growth caused by **narrow-bandwidth, low-photon-energy** radiation has not yet been adequately examined. For the wavelength range ≥30 nm where Al and Si are effective retrorereflectors, it is expected that that the rate of contamination will be much less severe than experienced with broad-band, higher-energy synchrotron radiation photons.

It is possible that an initial clearing of the UHV system of carbonaceous compounds by using the rf-discharge technique developed by Johnson et al.,[68,69] (followed by *in situ* deposition of the reflecting films) will yield adequately long lifetimes with high reflectance. Another option is to periodically overcoat the contaminated metallic films with fresh material to offset the effects of gradual deterioration that may occur while in the FEL resonator. Of course, the total thickness of the films must not become too great lest surface roughness increase. An ion gun, mounted in the vacuum chamber and used with specific beam parameters, may be an effective way to periodically sputter away aged films prior to evaporation of a fresh film.

<u>Undulator beam tube</u>: During transit through the narrow beam tubes within small-gap undulators, it is possible that inadequate vacuum, due to poor pumping conductance, might cause beam emittance growth

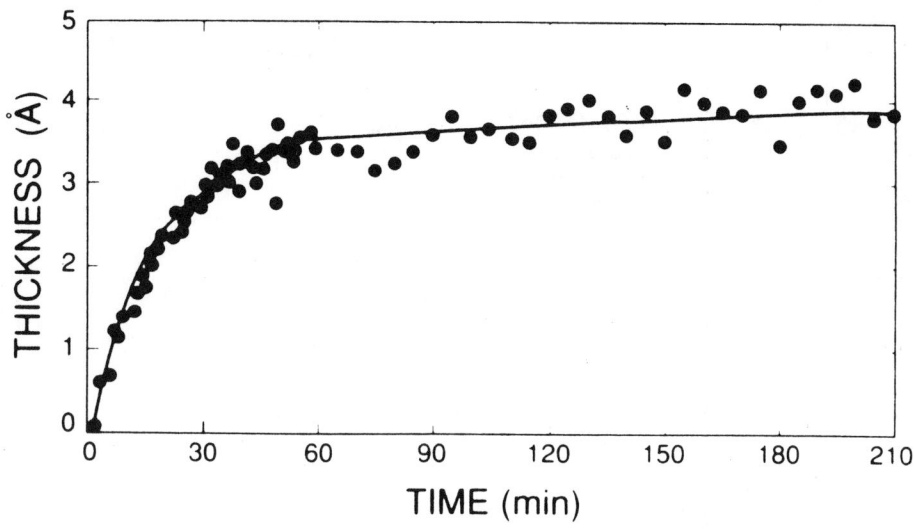

Fig.8. Ellipsometer measurement of the formation of a single monolayer of surface oxide on a fresh aluminum film exposed to 2 x 10^{-8} Torr of oxygen. After Scott et al.[64]

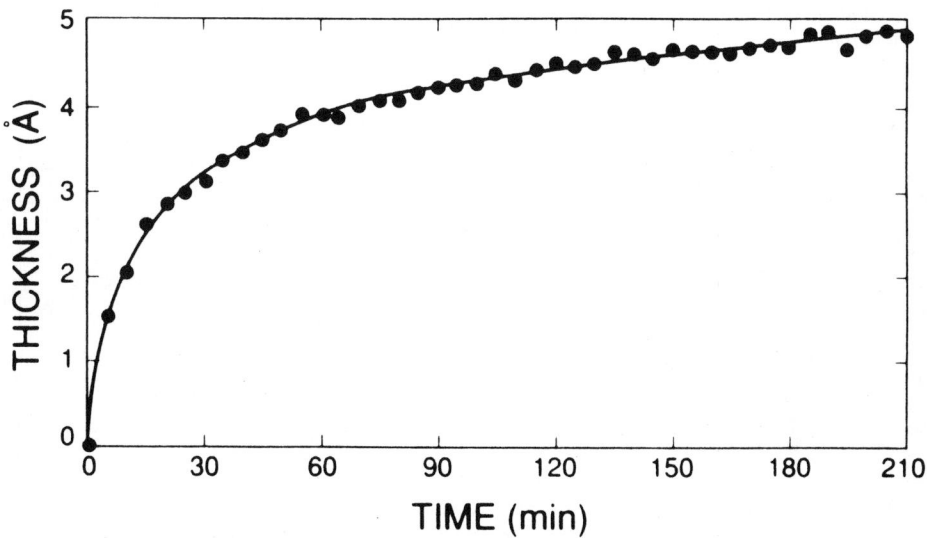

Fig.9. Ellipsometer measurement of the formation of a single monolayer of surface oxide on a fresh silicon film exposed to 10^{-7} Torr of oxygen. After Scott et al.[64]

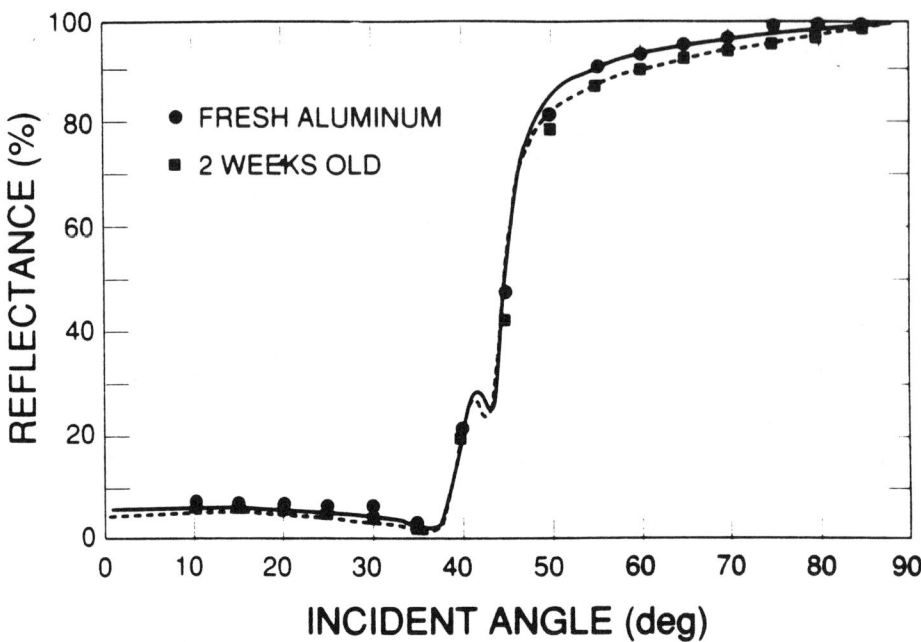

Fig.10. Reflectance vs angle of incidence for a fresh aluminum film overcoating a previously deposited and oxidized Al film (solid line). This same film was measured again after two weeks in a UHV system at a helium pressure of ≥2 x 10^{-9} Torr (dotted line). The interference effect between 35° and 45° is due to subsurface reflections. After Scott et al.[64]

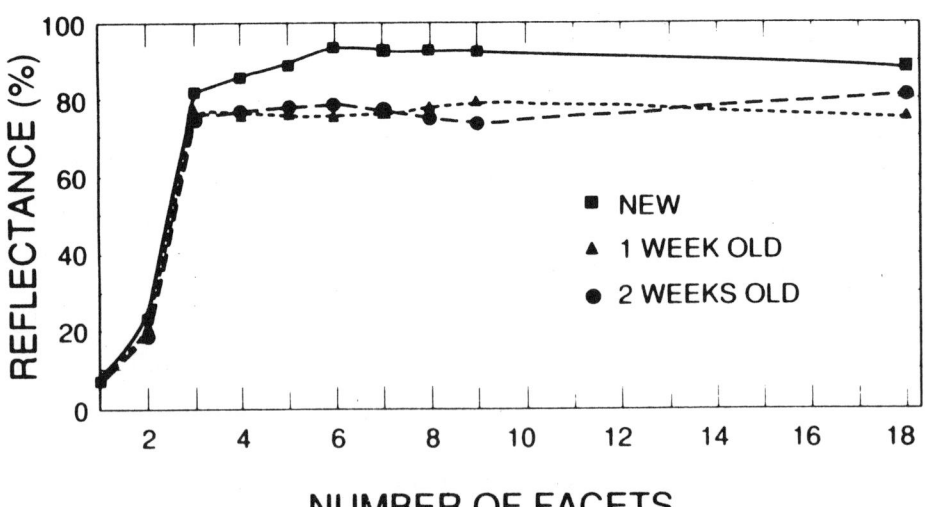

Fig.11. Net retroreflectance vs number of facets in a multifacet mirror calculated from the data of Fig. 10. No change is seen between one and two weeks. After Scott et al.[64]

via ion trapping. This could lead directly to restricted FEL gain in both single- and especially multiple passes of the same electron beam through the undulator, as in a storage-ring FEL configuration. Generally, small-gap undulators are not used in storage rings to avoid beam-scraping losses that restrict the beam lifetime.

Evidently, no problem is experienced with large-bore, short undulators such as the 1.3-m undulator used in the University of Paris, ACO-storage-ring FEL. Likewise, with magnet gaps of 4 to 9 mm and undulator lengths of 1 to 5 m, no beam deterioration problems have yet been experienced by developers of single-pass, rf-linac-driven FELs. For example, no gross emittance growth was observed during a Los Alamos test using 10^{-3}-T vacuum in a short, 1-m undulator.[71] However, at grossly higher pressures ≥ 1 Torr of hydrogen gas, beam degradation was observed in the Stanford University rf-linac FEL.[72] Even in this special case, where the gas was intentionally used to tune the FEL to short wavelengths, addition of a small amount of electron-attaching gas eliminated the beam scattering of the 4-ps pulses. Alternately, adjustment of the beam focal position compensated for the gas.[73] One may conclude that moderately poor vacuum conditions in a small-gap undulator will only be a problem for FELs in recirculating accelerators such as storage rings. The possible impact on an electron beam during its single-transit through a multiple-FEL system driven by an rf-linac, such as the proposed Los Alamos facility, is not likely to be significant.

SUMMARY

Free-electron lasers represent the next generation of coherent-radiation sources with peak- and average-power output capabilities that should surpass those of any existing, continuously tunable photon source by many orders-of-magnitude. Recent advances of FEL component technologies (electron linac, resonator mirrors, and magnetic undulator) should enable these devices to operate in the XUV, from <4 nm to 100 nm, as well as at more easily achieved longer wavelengths. Important developments include a high-brightness photo-injector, multifaceted resonator mirrors with retroreflectance >40%, and high-precision undulators with >200 periods. An rf-linac-based, multiple-FEL user facility, spanning wavelengths from 1 nm to 100 μm, has been designed and proposed by Los Alamos National Laboratory for scientific research and those industrial applications requiring expanded photon-parameter ranges. To enable such a facility to operate without significant degradation over long periods, contamination of the particular FEL components listed above must be prevented or controlled. Provision for UHV environments plus glow-discharge cleaning treatments should greatly mitigate contamination problems in future XUV FELs.

ACKNOWLEDGMENTS

Discussions with members of the Los Alamos Free-Electron Laser team were invaluable in preparing this manuscript. Particular thanks go to Scott Volz and Richard Sheffield regarding photocathode life-measurements, Donald Feldman and Roger Warren for characterization of FEL operation, Marion Scott on reflector contamination, and John Goldstein on XUV-FEL numerical simulations.

REFERENCES

1. I. B. Drobyazko, G. N. Kulipanov, V. N. Litvinenko, I. V. Pinayev, V. M. Popik, I. G. Silvestrov, A. N. Skrinsky, A. S. Sokolov, and N. A. Vinokurov, in *Free-Electron Lasers II*, Y. Petroff, Ed., SPIE Vol. 1133, pp. 2-10, (1989).
2. G. N. Kulipanov, V. N. Litvinenko, I. V. Pinayev, V. M. Popik, A. N. Skrinsky, A. Sokolov, and N. A. Vinokurov, Nucl. Instr. and Methods in Phys. Res. $\underline{A296}$, 1-3 (1990).
3. R. Prazeres, J. M. Ortega, C. Bazin, M. Bergher, M. Billardon, M. E. Couprie, M. Velghe, and Y. Petroff, Nucl. Instr. and Methods in Phys. Res. $\underline{A272}$, 68-72 (1988).
4. R. Prazeres, P. Guyot-Sionnest, D. Jaroszynski, J. M. Ortega, M. Billardon, M. E. Couprie, and M. Velghe, Nucl. Instr. and Methods in Phys. Res., to be publ., 1991.
5. J. M. J. Madey, in *Free Electron Generation of Extreme Ultraviolet Coherent Radiation*, J. Madey and C. Pellegrini, Eds., AIP Conf. Proc. No. 118, (Amer. Inst. Phys., NY, 1984), pp. 12-43.
6. J. C. Goldstein, in Proc. of the *ICFA Workshop on Low Emittance e^- - e^+ Beams*, J. B. Murphy and C. Pellegrini, Eds., (Brookhaven Natl. Lab., Upton, NY) BNL Rept. 52090, pp. 180-196, (1987).
7. H. Wiedemann, in *Int'l. Conf. on Insertion Devices for Synchrotron Sources*, R. Tatchyn, and I. Lindau, eds., Proc. SPIE Vol. 582, pp. 110-117, (1986).
8. J. E. La Sala, D. A. G. Deacon, and J. M. J. Madey, Nucl. Instr. and Methods in Phys. Res. $\underline{A250}$, 262-273 (1986).
9. J. E. La Sala, D. A. G. Deacon, and J. M. J. Madey, in *Int'l. Conf. on Insertion Devices for Synchrotron Sources*, op. cit., pp. 156-162, (1986).
10. J. M. J. Madey, S. V. Benson, and B. Burnham, Nucl. Instr. and Methods in Phys. Res., to be publ., 1991.
11. S. V. Benson, J. M. J. Madey, and W. Ying, Nucl. Instr. and Methods in Phys. Res., to be publ., 1991.
12. N. Marquardt, Proc. of *1989 IEEE Particle Accelerator Conf.*, F. Bennett and J. Kopta, Eds., IEEE Cat. No. 89CH2669-0 (IEEE Serv. Center, Piscataway, NJ, 1989), Vol.2, p. 780-782.
13. D. Nolle, F. Brinker, M. Negrazus, D. Schirmer, and K. Wille, Nucl. Instr. and Methods in Phys. Res. $\underline{A296}$, 263-269 (1990).
14. B. E. Newnam, J. C. Goldstein, J. S. Fraser, and R. K. Cooper, in *Free-Electron Generation of Extreme Ultraviolet Coherent Radiation*, op. cit., pp. 190-202.
15. J. C. Goldstein, B. E. Newnam, R. K. Cooper, and J. C. Comly, Jr., in *Laser Techniques in the Extreme Ultraviolet*, S. E. Harris and T. B. Lucatorto, Eds., AIP Conf. Proc. No. 119 (Amer. Inst. Phys., NY, 1984), pp. 293-303.
16. J. C. Goldstein, B. D. McVey, B. E. Newnam, in *Short Wavelength Coherent Radiation: Generation and Applications*, D. T. Attwood and J. Bokor, Eds., AIP Conf. Proc. No. 147, (Amer. Inst. Phys., NY, 1986), pp. 275-290.

17. J. C. Goldstein, B. D. McVey, and B. E. Newnam, in *Int'l. Conf. on Insertion Devices for Synchrotron Sources*, op. cit., pp. 350-360, (1986).
18. J. C. Goldstein and B. D. McVey, Nucl. Instr. and Methods in Phys. Res. A259, 203-209 (1987).
19. J. C. Goldstein, B. D. McVey and C. J. Elliott, Nucl. Instr. and Methods in Phys. A272, 177-182 (1988).
20. B. E. Carlsten and K. C. D. Chan, Nucl. Instr. and Methods in Phys. Res. A272, 208 (1988).
21. B. E. Carlsten, Nucl. Instr. and Methods in Phys. Res. A285, 313-319 (1989).
22. B. E. Newnam, Nucl. Instr. and Methods in Phys. Res. B40/41, 1053-1057 (1989).
23. B. E. Newnam, *1988 Linear Accelerator Conf. Proc.*, C. Leemann, Ed., CEBAF-Rept. 89-001, pp. 290-294, June, 1989.
24. J. C. Goldstein, B. D. McVey, and B. E. Newnam, Nucl. Instr. and Methods in Phys. Res. A296, 288-291 (1990).
25. J. C. Goldstein, R. W. Warren, and B. E. Newnam, Nucl. Instr. and Methods in Phys. Res., to be publ., 1991.
26. I. Ben-Zvi, L. H. Yu, S. Krinsky, and M. White, Nucl. Instr. and Methods in Phys. Res., to be publ., 1991.
27. D. W. Feldman, R. W. Warren, B. E. Carlsten, W. E. Stein, A. H. Lumpkin, S. C. Bender, G. Spalek, J. M. Watson, L. M. Young, J. S. Fraser, J. C. Goldstein, H. Takeda, T. F. Wang, R. B. Feldman, R. K. Cooper, W. J. D. Johnson, and C. A. Brau, IEEE J. Quantum Electron. QE-23, 1476-1488 (1987).
28. R. W. Warren, J. E. Sollid, D. W. Feldman, W. E. Stein, W. J. Johnson, A. H. Lumpkin, and J. C. Goldstein, Nucl. Instr. and Meth. in Phys. Res. A285, 1-10 (1989).
29. D. W. Feldman, in *Free-Electron Lasers II*, op. cit., pp. 36-53.
30. D. W. Feldman, S. C. Bender, B. E. Carlsten, J. Early, R. B. Feldman, W. J. D. Johnson, A. H. Lumpkin, P. G. O'Shea, W. E. Stein, R. L. Sheffield, and K. McKenna, Nucl. Instr. and Meth. in Phys. Res., to be publ., 1991.
31. S. V. Benson, J. Schultz, B. A. Hooper, R. Crane, and J. M. J. Madey, Nucl. Instr. and Meth. in Phys. Res. A272, 22-28 (1988).
32. S. V. Benson, W. S. Fann, B. A. Hooper, J. M. J. Madey, E. B. Szarmes, B. Richman, and L. Vintro, Nucl. Instr. and Meth. in Phys. Res. A296, 110-114 (1990).
33. R. L. Tokar, L. M. Young, A. H. Lumpkin, B. D. McVey, L. E. Thode, S. C. Bender, K. C. D. Chan, A. D. Yeremian, D. H. Dowell, A. R. Lowrey, and D. C. Quimby, Nucl. Instr. and Meth. in Phys. Res. A296, 115-126 (1990).
34. D. H. Dowell, M. L. Laucks, A. R. Lowrey, M. Bemes, A. Currie, P. Johnson, K. McCrary, J. Adamski, D. Pistoresi, D. R. Shoffstall, M. Bentz, R. Burns, R. Hudyma, K. Sun, W. Mower, R. Tokar, A. H. Lumpkin, S. Bender, B. McVey, J. Goldstein, and D. Shemwell, Nucl. Instr. and Meth. in Phys. Res., to be publ., 1991.
35. J. B. Murphy and C. Pelligrini, J. Opt. Soc. Am.-B 2, 259-264 (1985).
36. T. F. Wang, J. C. Goldstein, B. E. Newnam, and B. D. McVey, Intl. J. Electronics 65, 589-595 (1988).

37. S. D. Conradson and B. E. Newnam, in *Free-Electron Lasers and Applications*, D. Prosnitz, Ed., Proc. SPIE Vol. 1227, 134-144 (1990).
38. *Free-Electron Laser Applications in the Ultraviolet*, OSA Tech. Digest Series, Vol. 4, D. A. G. Deacon and B. E. Newnam, Eds., (Optical Soc. Am., Washington, D.C.), March 2-5, 1988.
39. B. D. McVey, Nucl. Instr. and Methods in Phys. Res. A250, 449 (1986).
40. B. D. McVey, J. C. Goldstein, R. D. McFarland, and B. E. Newnam, in *Laser Induced Damage in Optical Materials: 1990*, H. E. Bennett, A. H. Guenther, L. L. Chase, B. E. Newnam, and M. J. Soileau, Eds., (Natl. Inst. Stand. and Tech. U.S) Spec. Publ., SPIE Vol. 1441, to be publ., 1991.
41. *Report of the ALS/SSRL Users Workshop, May 9-11, 1983*, A. I. Bienenstock, T. Elioff, and E. E. Haller, co-chairmen, Lawrence Berkeley Laboratory Pub-5095, 1983.
42. *An ALS Handbook*, Lawrence Berkeley Laboratory PUB-643 Rev.2, April, 1989.
43. J. S. Fraser and R. L. Sheffield, IEEE J. Quantum Electron. QE-23, 1489-1496 (1987).
44. R. L. Sheffield, E. R. Gray, and J. S. Fraser, Nucl. Instr. and Methods in Phys. Res. A272, 222-226 (1988).
45. R. L. Sheffield, in *Proc. of 1989 IEEE Particle Accelerator Conf.*, op. cit., pp. 1098-1102 (1989).
46. R. L. Sheffield, W. D. Cornelius, D. C. Nguyen, R. W. Springer, B. C. Lamartine, E. R. Gray, J. M. Watson, and J. S. Fraser, in *1988 Linear Accelerator Conf. Proc.*, op. cit. pp. 520-522.
47. K. Batchelor, I. Ben-Zvi, R. Fernow, J. Gallardo, H. Kirk, C. Pelligrini, and A. van Steenbergen, Nucl. Instr. and Methods in Phys. Res. A296, 239-243 (1990).
48. M. Curtin, G. Bennett, R. Burke, A. Bhowmik, P. Metty, S. Benson, and J. M. J. Madey, Nucl. Instr. and Methods in Phys. Res. A296, 127-133 (1990).
49. S. K. Volz, Los Alamos Natl. Laboratory, priv. commun., 1990.
50. T. Kobari and H. J. Halama, J. Vac. Soc. Tech. A5, 2355 (1987).
51. A. G. Mathewson, Synchrotron Radiation News 3, 13-17 (1990).
52. H. C. Hseuh, T. S. Chou, and C. A. Christianson, J. Vac. Sci. Technol. A3, 518-522 (1985).
53. P. G. O'Shea, Los Alamos Natl. Lab., priv. commun., 1990.
54. D. W. Feldman, W. D. Cornelius, S. C. Bender, B. E. Carlsten, P. G. O'Shea, and R. L. Sheffield, in *Free Electron Lasers and Applications*, op. cit., pp. 2-13.
55. W. J. Choyke, and E. D. Palik, in *Handbook of Optical Constants*, E. D. Palik, Ed., (Academic Press, NY, 1985), pp. 587-596.
56. R. A. Roberts and W. C. Walker, Phys. Rev. 161, 730-735 (1967).
57. D. Y. Smith, E. Shiles and M. Inokuti, in *Handbook of Optical Constants of Solids*, op. cit., pp. 369-406.
58. D. Roussel-Dupre, Los Alamos Natl. Laboratory, verbal report on measurements of Mo/Si mirrors supplied by Ovonics, Inc., 1989.
59. B. E. Newnam, in *Laser Induced Damage in Optical Materials: 1985*, H. E. Bennett, A. H. Guenther, D. Milam, and B. E. Newnam, Eds., (Natl. Bur. Stand., Wash., DC), NBS Spec. Publ. 746, 261-269 (1988).

60. B. E. Newnam, U.S. Patent No. 4,917,447 issued April 17, 1990.
61. A. V. Vinogradov, I. V. Kozhevnikov, and A. V. Popov, Opt. Commun. 47, 361-363 (1983).
62. T.-Y. Hung and P. L. Hagelstein, Paper TuB4 presented at the *1990 Annual Mtng. of the Opt. Soc. America,* Nov. 4-9, 1990, Boston.
63. M. L. Scott, in OSA Proc. on *Short Wavelength Coherent Radiation: Generation and Applications,* R. Falcone and J. Kirz, Eds., (Opt. Soc. Am., Washington, DC, 1988), Vol. 2, pp. 322-324.
64. M. L. Scott, P. N. Arendt, B. J. Cameron, J. M. Saber and B. E. Newnam, Appl. Opt. 27, 1503-1507 (1988).
65. M. L. Scott, P. N. Arendt, B. Cameron, R. Cordi, B. Newnam, D. Windt, and W. Cash, in *X-Ray Imaging II (1986),* L. V. Knight, and D. K. Bowen, Eds., Proc. SPIE Vol. 691, pp. 20-27, (1986).
66. M. L. Scott, P. N. Arendt, B. J. Cameron, and B. E. Newnam, in *Soft X-Ray Optics and Technology,* E.-E. Koch, and G. Schmahl, Eds., Proc. SPIE Vol. 733, pp. 156-162, (1987).
67. M. L. Scott, Los Alamos Natl. Lab., priv. commun., 1988.
68. E. D. Johnson, S. L. Hulbert, R. F. Garrett, G. P. Williams, and M. L. Knotek, Rev. Sci. Instrum. 58, 1042-1045 (1987).
69. E. D. Johnson and R. F. Garrett, Nucl. Instr. and Methods in Phys. Res. A266, 381-385 (1988).
70. V. Rehn, in *Topical Conf. on Vacuum Design of Synchrotron Light Sources,* S. Bader, Ed., AIP Conf. Proc. (Amer. Inst. Phys., NY), to be publ. in 1991.
71. D. W. Feldman, Los Alamos Natl. Lab., priv. commun., 1990.
72. A. S. Fisher, R. H. Pantell, J. Feinstein, T. L. Deloney, M. B. Reid, and W. M. Grossman, Nucl. Instr. and Methods in Phys. Res. A250, 337-341 (1986).
73. A. S. Fisher, R. H. Pantell, M. B. Reid, J. Feinstein, A. H. Ho, M. Ozcan, and H. D. Dulman, Nucl. Instr. and Methods in Phys. Res. A272, 89-91 (1988).

REVIEW OF VACUUM SYSTEMS FOR X-RAY LITHOGRAPHY LIGHT SOURCES

J. C. Schuchman
Brookhaven National Laboratory, NSLS - Bldg. 725C
Upton, N.Y. 11973

ABSTRACT

This paper will review and give a status report on vacuum systems for X-Ray lithography light sources. It will include conventional machines and compact machines (machines using superconducting magnets). The vacuum systems will be described and compared with regard to basic machine parameters, pumping systems, types of pumps, chamber design and material, gauging and diagnostics, and machine performance.

INTRODUCTION

X-Ray Lithography makes use of soft X-Rays from synchrotron radiation to shadow print an image of a mask on a semiconductor wafer to produce integrated circuits. Electron storage rings are used as the X-Ray source. This paper will focus on the vacuum systems for these electron storage rings. Figure 1 shows schematically an electron storage ring, a photon beam line, a mask, and finally the wafer which is used to produce the high density chips.

Electron storage rings, as sources for X-Ray lithography, are generally small in size. Those with room temperature magnets have circumferences in the neighborhood of 30-50 meters, while those incorporating super-conducting magnets are in the 8-10 meter range. The latter machines are called compact light sources because of their small size. Ultra-high vacuum beam lines that are contiguous with the storage ring, and an electron injector make up the basic X-ray lithography facility.

Mistry[1] and Wiedemann[2] are two comprehensive references that apply to designing synchrotron light sources. X-ray lithography light sources (XLS) must satisfy the requirements mentioned in these references as well as beam power, wavelength, emittance lifetime and other specific parameters.

From a vacuum engineering standpoint, providing sufficient pumping speed in the dipole region and reducing gas desorption are paramount. Also, since an XLS will be operated as a production tool, automatic operation and high reliability[3] must be considered throughout the design. Figure 2 is an artist's rendering of the compact light source "HELIOS". It is typical of a racetrack style storage ring. It shows an electron beam injected from a linac, two

*Work performed under the auspices of U.S. DOE under contract DE-AC02-76Ch00016 and funded by DOD.

superconducting dipole magnets connected with room temperature components, and a number of lithography ports emanating from the dipoles.

Table I is a list of new light sources specifically designed for x-ray lithography,[4,5]. Table 1A shows conventional (room temperature) designs and Table IB compact designs. They are all-metal ultra-high vacuum systems; however, the approach taken to pumping differs for some of the machines. Mistry[1], Wiedemann[2], and Mathewson[6] discuss basic guidelines for all storage ring vacuum systems; for example, operating pressures in 10^{-10} Torr without beam and 10^{-9} Torr with beam, high gas loads due to photon stimulated desorption (PSD), "smooth" beam pipe to minimize instabilities, cleaning and conditioning.

CHAMBER MATERIAL

Stainless steel is used for the vacuum chambers of all the machines listed in Table I. For the compact rings, the advantages of aluminum extrusions, (such as-low cost, ease of machining, etc.) are lost when the density of component per unit length of ring is considered. There are beam monitors, striplines, clearing electrodes, injection chambers, flags, pump ports, beam ports all of which must fit into a minimum space. The result is much machining and welding, many ports and feedthrus, in essence a design where a material such as stainless steel is a natural choice. The larger conventional rings could be made from either stainless steel or aluminum depending on the overall design and economics.

When stainless steel is used within a changing magnetic field, i.e., dipole magnet during ramping, the alloy chosen must not undergo a phase change from austenitic to ferromagnetic martensitic due to forming, machining, or welding which could perturb the magnetic field. This naturally also applies to the superconducting coil support structure.

WARM VS COLD BORE

The use of superconducting magnets does not necessarily mean a cold bore (LHe temperature) chamber. Of the compact rings, COSY, HELIOS, NIJI-III, SIBERIA-SM and SIBERIA-AS have cold bore vacuum chambers, while AURORA, SUPER ALIS, and SXLS have room temperature chambers. Some advantages and disadvantages of a cold bore vacuum chamber are listed in Table I, Ref.[22]. Probably the most serious disadvantage is the effects of leaks during fabrication and during operation. Also, not too much is known about PSD from cold surfaces (see H. Jostlein et. al.[7]).

PUMPS

There is no one best pump for a synchrotron light source. A combination of pumps are generally used for optimum pumping.

Certainly, some form of cryopumping within a superconducting magnet is very appealing. This design would provide maximum pumping speed where it is needed. However, each magnet warm-up would re-emit previously cryopumped gases and re-contaminate conditioned surfaces. Experience gained on HELIOS will shed light on this approach.

NEG pumps have proven to be reliable and their use is continually expanding. However, for each use there are issues which must be decided. For example, which grade of NEG to use, the geometry and orientation of the NEG pump in the chamber, the activation and regeneration temperature and procedure, etc. SUPER-ALIS incorporates a NEG strip (St-707) as the main chamber pump. AURORA and SXLS use some standard NEG modules for supplemental pumping.

Titanium sublimation pumps continue to be used for reliable high speed pumping in ultra-high vacuum. However, the supply of titanium must eventually be replaced requiring venting and possible re-conditioning of the chamber surfaces. An advantage of the NEG pump is that a regeneration causes the sorbed gases to diffuse (except H_2) into the NEG material, however, the regeneration temperature does cause an increase in local thermal desorption of the heated surfaces.

HIGHLIGHTS OF XLS VACUUM SYSTEMS

HELIOS

HELIOS is a racetrack design XLS with superconducting dipole magnets, Figure 2. A schematic cross section of the vacuum chamber is shown in Figure 3. It is a cold bore machine with a continuous open slot for the photon beam to leave the magnet. A copper water cooled absorber is located on the outer vacuum vessel. Discrete sputter ion pumps supplement the cryogenic pumping. The cold surfaces provide an effective pumping speed of 2×10^5 ℓ/sec.

After assembly the vacuum system was conditioned by bakeout and glow discharge. As of mid July, after a short commissioning time, HELIOS achieved 50 mA of beam current at full energy of 700 MeV, with a lifetime in excess of one hour. At this point in the machine conditioning, the beam lifetime is determined almost entirely by the residual gas pressure [8].

AURORA

AURORA is a completely circular super conducting machine that incorporates a cryosorption pump within a warm vacuum chamber (see Figure 4). A NEG pump supplements the 40,000 ℓ/sec of cryosorption pumping speed. The system can achieve 3×10^{-10} Torr in one day and reaches 6×10^{-10} Torr without a bakeout [9]. Beam was successfully injected, ramped and stored at full energy early in 1990. This XLS is now offered for sale commercially.

SORTEC

The 1 GeV-200 mA Synchrotron Radiation Facility with critical wavelength of 15 Å is very well constructed and is ready to be used primarily for X-Ray Lithography and the development of necessary components, i.e., steppers, masks, cameras, photo resist etc. The vacuum chamber is electro-polished stainless steel. The limit on beam lifetime is imposed by Coulomb scattering and trapped ions. Clearing electrodes, 10 cm diameter discs are installed in straight sections and strips are mounted on the side walls in bending magnets [10].

Each chamber was conditioned by pre-baking at 250° C for forty-eight hours prior to assembly, followed by an in-situ bakeout at 150°C for forty-eight hours.

Sputter ion pumps of 300 ℓ/sec and 400 ℓ/sec, distributed ion pumps and 1000 ℓ/sec titanium sublimation pumps are used, giving a total pumping speed of 26,000 ℓ/sec. The pressure is 2×10^{-9} Torr with 200 mA of stored beam and less than 10^{-10} Torr without beam.

Figure 5 is a plot of beam current, lifetime and pressure versus time for the SORTEC ring, recorded one month after start of commissioning [11]. It not only shows that the design goals were met (lifetime of 4 hours at 200 ma), but also the relationship between stored beam pressure and lifetime for a conditioned machine. Typically, when beam is first injected, the pressure rises dramatically, increasing by a few decades, and then decreases with beam conditioning.

LUNA

LUNA is a room temperature 800 MeV conventional storage ring using four room-temperature 90° dipole magnets. The chambers are stainless steel. 400 ℓ/sec sputter ion pumps (SIP) are installed on each beam line take-off port. The design provides for future installation of non-evaporable getter (NEG) pumps and titanium sublimation pumps (TSP) at these locations. Pressure without beam is 4×10^{-10} Torr and 1×10^{-8} Torr with beam. The current lifetime of 10 mA for fifteen minutes is imposed by Coulomb scattering and ion trapping. One-meter-long stainless steel clearing electrodes are installed in the straight sections[10].

NAR

Figure 6 shows both the NAR and the SUPER-ALIS rings at NTT. Titanium sublimation pumps, distributed sputter ion pumps, and appendage sputter ion pumps are used. They provide an installed pumping speed of 2400 ℓ/sec [12]. The pumps are placed at high gas desorption locations.

SUPER-ALIS

SUPER-ALIS is a racetrack designed superconducting storage

ring with a warm chamber [13]. NEG strip pumps are installed on the inner side of the dipole chamber. Kobari, et.al. [14] describes the NEG performance under synchrotron radiation. High purity silicon beam absorbers were installed to reduce the photodesorption yield[15].

SXLS

The SXLS is a racetrack shaped storage ring using superconducting bending magnet with a room temperature vacuum chamber. Distributed ion pumps (DIP) and distributed NEG pumps will be built into each dipole chamber. Development is required for the DIP which operates in a four tesla field. Discrete SIP and TSP are installed in the straight sections and on the copper RF cavity. The surfaces in the dipole chambers will be treated to reduce PSD. Roughing and pumping during vacuum conditioning is via an oil-less combination turbo-molecular drag pump backed with a diaphragm fore pump [16].

CAMD

The vacuum chamber for the machine at LSU will be constructed of stainless steel, welded to form a smooth chamber. Lumped sputter-ion pumps will be used at radiation absorber locations. Provisions will be made for the addition of NEG pumps in the future if needed [17].

NIJI-III

The NIJI-III ring is shown in Figure 7. A cross section of the chamber is shown in Figure 7. The chamber is made from type 304L stainless steel pipe bent to the 0.5M radius by high frequency induction heating [18]. A liquid nitrogen cooled absorber (S/S type 304L) is installed inside the main dipole chamber. The absorber is finished by buffing, as are the dipole chambers. Electropolishing is used for complicated shapes found in the straight sections.

SIP and TSP, 230 ℓ/sec and 800 ℓ/sec respectively, with NEG pumps are installed in the straight sections. Two valves are used to separate the ring into halves.

COSY

COSY is a racetrack design with superconducting dipole magnets and a cold vacuum chamber. The chamber is stainless steel type 304 and the RF cavity is OFHC copper.

Pumping is via cryopumps (cold beam chamber), TSP and SIP. A four day bakeout results in a pressure of 2×10^{-10} Torr. Machine performance is limited by dipole field distortions due to ferromagnetic behavior of the stainless steel coil support structure [19].

SIBERIA-SM

The SIBERIA-SM project is described [20] as a 600 MeV machine with eight 45° superconducting dipole magnets. A plan view of the vacuum chamber-magnet assembly is shown in Figure 8. The chamber is made of stainless steel and is cooled to liquid helium temperature. A copper screen at liquid nitrogen temperature surrounds the chamber.

CONCLUSION

As more machines are built and become operational, technical problems, such as low energy injection, ion clearing, and superconducting magnet technology, will be better understood. Further work must still be done in designing light sources, which have generally been designed by R & D people, into production tools capable of automatic and highly reliable operation. Compact machines should be small and truly portable, with their design tending towards simple machines with easy access for maintenance.

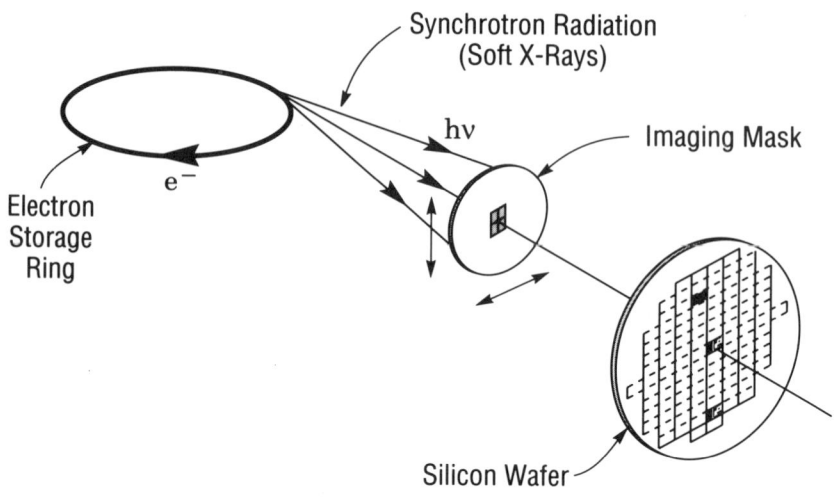

Fig. 1 X-ray lithography process-schematic, Ref. 21.

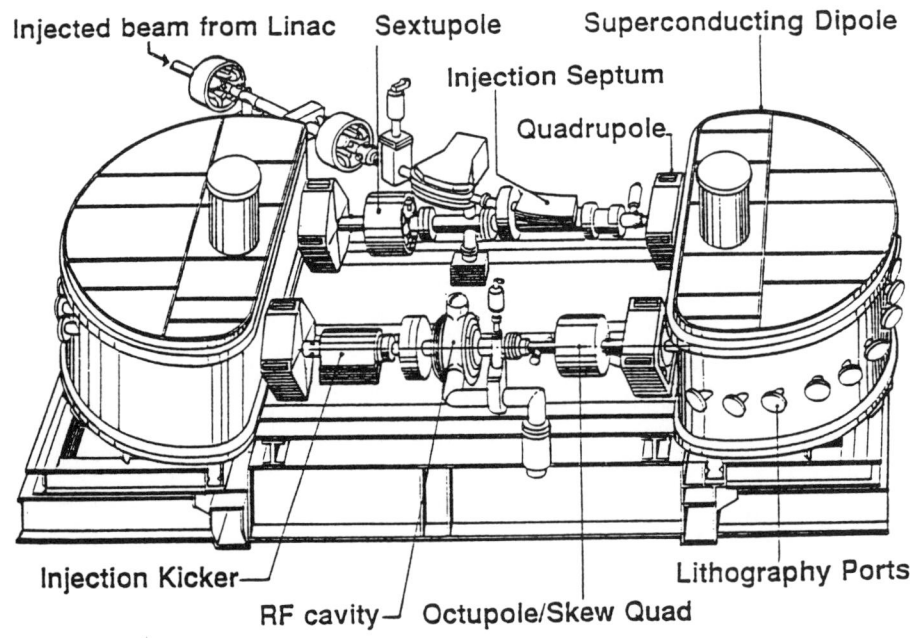

Fig. 2. Artist's rendering of HELIOS, Ref. 8.

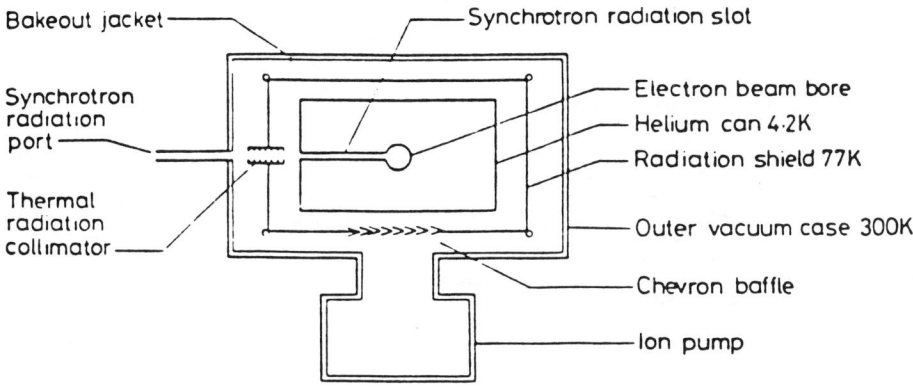

Fig. 3 Cross-section of HELIOS vacuum chamber-schematic, Ref. 23

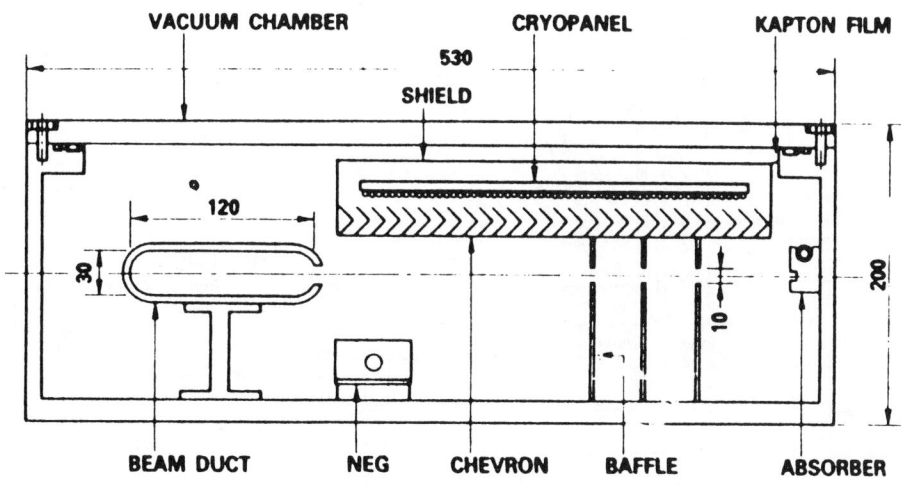

Fig. 4 Cross-section of AURORA vacuum chamber, Ref. 9

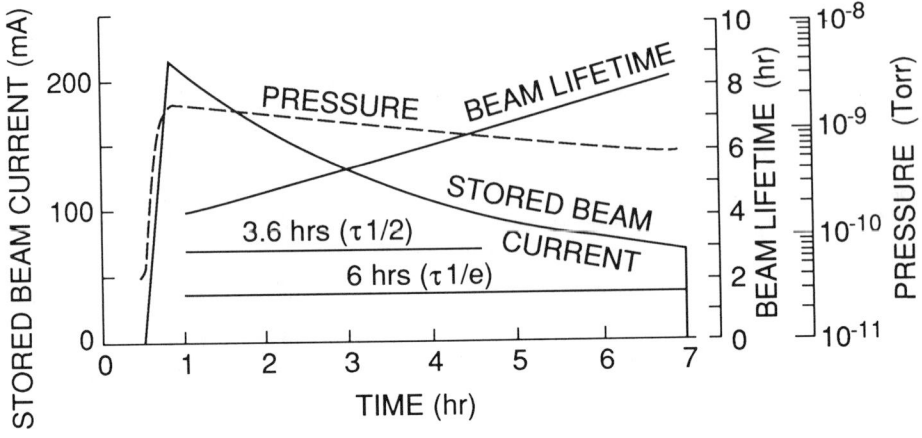

Fig. 5 Plot of beam current, lifetime and pressure versus time for SORTEC ring, Ref. 11.

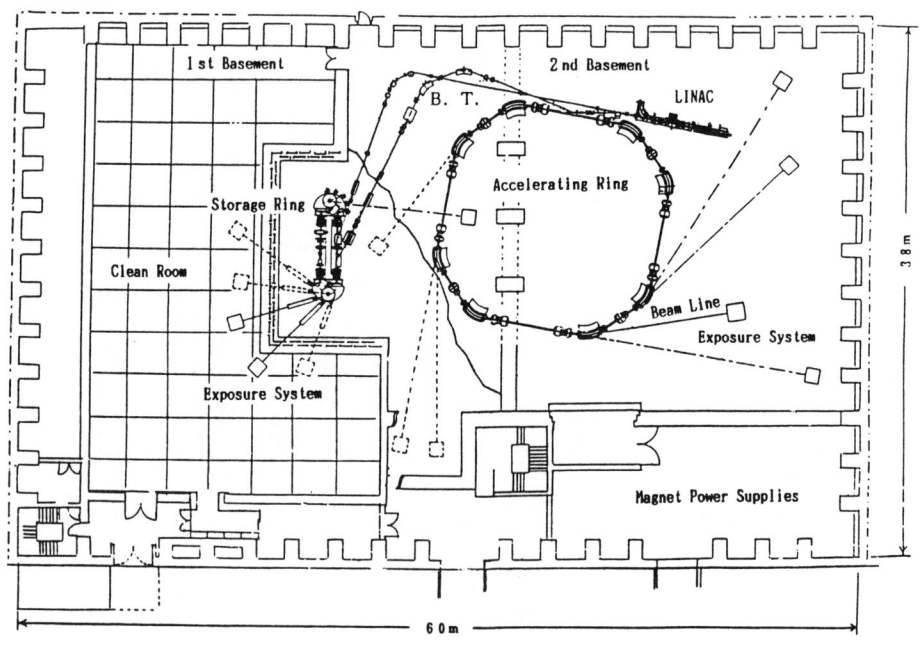

Fig. 6 Plan view of NAR and SUPER-ALIS facility, Ref. 12

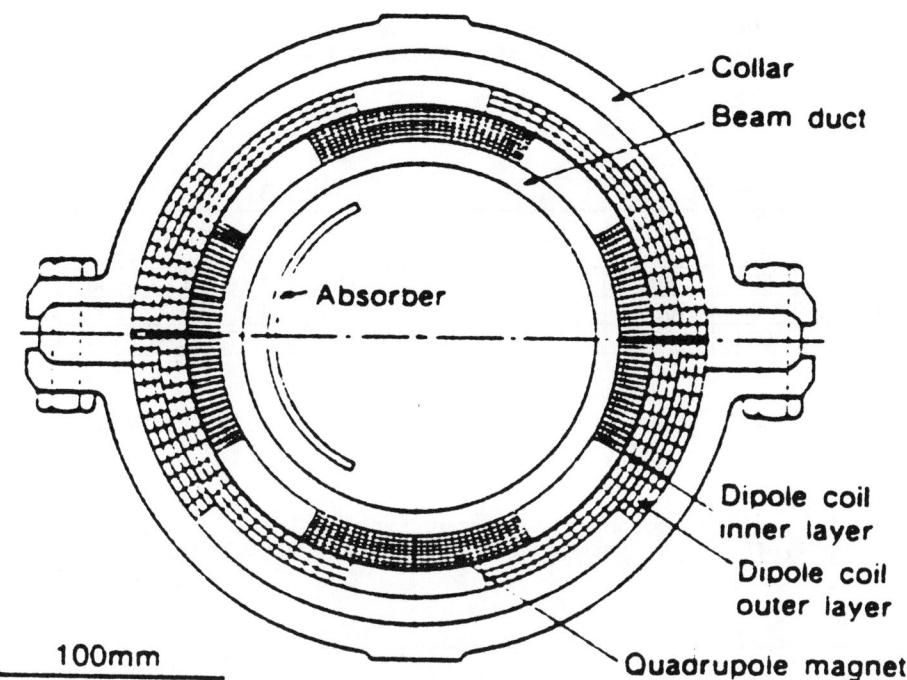

Fig. 7 Cross-section NIJI-III, vacuum chamber, Ref. 9

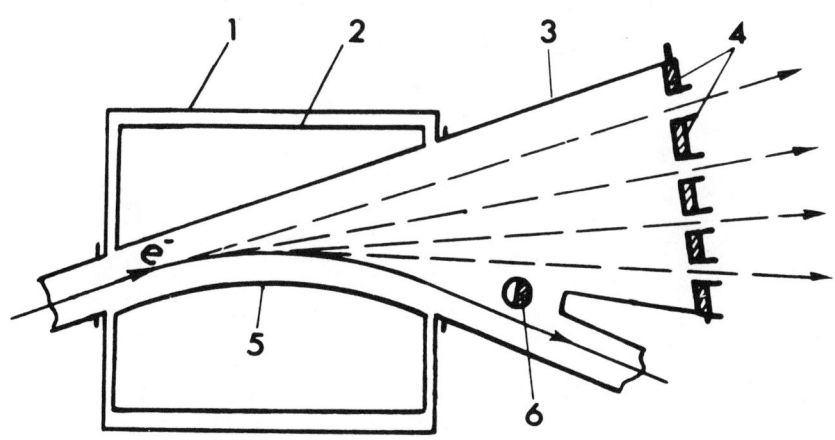

Fig. 8 Schematic of SIBERIA-SM vacuum chamber, Ref. 20.
(1) cryostat body, (2) nitrogen screen (3) SR extraction chamber
(4) radiation receiver, (5) vacuum chamber in a bending magnet,
(6) pinlike radiation receiver.

Table Ia. New conventional magnet rings being built for x-ray lithography, (L=linac, B=booster & M=racetrack microtron), Ref. 4 & 5.

Machine	C (m)	E (MeV)	λ_c (Å)	Location	E_{inj} (MeV)	B (T)	ρ (m)	I (ma)
NIJI II	17	600	37	ETL	200 L	1.4	1.4	N.A.
LUNA	23.5	800	22	IHI	45 L	1.33	2.0	50
NTT	52	800	20	Atsugi	15 L	1.46	1.83	500
SORTEC	46	1000	15.5	Tsukuba	1000 LB	1.2	2.77	200
CAMD	52.8	1200	9.5	LSU	150 L	1.37	2.93	400

Table Ib. New superconducting magnet rings being built for x-ray lithography, (L=linac, B=booster & M=racetrack microtron), Ref.4,5.

Machine	C (m)	E (MeV)	λ_c (Å)	Location	E_{inj} (MeV)	B (T)	ρ (m)	Beam Tube	Magnet Core	I (ma)
AURORA	3.14	650	10	SHI	150 M	4.34	0.5	warm	iron	300
COSY	9.6	590	12	BESSY	50 M	4.47	0.44	cold	air	100
SUPER ALIS	16.8	600	17.3	NTT	15 L, 600 B	3.0	0.66	warm	iron	500
HELIOS	9.6	700	8.5	OXFORD	<200 L	4.5	0.52	cold	air	200
SXLS	8.5	700	10	BNL	<200 L	3.85	0.6	warm	air	500
NIJI III	16	620	11.7	SEI	<200 L	4.13	0.5	cold	iron	200
*SIBERIA-SM	10	600	8.6	NPI	50-60 L	6	3.3	cold	iron/air	300
*SIBERIA-AS	8.5	600	13.6	NPI	50	3.8			iron/air	N.A.

* Proposed

REFERENCES

1. N. B. Mistry, AIP Conf. Proc. No. 171, BNL, Upton, NY, May 1988, p.1.
2. H. Wiedemann, AIP Conf. Proc. No. 171. BNL, Upton, NY, May 1988, p.10.
3. BNL 52005 UC-28, Workshop on Compact Storage Ring Technology:Applications to Lithography, BNL, Upton, NY, May 1986, p.12.
4. J. B. Murphy, Proc. of 1989 IEEE Particle Accel. Conf., Chicago Il., Mar. 1989, p.757.
5. V. V. Anashin, et.al., Compact Storage Rings Siberia-AS and Siberia-SM Synchrotron Radiation Sources for Lithography, Rev. Sci. Instrum. 60 (7), July, 1989, p.1767.
6. A. G. Mathewson, Ultra High Vacuum Technology for Synchrotron Light Sources, Synch. Rad. News, Vol.3, No.1, 1990, p.13.
7. H. Jostlein, et.al., BNL-51947 UC-13, NSLS Annual Report, Oct. 1985, p.124.
8. M.N. Wilson, et.al. Update on the Compact Synchrotron X-Ray Source HELIOS, presented at 34rd Microprocess Conf., Chiba, Japan, Jul. 1990.
9. T. Takayama, N. Yosumitu and H. Yamada, Superconducting Electron Storage Ring for a Synchrotron Light Source AURORA, SHI Ltd., Tanashi-city, Tokyo, Japan.
10. Private communication with H. Halama.
11. S. Nakamura, et.al., Status of the 1 GeV Synchrotron Radiation Source at SORTEC, presented at 3rd Microprocess Conf., Chiba, Japan, July 1990.
12. A. Shibayama, et.al., Rev. Sci. Instrum. 60 (7), July 1989, p.1779.
13. T. Hosokawa, et.al, Rev. Sci. Instrum. 60 (7), July 1989, p.1783.
14. T. Kobari, et.al., AIP Conf. Proc. No. 171, BNL, Upton, NY, 1988, p.100.
15. M. N. Wilson, Compact Synchrotron X-Ray Sources, presented at Eur. Particle Accel. Conf., Nice France, June 1990.
16. J. C. Schuchman, Vacuum System Design For a Superconducting X-Ray Lithography Light Source, presented at the 11th International Vac. Congress, Cologne FRG, Sept, 1989.
17. Private communications with V. Saile.
18. F. Miura, Y. Tsutsui and H. Takada, Proc. on Accel. Sci. and Tech., Osaka, Japan, Dec. 1989, p.131.
19. E. Weihreter, et.al, Status of the Superconducting Compact Storage Ring, COSY, Presented at Eur. Particle Accel. Conf., Nice, France, June 1990.
20. V.V. Anashin, et.al., Nuc. Instr. and Methods in Physics Res. A282 (1989), p.386.
21. BNL 52230, Report of the Fifth Workshop on Synchrotron X-Ray Lithography, BNL, Upton, NY, Nov. 1989.

22. J. C. Schuchman, Vacuum Design and Fabrication of Electron Storage Rings, J. Vac. Sci. Technol. A8 (3) May/Jun 1990. 1990.
23. D. E. Andrews, J. V. Worth, P. M. Williams, M. N. Wilson, AIP Conf. Proc. No. 171, BNL Upton, NY, May 1988, p. 135.

COMPARISON OF SYNCHROTRON RADIATION INDUCED GAS DESORPTION FROM Al, STAINLESS STEEL AND Cu CHAMBERS

A. G. Mathewson, O. Gröbner, P. Strubin, C.E.R.N., Geneva, Switzerland
P. Marin, R. Souchet, L.U.R.E., Orsay, France

ABSTRACT

The synchrotron radiation induced neutral gas desorption coefficients for Al, stainless steel and Cu have been measured by exposing test vacuum chambers made from those materials to 2.95 keV critical energy synchrotron radiation in a dedicated beam line on the DCI positron storage ring at the LURE Laboratory, Orsay, France.

In addition to the initial desorption coefficients, the decrease of the gas desorption with photon dose is also described. Al was found to have the largest initial desorption coefficients, with Cu a factor of about 10 lower and the lowest desorption coefficients were from stainless steel, where a 950°C vacuum degassing treatment had little effect.

Integration of the desorption coefficients with respect to photon dose gave the total quantity of gas desorbed. For a photon dose of $3 \; 10^{21}$ photons/m, almost 1 Torr l/m of H_2, CH_4, CO and CO_2 were desorbed from Al, 10^{-1} Torr l/m from Cu and $5 \; 10^{-2}$ Torr l/m from stainless steel.

In parallel, measurements of the photoelectron currents, believed to be the primary cause of the neutral gas desorption, were carried out. The highest photoelectron currents were measured in the Al chamber, followed by Cu with the lowest in the stainless steel chambers.

Using the results from the Al test chamber, the predicted pressure increases in the LEP vacuum system, when exposed to synchrotron radiation, agreed well with those actually measured.

INTRODUCTION

In synchrotron radiation light sources and high energy electron or positron storage rings the vacuum chambers are usually made from Al, stainless steel or Cu. The choice of the material is usually determined by cost, fabrication complexity and heat load from the synchrotron radiation. Because of its high thermal conductivity and ease of brazing, copper is also widely used as an absorber of synchrotron radiation. With the proper chemical cleaning and careful handling, the static thermal outgassing from these three metals is very similar. However, when subjected to synchrotron radiation, the photon induced neutral gas desorption can exceed the thermal outgassing by several orders of magnitude and it is this effect which determines the vacuum performance of the machine and decides the installed pumping capacity. With continued exposure to synchrotron radiation, the gas desorption decreases, i.e. the surface cleans up.

The initial photon induced gas desorption is important since it is this which determines the beam gas lifetime when the machine is started for the first time. The rate at which this desorption decreases with machine running time is more important since this determines the time for which the machine must be operated before acceptable beam-gas lifetimes can be obtained. Also the total quantity of gas desorbed is an important figure since it has consequences for the choice of the installed pumping capacity.

To obtain reliable figures for the desorption coefficients, the total amount of desorbed gas and the clean-up time for the vacuum system of the 27 km, 55 GeV electron positron storage ring LEP, test vacuum chambers made from Al, stainless steel and Cu were exposed to synchrotron radiation in a dedicated photon beam line on the DCI positron storage ring at the LURE Laboratory in Orsay, Paris[1]. In addition, measurements of the photoelectron currents produced by the photons in these chambers were made.

EXPERIMENTAL DETAILS

A. Vacuum system

Synchrotron radiation of 2.95 keV critical energy from a bending magnet of the DCI storage ring is extracted from the machine and passes through a collimator which limits the angular divergence of the radiation to 4.7 mrad in both the horizontal and vertical directions. This vertical collimation attenuates the continuous photon spectrum below about 6 eV. The test chambers could be pivoted in the horizontal plane so that the photons were incident on the side wall of the chamber at 11 mrad glancing angle of incidence where 3.12 m were irradiated.

The test chambers were pumped via a 72.5 l/s conductance for N_2 by a combination of a 400 l/s ion pump and two Ti sublimation pumps of about 1000 l/s each.

Pressure measurements were made with calibrated Bayard-Alpert gauges and a quadrupole gas analyser.

Data taking and analysis were performed by a specially developed minicomputer system.

B. Test chamber treatment

The aluminium test chamber was elliptical in cross section (131 mm x 70 mm) and manufactured from extruded tube. The material is a precipitation hardenable alloy of type ISO Al Mg Si 0.5. The chamber was chemically cleaned by spraying with an alkaline detergent at 60°C and about 100 bar, rinsed with cold demineralized water and dried with filtered hot air at 90°C[2].

The copper chamber was manufactured from 3 mm thick OFHC copper plate rolled in four 90 cm sections and electron beam welded to form a tube 131 mm in diameter and 3.6 m long. ConFlat flanges were vacuum brazed to each end. Before welding, each 90 cm section was chemically cleaned by first degreasing in perchloroethylene vapour at 121°C, then deoxidising by immersion in an alkaline detergent, pickling in HCl, passivation in a chromic acid/H_2SO_4 bath, rinsing in demineralized water and finally drying with N_2 gas. To ensure that the complete chamber was subjected to all stages of all treatments, the four sections were placed in the vacuum brazing oven although it was only the two end sections which were actually brazed to flanges. Once the chamber had been finally welded together it was given a last chemical cleaning in the deoxidising alkaline detergent.

Two 3.6 m long, 131 mm diameter, stainless steel chambers (SS 1 and SS 2) were manufactured from 316 L+N stainless steel tube. They were chemically cleaned first by immersion in perchloroethylene vapour at 121°C, then by immersion in an alkaline detergent at 65°C followed by rinsing in cold demineralized water and finally drying in a vacuum oven at 150°C. As an additional treatment, one chamber, SS 1, was degassed in a vacuum oven at 950°C.

Each test chamber was then vacuum tested by baking at 150°C (Cu and Al) or 300°C (stainless steel) for 24 h. When installed in the synchrotron radiation beam line, all test chambers were baked at 150°C for 24 h after which the base pressure was typically 2 10^{-8} Pa.

For ease of reference the different treatments are summarized in Table I.

Table I

TEST CHAMBER	TREATMENT
Aluminium	Alkaline etch
Copper	Vapour degreasing
	Alkaline deoxidation
	Acid etch
	Vacuum braze
	Alkaline deoxidation
Stainless steel (SS 1)	Vapour degreasing
	Alkaline detergent
	950°C vacuum degas
Stainless steel (SS 2)	Vapour degreasing
	Alkaline detergent

RESULTS

A. Gas desorption coefficients

The photon induced neutral gas desorption coefficients for Al as a function of the photon dose are shown in Fig. 1. There it can be seen that the gases desorbed are firstly H_2 followed closely by CO and CO_2, all in the 1 10^{-2} molecules/photon range, with CH_4 about a factor of 10 lower. With increasing photon dose all desorption coefficients decrease but the CH_4 decreases much faster than the other gases. After a dose of 2 10^{21} photons/m the H_2 has decreased by a factor of 13, CO and CO_2 by a factor of 50 and CH_4 by a factor of 300.

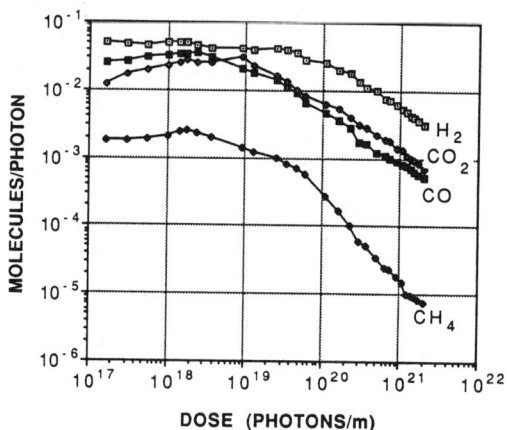

Fig. 1. The photon induced neutral gas desorption coefficients for Al as a function of the photon dose.

In Fig. 2 the corresponding results for Cu are shown. It can be seen immediately that, compared to Al, the initial desorption coefficients are all about a factor of 10 lower. However the rate of decrease with photon dose is also lower, in that after a dose of 2 10^{21} photons/m the desorption coefficients have only decreased by a factor of 10. Again it is H_2 which has the largest desorption coefficient followed closely by CO and CO_2 with CH_4 lower still.

The results for the stainless steel test chamber (SS 1) which was vacuum degassed at 950°C are shown

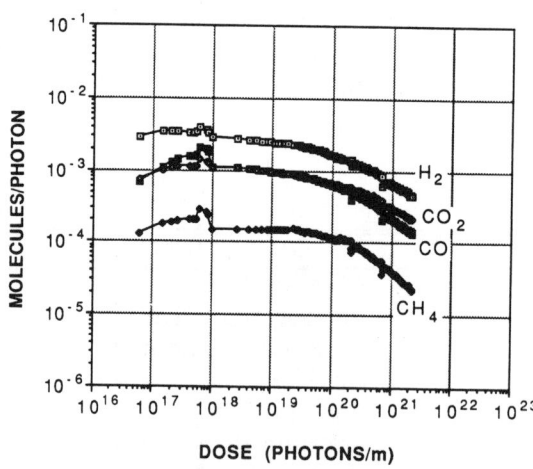

Fig. 2. The photon induced neutral gas desorption coefficients for Cu as a function of the photon dose.

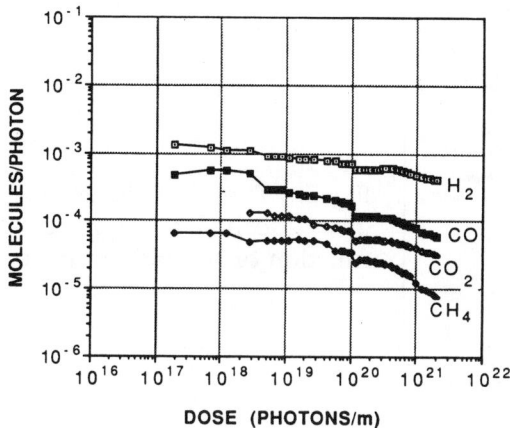

Fig. 3. The photon induced neutral gas desorption coefficients for stainless steel (SS 1) as a function of the photon dose.

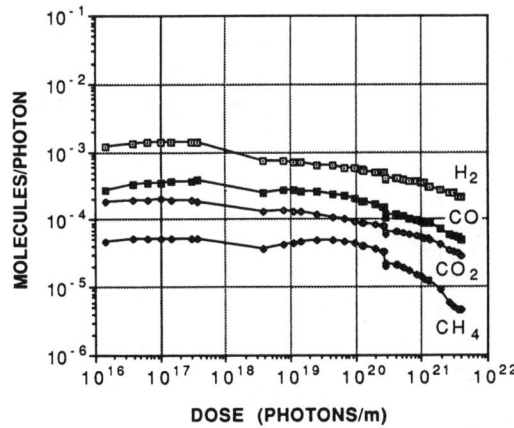

Fig. 4. The photon induced neutral gas desorption coefficients for stainless steel (SS 2) as a function of the photon dose.

in Fig. 3. The initial desorption coefficients are the lowest of all, with H_2 about $1\ 10^{-3}$ molecules/photon, CO and CO_2 in the 10^{-4} range and CH_4 in the 10^{-5} range. In common with Cu, the rate of decrease with photon dose is low and both H_2 and CO_2 have decreased by only a factor of about 3.5 and CO and CH_4 by a factor of 10.

Since the 950°C vacuum degassing may be difficult to carry out on large, complex vacuum chambers, this step was omitted from the treatment of test chamber SS 2. The results are shown in Fig. 4 and much to our surprise they were practically identical to those obtained from the vacuum degassed chamber SS 1.

With the advent of non-evaporable getters (NEG) the total quantity of gas desorbed by the synchrotron radiation is important in deciding how much NEG to install and in predicting how often it must be reconditioned during the operation of the machine[3]. By integration of the desorption coefficients with respect to photon dose, the total number of molecules desorbed per metre of chamber as a function of photon dose may be obtained. A more practical number is the amount of gas desorbed in Torr litres per metre of chamber, where 1 Torr litre = $3.3\ 10^{19}$ molecules. This integration has been carried out for Al, Cu and stainless steel (SS 2) and the results are shown in Fig. 5, 6 and 7 respectively.

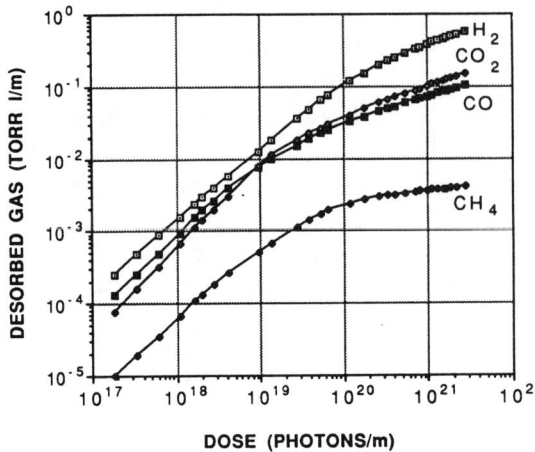

Fig. 5. The total quantity of gas desorbed from Al as a function of the photon dose.

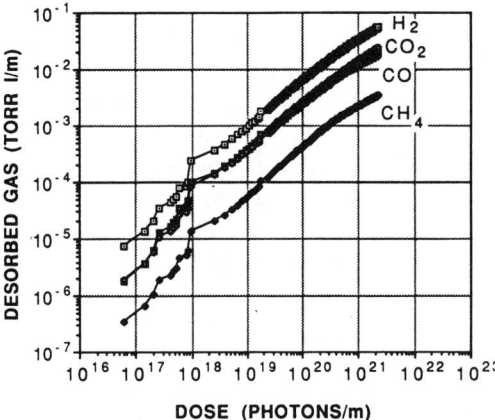

Fig. 6. The total quantity of gas desorbed from Cu as a function of the photon dose.

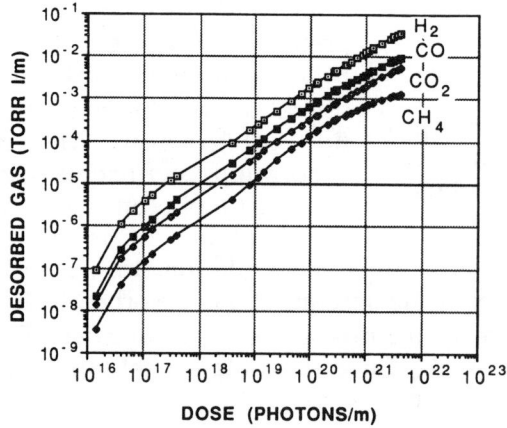

Fig. 7. The total quantity of gas desorbed from stainless steel (SS 2) as a function of the photon dose.

For the Al, which had the highest desorption coefficients, it can be seen that, after $3 \cdot 10^{21}$ photons/m, almost 1 Torr l/m of H_2 had been desorbed, 0.1 Torr l/m of CO and CO_2 but only $4 \cdot 10^{-3}$ Torr l/m of CH_4.

After $3 \cdot 10^{21}$ photons/m, the Cu test chamber had desorbed $5 \cdot 10^{-2}$ Torr l/m of H_2, $2 \cdot 10^{-2}$ Torr l/m of CO_2, $1.8 \cdot 10^{-2}$ Torr l/m of CO and $3 \cdot 10^{-3}$ Torr l/m of CH_4.

For the stainless steel which had the lowest desorption coefficients, after the same photon dose, $3 \cdot 10^{-2}$ Torr l/m of H_2 had been desorbed, $1 \cdot 10^{-2}$ Torr l/m of CO, $5 \cdot 10^{-3}$ Torr l/m of CO_2 and $1 \cdot 10^{-4}$ Torr l/m of CH_4.

Even after $3 \cdot 10^{21}$ photons/m, in all the test chambers the quantity of gas desorbed was still increasing and showing only a slight tendency to level off.

Another practical piece of information useful in the design of pumps for vacuum systems exposed to synchrotron radiation is how much gas has to be removed before the desorption coefficients drop below a certain value. These results are shown in Fig. 8, 9 and 10 for Al, Cu and stainless steel respectively.

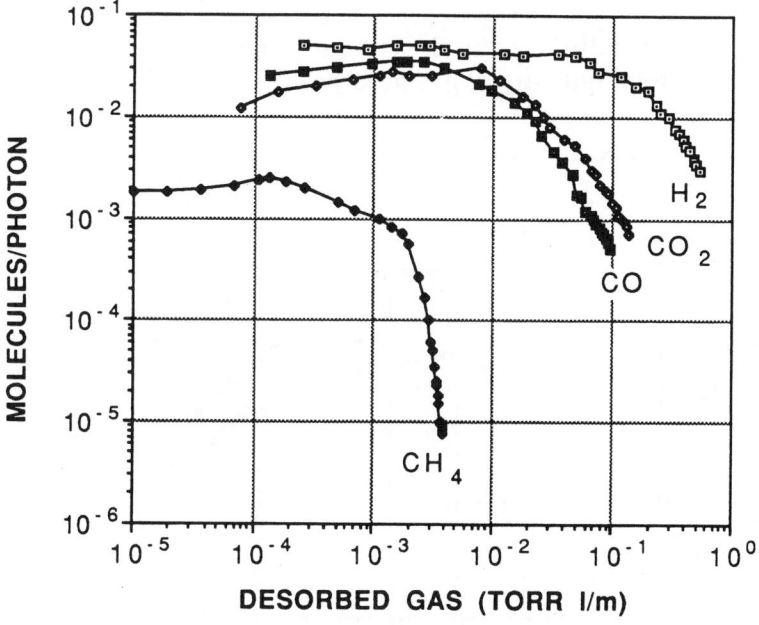

Fig. 8. The photon induced neutral gas desorption coefficients for Al as a function of the total quantity of gas desorbed.

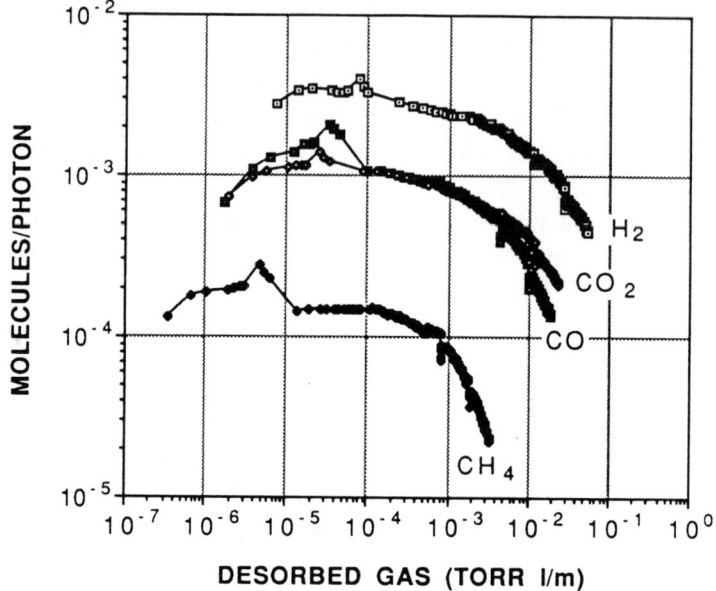

Fig. 9. The photon induced neutral gas desorption coefficients for Cu as a function of the total quantity of gas desorbed.

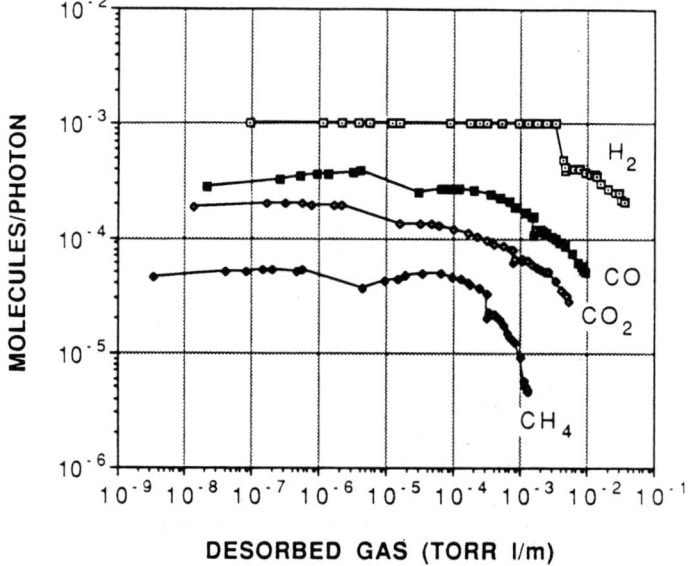

Fig. 10. The photon induced neutral gas desorption coefficients for stainless steel as a function of the total quantity of gas desorbed.

B. Photoelectrons

In parallel with the measurements on the photon induced gas desorption, the photoelectron currents in the test chambers were also measured. In the central part of the chamber, a 20 cm long, 1 mm diameter stainless steel wire, running along the axis of the chamber, was suspended from an insulated feedthrough and photoelectrons produced in the vicinity were collected by applying + 1000 V to the wire. Applying higher voltages simply collected more photoelectrons from further along the chamber, thus all measurements are relative, but nevertheless give an indication of the photoelectron currents to be expected.

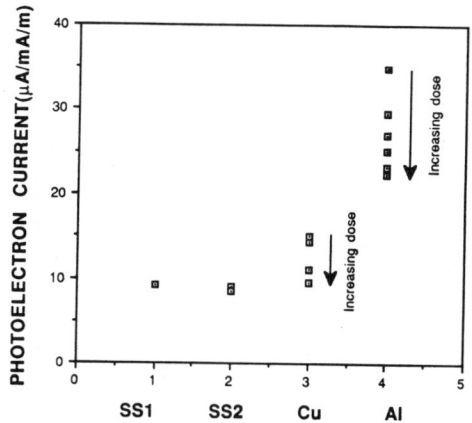

Fig. 11. The measured photoelectron currents in the four test chambers.

The results are shown in Fig. 11 for the four test chambers where it can be seen that the highest photoelectron currents were produced in the Al chambers, followed by the Cu then the stainless steel. With increasing photon dose, as shown by the arrows in the figure, the photoelectron currents decreased. In the stainless steel chambers there was almost no change with photon dose.

C. Extrapolation to LEP

The injection energy of the LEP machine is 20 GeV which corresponds to a critical energy of 5.7 keV for the synchrotron radiation compared to 2.95 keV in DCI. Also, for a large part of the time, LEP would run at 46 GeV where the critical energy is 69.7 keV.

The expression for the desorption coefficient η for a given gas is given by:

$$\eta = \frac{3.3 \cdot 10^{19} \cdot \Delta P \cdot S \cdot 2 \cdot \pi \cdot R}{8.1 \cdot 10^{17} \cdot E \cdot I} \quad \text{(Molecules/photon)} \quad (1)$$

where: ΔP is the absolute pressure increase due to the photons (Torr)
S is the pumping speed for the gas (l/s/m)
E is the beam energy (GeV)
I is the beam current (mA)
R is the bending radius (m).

The vacuum chamber in LEP is made from Al thus we have taken the desorption coefficients from Fig. 1, applied the appropriate LEP pumping speed for each gas, the beam current and the beam energy (either 20 GeV or 46 Gev) and calculated the pressure increase for each of the four gases from equation (1). These absolute pressures were then converted to N_2 equivalent to be able to compare them with the gauges installed in the LEP machine.

No corrections were made for the effect of angle of incidence -11 mrad in DCI and 7 mrad in LEP- nor for the higher critical energy of the photons in LEP. Both these effects would tend to increase the desorption coefficients.

In the LEP machine, H_2, CO and CO_2 are pumped by NEG strip running practically the whole length of the machine thus the linear pumping speed is well defined. But CH_4 is pumped only by 60 l/s ion pumps placed every 20 m giving a linear pumping speed which changes with distance along the vacuum chamber. The calculations were made with a linear pumping speed for CH_4 of 1 l/s/m, a figure which could easily be in error by more than a factor of two up or down, and, since CH_4 dominates the residual gas due to its low pumping speed compared to the other gases,1 l/s/m as opposed to about 500 l/s/m, this uncertainty is directly reflected in the total pressure calculation.

The results are shown in Fig. 12 where it can be seen that, despite not having corrected for all the effects mentioned above, the agreement between the extrapolated and measured total pressures is good and gives confidence that the results from the test chambers may be used to predict the vacuum behaviour in an actual machine with a reasonable degree of accuracy. An idea of the precision in the measured LEP data can be gained from the points just above a dose of 10^{19} photons/m in Fig. 12. A series of four measurements gives a factor of two between the lowest and the highest.

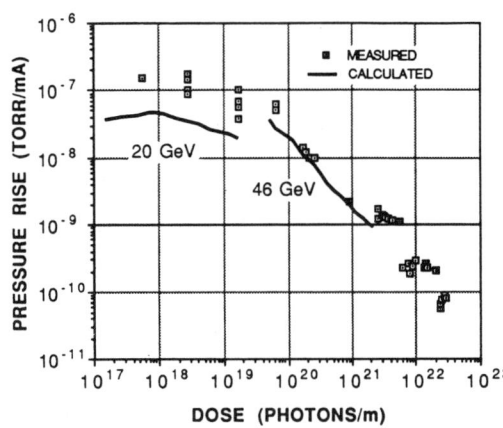

Fig. 12. A comparison between the predicted and measured pressure increases in the LEP machine.

CONCLUSIONS

Under photon bombardment the gases desorbed from all test chambers were, in order of importance, H_2, closely followed by CO and CO_2, with CH_4 last. Also, with increasing photon dose, the H_2, CO and CO_2 desorption coefficients in a given chamber all decreased at about more or less the same rate but the CH_4 decreased faster.

The highest initial desorption coefficients were measured in the Al chamber followed by Cu then stainless steel. The results show that the higher the initial desorption rate the higher is the decrease in the desorption rate. The impression is that, at sufficiently high photon dose, the desorption coefficients for all three metals will tend to similar low values.

Using both 4 keV and 500 eV critical energy photons to measure photon induced gas desorption from Al, Cu and stainless steel, similar results have been found[4,5].

The high temperature (950°C) vacuum degassing of stainless steel had little effect on the photon-induced desorption coefficients despite the fact that such a treatment produces a clean surface as seen by Auger surface analysis and greatly reduces the electron induced gas desorption coefficients[6].

The photoelectron currents measured in the test chambers showed the same tendency as the desorption coefficients. The highest currents were measured in the Al chamber followed by Cu and then stainless steel. In addition, the photoelectron currents were the same in the two stainless steel chambers. Also in the Al and Cu chambers the photoelectron currents decreased with increasing photon dose. Since electrons contribute to the gas desorption process[6], the form of the desorption versus photon dose curves reflects both the cleaning of the surface and the reduction in the photoelectron currents.

That the data obtained from test chambers can be used with a reasonable degree of accuracy to predict the vacuum behaviour in a real machine has been verified in the LEP electron positron storage ring. Despite differences in photon critical energy and some uncertainties in pumping speed, the measured and calculated pressures in LEP agreed to within a factor of two. Thus we are confident that our data can be used to better design the vacuum systems of future electron or positron storage rings.

REFERENCES

1. O. Gröbner, A. G. Mathewson, H. Stori, P. Strubin and R. Souchet, Vacuum 33 (7), 397, (1983).

2. A. G. Mathewson, Vuoto, Vol. XVII, No. 3, 102, (July-September 1987).
3. C. Benvenuti and F. Francia, CERN Report, CERN-LEP-VA/89-61, (1989).
4. S. Ueda, M. Matsumoto, T. Kobari, T. Ikeguchi, M. Kobayashi and Y. Hori, 11th Int. Vac. Congress and 7th Int. Conf. on Solid Surfaces, 25th-29th Sept. (1989).
5. H. J. Halama and C. L. Foerster, 11th Int. Vac. Congress and 7th Int. Conf. on Solid Surfaces, 25th-29th Sept. (1989).
6. M-H. Achard, R. Calder and A. G. Mathewson, Vacuum 29(2), 53, (1979).

THE SEARCH FOR LOW PHOTODESORPTION COATINGS*

C. L. Foerster and G. Korn
Brookhaven National Laboratory, NSLS - Bldg. 725C
Upton, New York 11973

ABSTRACT

Low photo desorption (PSD) from surfaces of vacuum chambers increases the beam lifetime and reduces the cost of the pumping system of any storage ring. In compact rings where all radiated power (~10 kW) is incident on a few meters only, low PSD and good thermal conductivity of photon absorbers are of particular importance. An experimental chamber in which one meter long bars can be exposed to white photon beam with 500 eV critical energy has been built and installed on the U10B beamline in the VUV ring at the NSLS. Several reference bars made of high purity copper and a TiN coating on copper have been measured. Subsequent runs will include gold coating on copper, aluminum (200 C baked), hard carbon coating on copper and uncoated beryllium bars. In this paper the desorption coefficients will be measured and compared.

I. INTRODUCTION

Vacuum surfaces having low Photo Stimulated Desorption (PSD) yields are necessary for good beam lifetimes and for reasonable construction cost of storage rings where synchrotron radiation will exist. PSD is the dominant gas load during operations and increases with beam current. To date, stainless steel and aluminum have been acceptable materials for beam chamber fabrication. Copper due to its good thermal and electical properties has been successfully utilized as photon absorbers. As is well known, the preparation of the material and surface treatments is a primary consideration.

New applications of electron storage rings such as compact rings and B factories place severe demands on vacuum system design due to limited space for pumping and extremely high currents. Much more careful and extensive studies of new materials and surfaces will be required to resolve both the photodesorption and heat dissipation. In addition the effect of angle of incidence, diffuse and specular reflection as well as photoelectron generation will have to be better understood to optimize the choice of materials. We have therefore built an experimental apparatus that can rapidly change the angle of incidence and measure all the above effects. At present it is operational on the U10B beam line having critical energy of 500 eV. Plans are being made to adapt it to the X28C, X-ray beam line with variable photon energy of 120 to 5000 eV.

Since a substantial amount of data exists on aluminum and stainless steel we will concentrate our efforts on copper, berillium and thin films of TiN, hard carbon and gold on copper.

*Work performed under the auspices of U.S. DOE under contract DE-AC02-76Ch00016

Stainless and aluminum will also be investigated for comparison. Preliminary results on 1.21m long Cu and TiN coated Cu bars will be presented in this paper. Two test set-ups were used for the measurements.

II. DESCRIPTION

A. Experimental Set-Up

The details of most of our previous experimental set-ups have been described [1,2,3] when aluminum, stainless steel, and copper coated chambers were tested. In present experiments, see Fig. 1 schematics, one meter long samples were exposed to white light with critical energy of 500 eV. The vertical collimator (C_v) was adjusted to 3.5 mrad and the horizontal collimator (C_h) was initially set to 10 mrad. The horizontal, C_H, collimator must be adjusted smaller when the sample is rotated for incident angles less than 50 mrad to keep the primary photons within the length of the sample. The samples receive photons directly from the source since there are neither mirrors nor monochromaters in this beam line.

The total incident photon flux per beam current I, per horizontal opening angle Θ, and per second t is given by [4] where E is the machine energy of

$$N/I\,\Theta t = 1.28 \times 10^{14}\, E \text{ photons } mA^{-1}\, mrad^{-1}\, s^{-1}, \quad (1)$$

0.75 GeV. Since the main gases desorbed are H_2, CH_4, CO and CO_2, the residual gas analyzer (RGA) was calibrated to yield their relative sensitivities. The absolute partial pressures have always been calculated from the calibrated BA gauge readings. The relative abundance of desorbed gases is obtained from RGA data. The RGA was recalibrated for each test run on each sample.

The specific molecular desorption yield [4] η_i is given by

$$\eta_i = \frac{GS_i\Delta P_i/I}{(N/I\Theta\, t)\, \theta_1} \text{ molecules photon}^{-1}, \quad (2)$$

where $\Delta P_i/I$ is the specific pressure rise of each gas in Torr mA^{-1}, Θ_1 is the horizontal opening angle in the experiment, $G = 3.2 \times 10^{19}$ molecules $Torr^{-1}\, l^{-1}$, S_i is the pumping speed in ls^{-1} for each gas species at orifice 01.

The beam line between the ring and valve, V3 in Fig. 1a was vacuum baked to 200 C for forty-eight hours and conditioned with V3 closed. The nitrogen conductance of orifice 01 is 47.5 l/s and can be considered the speed Si in Eq 2 when V3 is open. Orifice 01 is a rectangular duct and its calculated conductance was verified by in-situ nitrogen speed measurements using the pressure drop across a known conductance.

In test set-up #1, in Fig. 1b, the sample is mounted length wise across the chamber in the horizontal plane, making the photon incidence angle, 100 mrad. Three parallel pick up wires spaced 10mm apart are located opposite the sample bar to measure photo electrons and to sample diffusely reflected photons. The chamber was constructed of 304 stainless steel, vacuum baked to 300 C for forty-eight hours and argon-oxygen glow discharge conditioned just prior to measurements with installed samples. The window on the end was used for alignment prior to installation of the sample. The sample to be tested is secured in the test chamber to stainless steel plates welded inside each end of the chamber.

In test set-up #2 the sample was mounted horizontally length wise on the chamber wall (see Fig. 1c). This chamber was also fabricated with 304 stainless steel and was vacuum baked at 200 C for 72 hours. It has a horizontal pick up wire opposite the sample, and a pick up wire along the top of the chamber. At the down stream end of the test chamber is a water cooled, electrically insulated, photon stop. As with chamber #1 the ends of the sample are mounted to plates welded on the chamber wall. The chamber is mounted to a rotatable X-Y table with its center of rotation located relative to the horizontal center the exposed sample face.

B. Oxygen free Copper (OF-Cu)

A forty eight inch long by two inches wide and one half inch thick OF-Cu sample was exposed to photons at 100 mrad incidence angle. The sample was etched and solvent cleaned following standard NSLS procedures prior to installation in test set-up #1.

The test set-up #1 with sample was vacuum baked 175 C for 48 hours and pressure was less than 1×10^{-9} torr after cool down. The desorption yields versus dose are shown in Fig. 3. The sample was removed from test set-up # 1 and later installed in test set-up #2 with the unexposed side away from the wall for measurements. Test set-up #2 was vacuum baked with sample to 175 C for 72 hours.

C. Titanium Nitride Coated Copper (TiN-Cu)

A copper sample (48 x 2 x 1/2) was cleaned and sent out to a vendor for titanium nitride coating. The sample was first mounted in test set-up #1, baked to 175 C for forty eight hours. After cool down, pressure was less than 1×10^{-9} Torr. The desorption yields versus dose are shown in Fig. 2. After run #1 the sample was removed from test chamber one and conditioned at 200 C including an Argon glow discharge dose of 2×10^{18} ions/cm^2. It was then re-installed in test chamber #1 for measurement of its unexposed face. It was baked to 175 C for forty eight hours in situ. Cool down pressure was again less than 1×10^{-9} torr. The desorption yields versus dose are shown in Fig. 4.

III. DISCUSSION

A. Oxygen Free Copper

OF-Copper is commonly used in storage rings and beam lines to absorb unwanted power. It is normally used internally and usually water cooled. Our results are comparable to our previously run copper absorber[6], plated copper[2] and recent work[5] of Ueda et al. We did not see traces of high mass containments we have experienced with plated copper. Our NSLS copper cleaning procedure includes an etch.

The photon stop in test set-up #2 is constructed of OF-Cu and was conditioned after bake using direct photon from the source normal to its surface. The photo electron current was the same as the OF-Cu sample in test set-up #1. The desorption yield was almost the same as Fig. 2.

B. Titanium Nitride Coated OF-Copper

An OF-Cu sample was coated in vacuum by a vendor using his propriority process. This is the type coating that has been used in storage ring RF cavities to improve vacuum operations. The yield η from titanium nitride coated OF-Cu was almost the same as uncoated Of-Copper. (See Fig. 2 and 3) The coated sample is very slightly higher with clean up rates for H_2, CO, and CO_2 the same as uncoated.

Argon glow discharge conditioning of the TiN-Cu sample as expected, resulted in a large reduction of η yield. Initially, reduction of half an order of magnitude for H_2, an order of magnitude for CO, and one and a half order of magnitude for CO_2. The yield was seen to decrease significantly after an exposure of 10^{23} photons per meter.

CONCLUSION

Vacuum baked OF-Copper and TiN coated copper have approximately the same desorption characteristics. Our previous work and that of other has found vacuum baked copper and stainless to be almost the same. Therefore, copper, TiN coated copper and stainless steel have almost the same desorption coefficients after the same vacuum bake procedure.

Glow discharge conditioning of TiN coated copper results in reduced photo desorption. This was previously experienced to a greater degree with stainless steel. An Argon oxygen glow had been used for stainless.

Additional measurements of other materials are either planned or in process awaiting their turn. Among those in line are gold plated copper, berillium bars, aluminum bar ($200°C$ baked), and hard carbon coated copper. We have not yet developed a good homogeneous hard carbon process for our large sample.

ACKNOWLEDGEMENTS

The authors would like to thank H. Halama for fruitful discussions and suggestions, E. Gaudet, and the NSLS Vacuum Group for their excellent support. The assistance of the NSLS operations crew is also gratefully acknowledged.

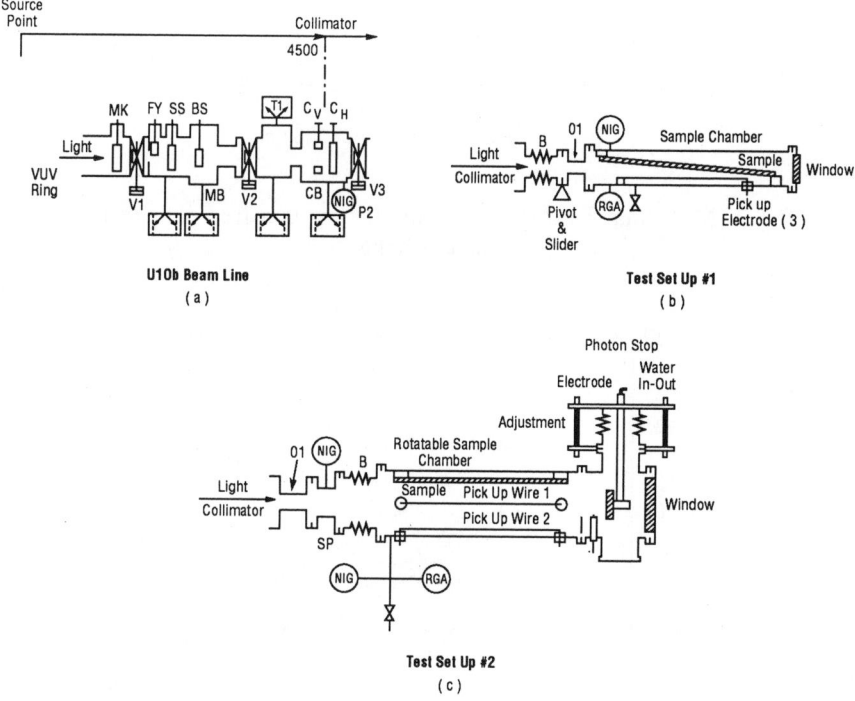

Fig. 1 Schematic diagram of the beamline and test set-ups. (a) Components: MK-mast for front end valve-VI; FV fast valve; SS-safety shutter; BS-beam stop; MB-mirror box; V2-U10B isolation valve; CB-collimator box with adjustable vertical-CV and horizontal collimators-CH; V3-isolation valve. (b) components; B-Bellows; 01-rectangular conductance. (c) 01-rectangular conductance; SP-spool, piece; set-up #2.

330 The Search for Low Photodesorption Coatings

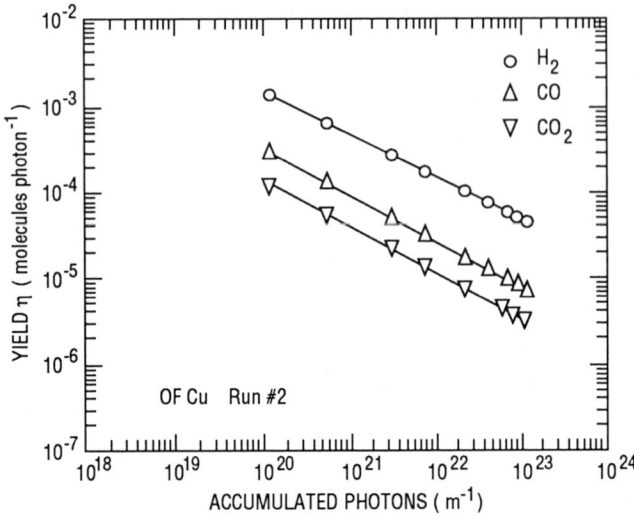

Fig. 2 Molecular Desorption Yields for oxygen free copper after vacuum bake

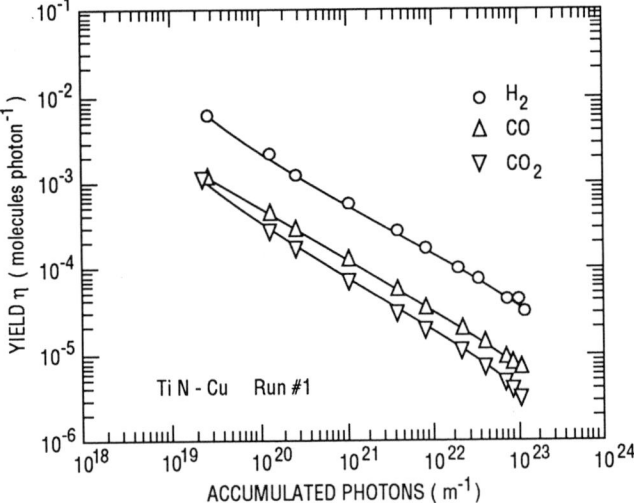

Fig. 3 Molecular desorption yields for Titanium nitride coated copper after vacuum bake

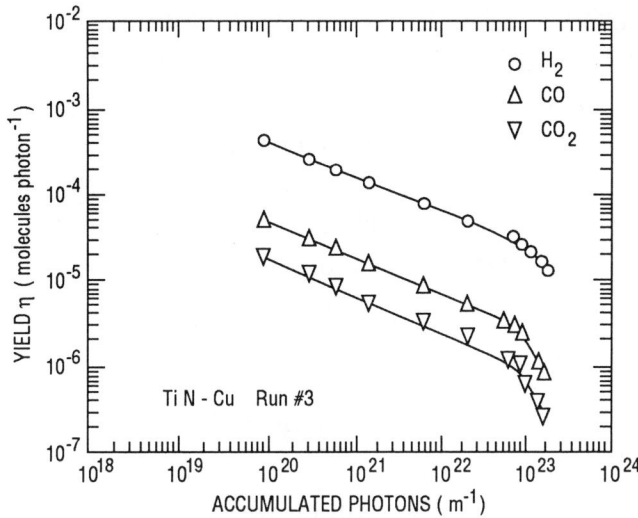

Fig. 4 Molecular desorption yields for glow discharge conditioned Titanium Nitride coated copper

REFERENCES

1. T. Kobari and H. Halama, J. Vac. Sci Techol. A5 (4), 2355 (1987).
2. C. L. Foerster, H. Halama, and C. Lanni, J. Vac. Sci. Tecnhol. A8 (3), 2856 (1990).
3. H. J. Halama and C. L. Foerster, Vacuum, in press.
4. O Gröbner, A.G. Mathewson, H. Stör, and P. Strubin, Vacuum 33, 397 (1983).
5. S. Ueda, M. Matsumoto, T Kobari, T.Ikeguchi, M. Kobayashi, Y. Hori, in press, 11th IVC 87th ICSS (Köln, 1989).
6. T. S. Chou, C.L. Foerster, H. Halama and C. Lanni, American Inst. of Physics, Conf. Proced. 171 P 334 (1988).

INVESTIGATION OF PHOTODESORPTION IN A CHAMBER OF THE ELECTRON STORAGE RING

Masanori Kobayashi

Photon Factory, National Laboratory for High Energy Physics,
Oho 1-1, Tsukuba Ibaraki 305,
Japan

ABSTRACT

A model of the photodesorption which occurrs in an electron storage ring and beam line experimental apparatus is presented. The inner surface of the chamber is divided into three regions, i.e., thermal desorption area; photodesorption area, irradiated by directly incident photons; and photodesorption area, irradiated by scattered photons. The model includes readsorption and diffusion. Calculated results are compared with experimental results. The simplest Henry's isotherm was not sufficient to describe photodesorption but a Langmuire type isotherm was more suitable.

INTRODUCTION

In the vacuum system of an electron storage ring, the main gas load is not thermal desorption but photodesorption by synchrotron radiation. The experimental results of photodesorption have been accumulated at PF-ring[1], Daresbury-ring[2], PF-BL21-test ducts[3], CERN-DCI-test ducts[4,5], NSLS-VUV-test ducts[6], etc. The results showed that photodesorption decreases with photon accumulation and that photodesorption depends on materials, surface treatment, angle of incidence and reflection of photons. Photodesorption caused by direct photon irradiation was considered as a main gas source at first. Photodesorption by scattered photon irradiation was recognized to be an effective gas source by the experiments using a uni-directional detector at PF[7]. Similar results were obtained at DCI by using long duct in which photons incident on the surface have a small angle of incidence[8]. Recently it was also pointed out by both groups mentioned above that diffusion was an important factor to explain the observed results[1,9].

Experimental results accumulated at many laboratories are, however, difficult to compare to each one of them. One of the reasons for the difficulty is that photodesorption has been evaluated by the photodesorption coefficient η (molecules/photon) which is an engineering factor to describe vacuum characteristics under photon irradiation. Another reason is that the physical process of photodesorption in a chamber is not clear and has not been discussed sufficiently. If we obtain good understanding of the physical process of photodesorption, we can then use the coefficient to describe vacuum characteristics under photon irradiation and apply it to new, challenging machines.

A model of photodesorption is now presented. The model has to consider the usual thermal evacuation process when a photon shutter is closed, that is, desorption and readsorption process should be taken into account in the model. The diffusion process should also be included in the model. Calculated results using the model will be compared with observed η coefficients. Sticking probability, rest time and surface concentration of adsorbed molecules, etc., are evaluated as to whether they apply to the usual vacuum process in a chamber.

MODEL

Before we present a model of photodesorption, we will start to consider the balance of molecules between the gas phase and the adsorbed phase. At the ultimate pressure region, we can describe the relation of the amount of adsorption on a unit surface by N_S (molecules/cm^2) and the gas molecular density corresponding to the pressure by N_g (molecules/cm^3) as follows,

$$N_S \approx (1/4)\bar{v} N_g \tau (s + A_p/A) \tag{1}$$

where v is thermal mean speed of gas molecules, s is sticking probability and τ is the rest time of molecules, and A is the inner surface area and A_p is the pumping port area taking account of Clausing factor for the pump. As it is well known, $\tau = \tau_0 \exp[E_d/kT]$ where E_d is activation energy for desorption and τ_0 is a constant and in many cases it is assumed to be 10^{-13} second[10]. At that region, we don't know exactly whether a practical surface of the vacuum chamber may be cleaned out and N_s is far less than a monolayer adsorption N_{sM}, or the surface may be covered with molecules in nearly monolayer. If the surface is in the former condition and sticking probability may be almost equal to 1 then the surface can act as a pump. When photons come in this vacuum system and photodesorption occurs, the increment of N_g will not be so high. On the other hand, if the surface is in the latter condition, the sticking probability at the equilibrium may be small, and increment of N_g will be so high. Many experimental results of photode-sorption indicate that the latter is the correct image of the vacuum system with practical surfaces. So we were taken account of the readsorption in the model of photodesorption process.

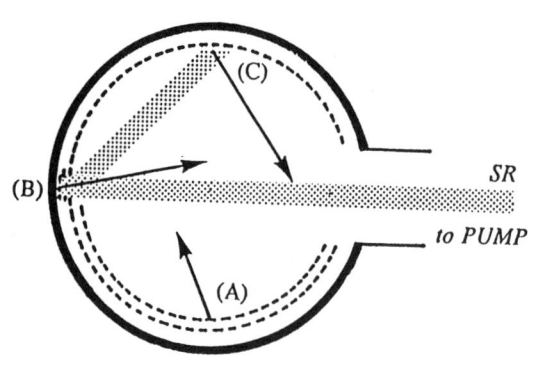

Fig.1 Schematics of photodesorption in a vacuum chamber. Inner surface is divided into (A) thermal desorption, (B) photodesorption by direct photons and (C) photodesorption by scattered photons. Molecules desorbed can readsorb on the surface with sticking probability s and rest time τ. Other parameters are shown in Table 1.

Next, we write the equations of N_g and N_s in photodesorption process. Figure 1 shows schematically a vacuum chamber in which photodesorption occurs. Photons come and enter the chamber through an orifice, and the chamber is evacuated through the orifice. The inner surface of the chamber is divided in three regions, that is, in the region A, desorption occurs in usual thermal process, in the region B, desorption occurs by irradiation of photons which directly incident on the area, and in the region C, desorption occurs by irradiation of scattered photons. The surface areas are designated by A_A, A_B and A_C which correspond to the three regions, respectively. Volume of the chamber is designated by V. Molecules in the three regions and in the gas phase can be described as follows,

region A: $\quad dN_{sA}/dt=(1/4)\bar{v}s_A N_g - N_{sA}/\tau_A \quad$ (2)

region B: $\quad dN_{sB}/dt=(1/4)\bar{v}s_B N_g - N_{sB}(1/\tau_B + N_{pB} Y_e \sigma) \quad$ (3)

region C: $\quad dN_{sC}/dt=(1/4)\bar{v}s_C N_g - N_{sC}(1/\tau_C + N_{pC} Y_e \sigma) \quad$ (4)

gas phase: $dN_g/dt =$
$-(S/V)N_g - (A_A/V)dN_{sA}/dt - (A_B/V)dN_{sB}/dt -$
$(A_C/V)dN_{sC}/dt + (A_A D_A/V)\partial N_{cA}/\partial x]_{x=0} -$
$(A_B D_B/V)\partial N_{cB}/\partial x]_{x=0} - (A_C D_C/V)\partial N_{cC}/\partial x]_{x=0} \quad$ (5)

where S is the pumping speed of the chamber. It can be expressed as $(1/4)\bar{v}A_p K_C$ where A_p is the orifice aperture and K_C is Clausing factor of the pump.

Photon intensity F_p can be calculated by the relation, F_p (photon/sec) = $8.08 \times 10^{17} I_b E_b/(2\pi) \times \Delta\Theta\Delta\psi$, where I_b is stored beam current (mA) and E_b is beam energy (GeV), and $\Delta\Theta$ and $\Delta\psi$ are angular divergence limited by the orifice in horizontal and vertical, respectively[4]. Photon flux intensity N_p on a unit surface area, which is effective for photodesorption, is defined by using reflectivity R at the first incidence point of photons,

region B: $N_{pB}=F_p(1-R)/A_B$ (6)
region C: $N_{pC}=F_pR/A_C$ (7)

As is shown by Eqs. (3) and (4), the photodesorption rates are expressed as $N_S N_p Y_e \sigma$ where Y_e is photoelectron yield (photoelectron/photon) and σ is cross section (cm^2/photo-electron) of electron stimulated desorption.

When we take account of diffusion, we assume that molecules which diffuse out from the bulk quickly emitted from the surface with emission coefficient α_A, α_B and α_C. The molecules do not accumulate at the surface, so the emission is proportional to the tangent of bulk density at the surface. We can express the emitted molecules caused by diffusion by using usual diffusion equation ($D\partial^2 N_c(t)/\partial x^2 = \partial N_c(t)/\partial t$) in semi-infinite bulk[11], and the equations are as follows,

$$D_A \partial N_{cA}(t)/\partial x |_{x=0} = N_{cA} d/\tau_A \times \exp[(\alpha_A^2/D_A) \times t] \times \text{erfc}[\alpha_A \sqrt{t}/D_A] \quad (8)$$

$$D_B \partial N_{cB}(t)/\partial x |_{x=0} = N_{cB} d(1/\tau_B + N_{pB} Y_e \sigma) \times \exp[(\alpha_B^2/D_B) \times t] \times \text{erfc}[\alpha_B \sqrt{t}/D_B] \quad (9)$$

$$D_C \partial N_{cC}(t)/\partial x |_{x=0} = N_{cC} d(1/\tau_C + N_{pC} Y_e \sigma) \times \exp[(\alpha_C^2/D_C) \times t] \times \text{erfc}[\alpha_C \sqrt{t}/D_C] \quad (10)$$

where d is surface thickness ($\sim 3 \times 10^{-8}$cm)[10], D_A, D_B and D_C are diffusion constant, $\alpha_A = d/\tau_A$, $\alpha_B = d(1/\tau_B + N_{pB} Y_e \sigma)$, and $\alpha_C = d(1/\tau_B + N_{pC} Y_e \sigma)$, and erfc[$\alpha_A \sqrt{t}/D_A$], erfc[$\alpha_B \sqrt{t}/D_B$] and erfc[$\alpha_C \sqrt{t}/D_C$] are error function complementary, which are correspond to the region A, B and C, respectively. We substituted the last terms in Eq.(5) by Eqs.(8)-(10) and solved the equations.

ANALYSES

A. Analytical expression (quasi-equilibrium approximation)

To solve these equations analytically, we assume that quasi-equilibrium condition is filled between gas phase and adsorbed phase, and that sticking probability and rest time are constants; i.e., Henry's isotherm was assumed. In three regions, the following conditions are assumed;

region A: $(1/4)\bar{v}sN_g$ and $N_{sA}/\tau_A \gg dN_{sA}/dt$ (11)
region B: $(1/4)\bar{v}sN_g$ and $N_{sB}(1/\tau_B+N_{pB}Ye\sigma) \gg dN_{sB}/dt$ (12)
region C: $(1/4)\bar{v}sN_g$ and $N_{sC}(1/\tau_C+N_{pC}Ye\sigma) \gg dN_{sC}/dt$. (13)

From Eqs.(11)~(13) and (2)~(4), we can define an effective volume V_e including re-adsorption as

$$V_e = V + (1/4)\bar{v}s\{A_A\tau_A + A_B/(1/\tau_B+N_{pB}Ye\sigma) + A_C/(1/\tau_C+N_{pC}Ye\sigma)\}. \quad (14)$$

Thus the differential equation (5) to be solved can be rewritten, and it is

$$dN_g/dt = -\gamma N_g + \gamma_A N_{cA} d/\tau_A \times \exp[(\alpha_A^2/D_A)\times t] \times \text{erfc}[\alpha_A\sqrt{t/D_A}]$$
$$+\gamma_B N_{cB} d(1/\tau_B+N_{pB}Ye\sigma) \times \exp[(\alpha_B^2/D_B)\times t] \times \text{erfc}[\alpha_B\sqrt{t/D_B}]$$
$$+\gamma_C N_{cC} d(1/\tau_C+N_{pC}Ye\sigma) \times \exp[(\alpha_C^2/D_C)\times t] \times \text{erfc}[\alpha_C\sqrt{t/D_C}] \quad (15)$$

where $\gamma = S/V_e$, $\gamma_A = A_A\alpha_A/V_e$, $\gamma_B = A_B\alpha_B/V_e$, and $\gamma_C = A_C\alpha_C/V_e$.

A similar but more simple equation was presented by K. Dimoff et.al. for modeling of impurity release rate of glow discharge conditioning of vacuum vessels[12]. The emission rate α was given as a constant and was equal to $(1/\tau_0)\exp[-E/kT]$. In the Eqs.(8)~(10), the emission factors α_B and α_C are more complicated, but formal solution of Eq.(15) could be obtained by the Laplace transform method. We set $N_g = N_g(0)$, $N_{cA} = N_{cA}(0)$, $N_{cB} = N_{cB}(0)$ and $N_{cC} = N_{cC}(0)$ at t=0 for initial condition. After setting $N_g(0)$ and $N_s(0)$, we choose τ or s. The chosen ones can be solved according to Eq.(1). The obtained solution is

$$N_g(t) = N_g(0)\exp[-\gamma t]$$
$$+ \gamma_A N_c A(0)/\{\gamma(1+\alpha_A^2/D_A\gamma)\} \times$$
$$\{\exp[(\alpha_A^2/D_A)\times t]\times \text{erfc}[\alpha_A\sqrt{t/D_A}] + \exp[-\gamma t](\alpha_A^2/i\sqrt{D_A\gamma_A})$$
$$\text{erf}[i\sqrt{\gamma t}]-1)\} + \gamma_B N_c B(0)/\{\gamma(1+\alpha_B^2/D_B\gamma)\} \times$$
$$\{\exp[(\alpha_B^2/D_B)\times t]\times \text{erfc}[\alpha_B\sqrt{t/D_B}] + \exp[-\gamma t](\alpha_B^2/i\sqrt{D_B\gamma_B})$$
$$\text{erf}[i\sqrt{\gamma t}]-1)\} + \gamma_C N_c C(0)/\{\gamma(1+\alpha_C^2/D_C\gamma)\} \times$$
$$\{\exp[(\alpha_C^2/D_C)\times t]\times \text{erfc}[\alpha_C\sqrt{t/D_C}] + \exp[-\gamma t](\alpha_C^2/i\sqrt{D_C\gamma_C})$$
$$\text{erf}[i\sqrt{\gamma t}]-1)\}. \tag{16}$$

This solution gives the gas density in the chamber including diffusion and readsorp-tion of molecules which desorbed in photodesorption process.

B. Direct analyses by computer simulation

In the analytical solutions including readsorption and diffusion under photon irradiation, we assumed that s and τ are constant, but calculated curves of η seemed to be different from the experimental values. When we want to get better fittings, we may handle s and τ as variables. Then the equations become non-linear, so we have to solve the Eqs.(2)~(5) with Eqs.(8)~(10) directly by a computer.

We assumed the sticking probability varies with the amount of adsorption, that is,

$$s = s_0(1-\Theta) \tag{17}$$

where s_0 is the sticking probability on a clean surface with no adsorption, and Θ is coverage. The rest time τ is assumed to vary with the activation energy of desorption,

$$E = E_0 - (E_0 - E_1)\Theta \tag{18}$$

where E_0 is the energy on a clean surface corresponding to s_0, and E_1 on the full coverage surface.

To solve the equations (2)~(4), we adopted the 4-dimensional Runge-Kutta method, and assumed that the surface coverages are same on the three regions at initial. The photodesorption coefficient η can be calculated from the difference of N_g when the photon shutter is open or close.

RESULTS

A. Results under quasi-equilibrium condition

To obtain photodesorption coefficient η, we first calculated N_g without photons as usual thermal process, then calculated N_g with photons, that is, both process were calculated independently. We defined η as,

$$\eta = \{N_g(t, N_p>0) - N_g(t, N_p=0)\} \times (1/4)\bar{v}A_p/F_p \qquad (19)$$

where $(1/4)\bar{v}A_p$ is pumping speed of the orifice, $N_g(t, N_p>0)$ indicates the number of gas molecules per cm^3 at time t when the photon shutter is open and $N_g(t, N_p=0)$ applies when the shutter is closed. Photon flux F_p (photon/s/slit) at BL-21 of PF, is

$$F_p = 8.08 \times 10^{17} I_b E_b / (2\pi) \times 5/13000 \times 40\% \qquad (20)$$

and typical values of I_b is 300 mA when beam lifetime was over 15 hours, and E_b was 2.5GeV. The 40% is a calculated transmission factor of the collimation slit (5mm in width and 5mm in height) taking account of all photon energies and the distance (13m) between the slit and the radiation source. Photon dose to the experimental system was defined as $\int F_p \times dt(s)$ for open time of the photon shutter. The experimental set-up and results were already presented[3,7]. The gas densities $N_g(0)$ were obtained from the observed pressures of about $0.8 \sim 3 \times 10^{-9}$ Torr. Photoelectron yield Y_e(photoelectrons/photon) and the cross section of electron stimulated desorption are chosen as 0.1 and 1×10^{-17} cm^2, respectively[10]. Surface density $N_s(0)$ is assumed to be 1×10^{15} molecules/cm^2 for three regions. The other parameters are listed in Table 1.

Real surfaces are not atomically flat and many photons can penetrate into such surfaces, so photons and secondary electrons interact with not only molecules on the most outer surface but in the deeper surface layer. We introduced a new parameter which is "effective surface thickness". It was assumed to be 10 for the region B where photons irradiate directly, and 3 for the region C where scattered photons (with lower energy) irradiate. Thus the desorption rate at region B is

Table 1. Parameters of the System Used in the Calculation

Parameter		unit	Region A	Region B	Region C
surface area	A	cm^2	$A_A = 280$	$A_B = 0.25$	$A_C = 1120$
reflectivity	R			0.5	
photon intensity	N_p	ph/s/cm^2	0	$F_p(1-R)/A_B$	$F_p R/A_C$
pumping speed	S	l/s	3		
diffusion constant	D	cm^2/s	1×10^{-17}	1×10^{-14}	5×10^{-16}

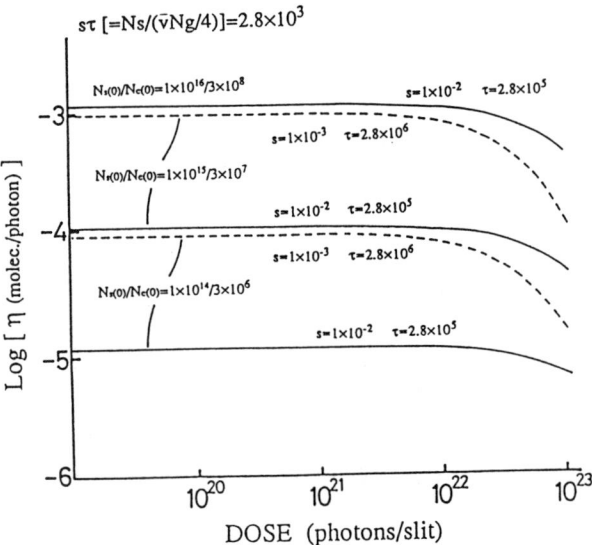

Fig.2 Photodesorption coefficient as a function of photon dose, where parameters are sticking probability and rest time. $s\tau = 2.8 \times 10^3 =$ constant, $N_S(0)/N_g(0)$ are as per the figure and $N_{cA}(0) = N_{cB}(0) = N_{cC}(0) = 5.6 \times 10^{19}$

$N_{SB}(1/\tau_B + 10 \times N_{pB} Y e \sigma)$ and the rate at region C is $N_{SC}(1/\tau_C + 3 \times N_{pC} Y e \sigma)$ are imployed in the calculation.

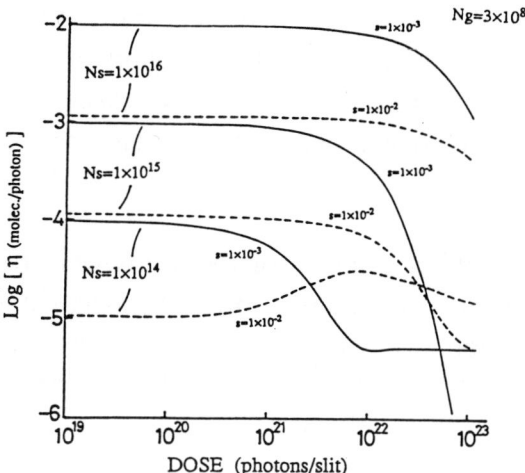

Fig.3 Photodesorption coefficient as a function of photon dose, where parameters are $N_S(0)$. Solid lines correspond to $s=10^{-3}$ and dashed lines $s=10^{-2}$.

Figure 2 and 3 show the photodesorption coefficient for carbon monoxide as a function of photon dose. In Fig. 2, parameters are sticking probability and rest time. In Fig.3, parameters are the amount of adsorption, $N_S(0)=10^{16}$, 10^{15} and 10^{14}, while $N_g=3\times10^8=$ constant. The procedure to obtain the η in Eq.(19) was different from the usual method, so the values of η with $s=10^{-2}$ and 10^{-3} at $N_S=10^{14}$ in higher photon dose in Fig.3 may have no physical meaning.

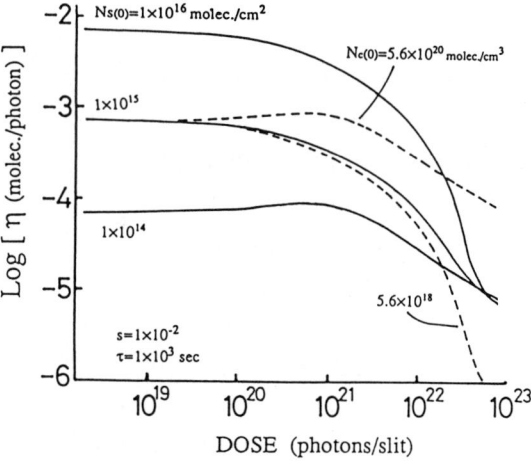

Fig.4 Photodesorption coefficient as a function of photon dose, in which parameters are $N_S(0)$ for solid lines and initial bulk concentration $N_C(0)$ for dashed lines.

Figure 4 shows the photodesorption coefficient for hydrogen as a function of photon dose. Parameters are initial amounts of adsorption, and the results are indicated by solid lines. Dashed lines indicate the results when initial concentrations of bulk are changed.

B. Results using variable s and τ

To make readsorption effect clear, diffusion terms were first excuded from the equations and only the readsorption process was taken account in the calculation. To get photodesorption coefficient η, we opened and closed the photon shutter intermittently and obtained single N_g curve. The coefficient is defined as,

$$\eta = \{N_g(t, N_p>0) - N_g(t\ latest, N_p=0)\} \times (1/4)\bar{v} A_p / F_p \quad (21)$$

where $N_g(t_{latest}, N_p=0)$ indicates the number of gas molecules per cm^3 at the latest time when the shutter was closed. Some results are shown in Fig.5, where sticking proba-bility varied in $1\times 10^{-4} \leq s \leq 0.5$,

Fig.5 Photodesorption co efficient against photon dose. Sticking probability changes from 1×10^{-4} to 0.5, and τ changes according to the amount of adsorption which is described in the text.

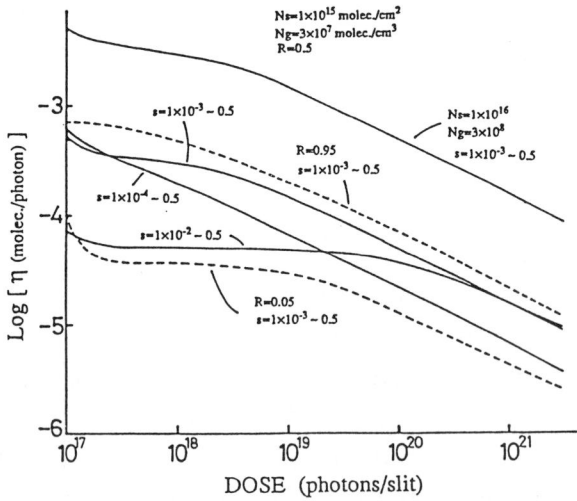

E_1 was defined from the equilibrium conditions and $E_0 = E_1 + 4 kcal/mol$. The η began to decrease at less photon dose, while the results calculated by analytical method show in Figs.2 and 3 were

constant to higher dose. The tangents of Log η against Log photon dose seemed to be between 0 and -1/2 and not between -1/2 and 1. The results with constant sticking probability not shown in the figure showed similar tendency as those obtained by analytical method. Thus the curve shown in Fig.5 depend not only on the calculation method but on the characteristics of the sticking probability.

DISCUSSION

A. Comparison with the observed results

Figure 6 shows the calculated results of photodesorption η as a function of photon dose (photons/slit). In the figure, dashed lines indicate the experimental results[3], and the solid lines indicate the calculated results. To fit the calculated results with the experimental values, the parameters were selected carefully.

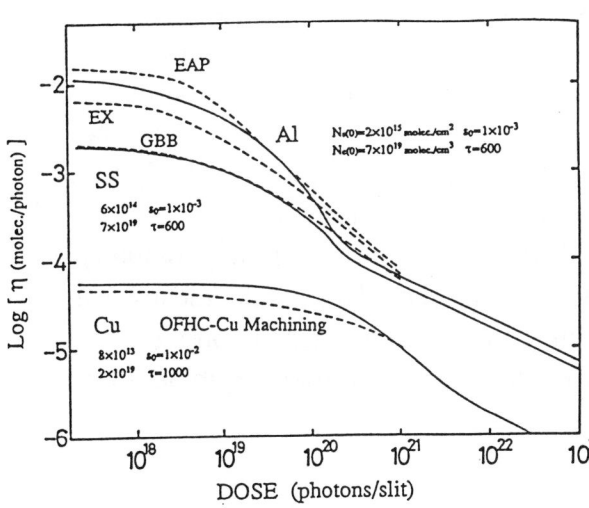

Fig.6 Photodesorption coefficients (molecules/photon) as a function of photon dose (photon /slit). Solid lines indicate calculated values and dashed lines show the experimental values. $N_S(0)=2\times10^{15}$, and $N_{cA}(0)=N_{cB}(0)=N_{cC}(0)=7\times10^{19}$ for alumi-num, $N_S(0)=6\times10^{14}$, and $N_{cA}(0)=N_{cB}(0)=N_{cC}(0)=7\times10^{19}$ for stainless steel and $N_S(0)=8\times10^{13}$, and $N_{cA}(0)=N_{cB}(0)=N_{cC}(0)=2\times10^{19}$ for OFHC copper.

B. Time variation of N_g and N_s

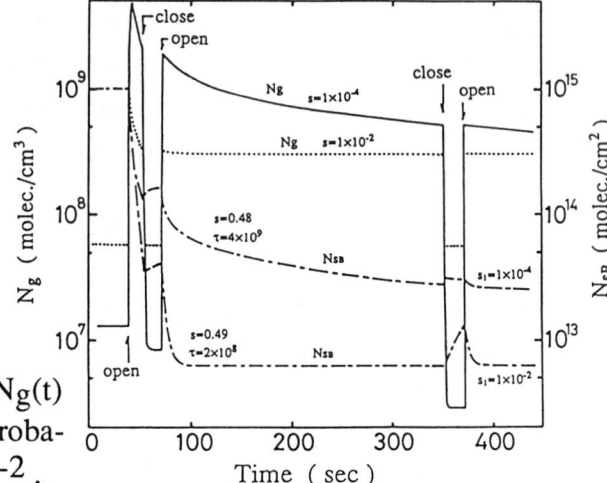

Fig.7 Time variations of $N_g(t)$ and $N_{SB}(t)$. Sticking probabilities are 10^{-4} and 10^{-2}.

An example of variation of $N_g(t)$ and $N_{SB}(t)$ were obtained by the simulation and they are shown in Fig.7, where the photon shutter was opened three times. For the system with small sticking probability $s=10^{-4}$, increments of N_g were larger and after photon irradiation bases became lower than that of initial. On the contrary, increment of N_g with $s=10^{-2}$ were smaller than the results mentioned above and the bases after irradiation were almost similar as the initial. In both of the examples initial amount of adsorption were $N_S(0)=2\times10^{15}$. They decreased with photon dose and they gradually increased when the photon shutter was closed. This results means that practical time constant of cleaning process did not depend on the outgassing from the directly irradiated region but those irradiated by scattered photons.

C. Effective η and intrinsic η

The photodesorption coefficient has been defined as $\eta=kS(p-p_b)/F_p$ where k is a conversion factor of pressure p to number of gas molecules per unit volume, p_b is pressure without photons, S is pumping speed of the system. This η which is called photodesorption coefficient has a unit of [molecules/photon]. In exactly meaning, however, its unit is not [*desorbed molecules*/photon] but [*pumped-out molecules*/photon] from the vacuum chamber to an outer system. The model shown above indicates that readsorption process can not be neglected but important to describe photodesorption in the chamber. This means that the vacuum system has the second pump, i.e., a surface pump, which pumping speed is designated by S_S and its surface is irradiated by photons. When we want to describe *intrinsic* desorption coefficient $\eta*$ with a unit of [*desorbed molecules* /photon], the coefficient should be defined as

$$\eta* = k(S+S_S)(p-p_b)/F_p = \eta(1+S_S/S) \tag{22}$$

This relation indicates that we can obtain $\eta=\eta*$, and η becomes independent of pumping speed only when $S_S \ll S$, i.e., $s \ll A_p/A$.

CONCLUSION

The model proposed here can be useful to understand photodesorption in the chamber. The Runge-Kutta method was better than the analytical equations in low photon dose, but it required extensive time for calculations with higher photon dose.

To reduce photodesorption, it is effective to reduce surface concentration of molecules in lower dose. In higher dose, reduction of bulk concentration of duct materials is effective. We have not yet calculate the pressures in a cold bore system such as SSC Magnet Beam Tube, but we believe this method will also be applicable to such a system.

It was straightforward to expand the 4-dimensional equations to 5-dimensional ones in order to describe pressures in the antechamber vacuum system. Preliminary results required an additional pump in the beam duct or at the slit between the antechamber and the beam duct to improve pressures in the beam duct.

ACKNOWLEDGEMENT

I thank Prof.Y.Kamiya for solving the equation, Mr.M.Matsumoto for checking the program and Dr.Y.Hori for discussing and preparing the figures.

REFERENCES

[1] M.Kobayashi, AIP conference proceedings No.171, New York (1988). (American Vacuum Society Series 5, Vacuum design of advanced and compact synchrotron light sources) pp155.
[2] B.A.Trickett, Vacuum **37**(1987)747.
[3] S.Ueda, M.Matsumoto, T.Kobari, T.Ikeguchi, M.Kobayashi and Y.Hori, 11th International Vacuum Congress and 7th International Conference on Solid Surfaces (held in Cologne, Germany, Sept. 22-29, 1989). This is printing in Vacuum (1990) as a special issue.
[4] O.Grobner, A.G.Mathewson, H.Stori, P.Strubin and R.Souchet, Vacuum **33**(1983)397.
[5] O.Grobner et.al., LEP Vacuum Technical Note 6/Jan/1986.
[6] T.Kobari and H.Halama, J.Vac.Sci.Technol. **A5**(4)(1987)2355.
[7] M.Kobayashi, M.Matsumoto and S.Ueda, J.Vac.Sci.Technol. **A5**(4)(1987)2417.
[8] O.Grobner,.G.Mathewson, P.Strubin, E.Alge and R.Souchet, J.Vac.Sci.Technol. **A7**(2)(1989)223.
[9] M.Andritshky, O.Grobner, A.G.Mathewson, F.Schumann, P.Strubin and R.Souchet, Vacuum 38 No.8-10(1988)933.
[10] P.A.Redhead, J.P.Hobson and E.V.Kornelsen, *Scientific Foundations of Vacuum Technique* 2nd.Ed. (Jon Wiley, 1962).
[11] H.S.Carslaw and J.C.Jaeger, *Conduction of Heat in Solid* (Oxford Univ.Press, London 1959).
[12] K.Dimoff, C.Boucher, J.Parbhakar and A.K.Viji, JVST A6(5) Sept/Oct(1988)2876.

PHOTOELECTRON EFFECT ON PHOTODESORPTION IN A CHAMBER IRRADIATED BY SYNCHROTRON RADIATION

T. Kobari, M. Matsumoto, T. Ikeguchi, and S. Ueda
MERL, Hitachi, Ltd. Tsuchiura, Ibaraki, 300, JAPAN
M. Kobayashi and Y. Hori
Photon Factory, KEK, Tsukuba, Ibaraki, JAPAN

ABSTRACT

Photodesorption in vacuum chambers exposed to synchrotron radiation were measured at the PF storage ring of KEK. Two types of wire electrodes, one is a cylindrical grid in an axial direction near the inner surface of the test chamber, and the other is a semi-spherical grid around a surface irradiated by synchrotron radiation, were installed in the test chambers. Desorption and photoelectron currents were measured under conditions of bias voltages applied to the electrodes. Photodesorption caused by synchrotron radiation is greatly influenced by bias voltage. Therefore, electron stimulated desorption due to photoelectron contributes considerably to the photodesorption. Photodesorption can be decreased by a negative electric field around the primary incident surface of the chamber.

INTRODUCTION

Attaining long beam lifetime and stability is one of the most essential functions of the electron or positron storage rings. Pressure increase caused by photodesorption shortens lifetime and induces various instabilities. It is very important for machine designers to reduce photodesorption. Photodesorption is affected by photon energy, the number of photons, angle of incidence, materials, and surface treatments.[1-6] There have been some reports on the photodesorption process in vacuum chambers exposed to synchrotron radiation. The desorption rate of the surface directly irradiated by synchrotron radiation decreases quickly in comparison with that of the surface irradiated by scattered photons or photoelectrons[7] and photoelectron mainly produces the desorption.[2,8,9] The dependence of the number of scattered photons on glancing angle[8] and the dependence of photodesorption on the oxide layer of a surface are also studied.[9]

The focus of this experiment has been to study the effect of the photoelectron on the photodesorption from test chambers made of various materials using grid type electrodes which produce the electric fields in the test chambers.

EXPERIMENTAL SETUP

The experimental setup, consisting of the apparatus for measurement and the test chamber, is shown in Figure 1. This test section, fixed to a movable base, is connected to the PF ring. Synchrotron radiation collimated by the slit enters the test chamber from the PF ring. The collimation angle is 0.385 mrad in both vertical and horizontal directions. The size of the photon beam is 5 mm in both directions. The incidence angle of synchrotron radiation to the test chamber is normal. The slit also works as an orifice which conductance is 3 l/s(N_2 equi.) and is located between two B-A gauges, IG_1 and IG_2. The outgas from the test chambers are determined by multiplying the difference of pressure between two gauges by the conductance. The test section can be moved vertically and can be rotated around the source point of synchrotron radiation. The movable base is used in order to change the flux and energy of photons coming into the test chamber and to keep the irradiated point fixed. Two electrodes are installed in the test chamber. One is a cylindrical grid (grid B) in the axial direction near the inner surface of the test chamber, which consists of eighteen longitudinal wires and four circular wires. Grid B has a length of 270 mm and has a diameter of 50 mm. The other is a semi-spherical grid (grid A) around the irradiated surface, which consists of six circular wires. The radius of semi-sphere, Grid A is 20 mm. These grids are made from tungsten wires which diameters are 0.3 mm. Bias voltage can be applied to these grids separately and photoelectron currents can be measured by ammeters. Before this experiment, these electrodes were installed and cleaned under synchrotron radiation with a positive voltage

Table I Test chamber

Chamber	Material	Surface treatment	History of SR exposure
Aℓ	Aluminum ally (A6063)	Extrusion	21Ahr (3.7×10^{21} photons)
Cu	OFC-Class1 (ASTM)	Machining	76Ahr (1.3×10^{22} photons)
SS	Type 316L SS (Vacuum melted)	Electro-polishing	NO

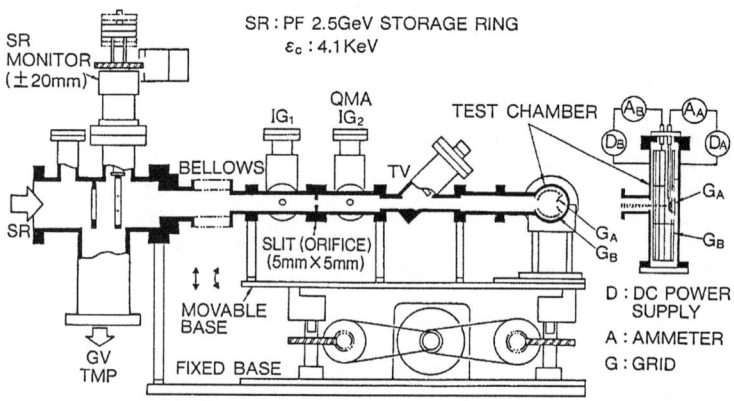

Fig. 1. Experimental setup and test chamber with grid type electrodes.

applied. All test chambers have diameters of 60 mm and are 300 mm long. Irradiated surfaces are cooled by water from outside. Test chamber specifications are shown in Table I. The test chambers made of type A6063 extruded aluminum alloy and high purity oxygen free copper (ASTM Class-1 OFC) had been exposed to synchrotron radiation just once before this experiment. The directly irradiated surfaces were moved 10 mm in the axial direction compared with the previous irradiated surfaces.[5]

RESULTS AND DISCUSSION

A. Desorption yields through experimentation.

Figure 2 shows the photodesorption yields η(molecules/photon) of the three kinds of materials based on nitrogen equivalent total pressures as a function of the beam dose. η is calculated from the outgas which is measured by B-A gauges and the incident photon flux[7]. The designations, Aℓ, SS, and Cu in the figure indicate the yields of aluminum alloy, stainless steel, and oxygen free copper. The designations of A1~A5, S1~S8, and C1~C3 indicate the experiments with bias voltage and, or change in the slit position from the center of synchrotron radiation. The desorption yield of Aℓ is the highest and that of Cu is the lowest under conditions of no bias voltage and no change in the slit position throughout the experiment. SS is a new stainless steel chamber, but its desorption yield is lower than that of the Aℓ chamber. The results reported in this paper were obtained under conditions in which the center of the slit was set in the center of the synchrotron radiation. Therefore, the highest possible photon flux was introduced into the test chamber.

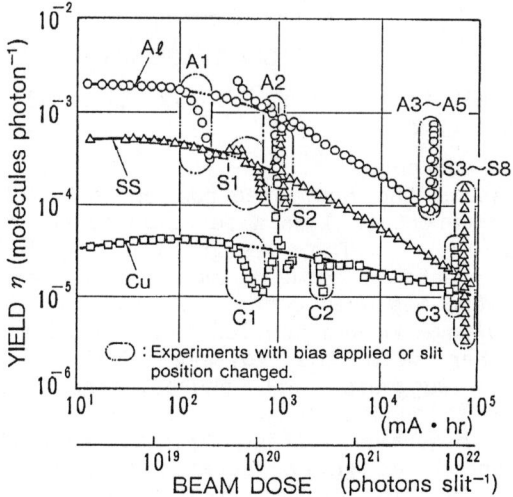

Fig. 2. Photodesorption yield η vs. beam dose. (N_2 equi. total yield)

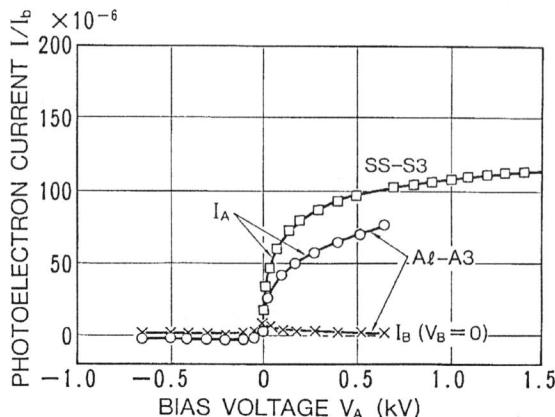

Fig. 3. Photoelectron current I/I_b vs. bias voltage V_A applied to the grid A.

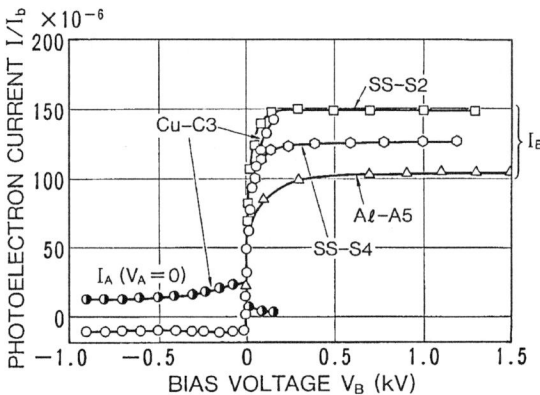

Fig. 4. Photoelectron current I/I_b vs. bias voltage V_B applied to the grid B.

B. Photoelectron currents.

Figure 3 shows the photoelectron currents (I) normalized by the stored beam current (I_b) as a function of the bias voltage (V_A) applied to grid A. The designations I_A and I_B are the photoelectron currents of grid A and grid B, respectively. Both SS-S3 and Aℓ-3 were measured under beam doses of more than 20 Ahr. When positive voltage V_A was applied and gradually increased, I_A/I_b increases together with the voltage rise and I_B/I_b slightly decrease. This suggests that photoelectrons produced on the chamber wall due to photon incidence are attracted by the positive grid A. When negative voltage V_A was applied and gradually increased, I_A and I_B slightly decrease to about -50V. Figure 4 shows I/I_b as a function of bias voltage (V_B) applied to grid B. I_B and I_A with the bias of V_B, behave the same as I_A and I_B of Figure 3, respectively. When positive bias was applied, photoelectron currents become constant at voltage lower than about 300 V. This is different from the photoelectron current I_A of Figure 3 with the bias V_A. It is presumed that the efficiency of trapping photoelectrons of grid B is higher than that of grid A because of the shapes and arrangements of the grids in the test chamber. The photoelectron current of SS-S4 at the beam dose of 80 Ahr is less than that of SS-S2 at the beam dose of 1 Ahr. The reduction in photoelectron current is probably due to the "beam cleaning" effect of exposure to synchrotron radiation. Assuming that most photoelectrons are trapped by grid B at a steady photoelectron current, photoelectron yields can be calculated from the current of grid B and the incident photon flux. For example, the calculated photoelectron yield of the aluminum alloy test chamber $A\ell$-A5 is 0.013. Normal incidence photoelectron yield Y(electrons/photon) of the aluminum oxide is expressed as $Y=5.71\varepsilon^{-0.82}$ (400 eV < ε < 10 ε_c eV),[8] where ε is photon energy and ε_c is the critical energy of synchrotron radiation. By substituting a photoelectron yield of 0.013 in the formula, the photon energy equivalent to the synchrotron radiation used here having critical energy of 4.1 keV is found to be 1.67 keV.

C. Effects of bias voltages on the desorption.

Applying bias voltage to the grids changes the photodesorption from the test chambers. Figure 5 shows an example of the outgas changes. When negative bias voltage V_A was applied and gradually increased, the outgas due to photodesorption decreased. The designation Q indicates desorption with no bias voltage, and ΔQ is a deviation due to the bias voltage from the outgas Q. In this case, Q is an approximate value attained from the values of Q before and after the bias experiment. The sign ΔQ is defined as negative when Q is increased.

Figure 6 and 7 show the ratio of outgas, $(Q-\Delta Q)/Q$, affected

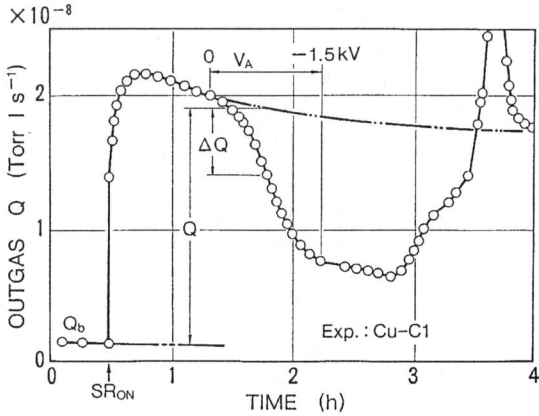

Fig. 5. Changes of outgas due to the bias voltage V_A.

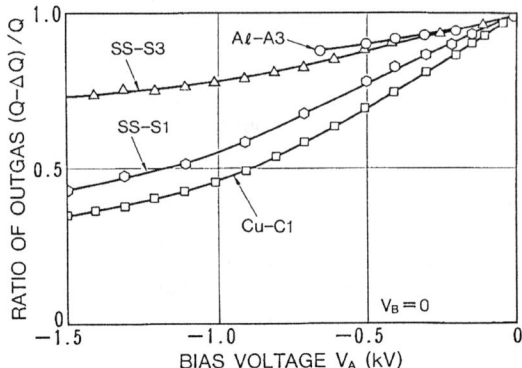

Fig. 6. Ratio of outgas as a function of the bias voltage V_A.

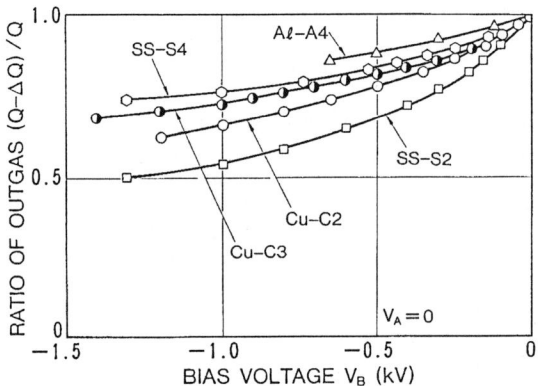

Fig. 7. Ratio of outgas as a function of the bias voltage V_B.

by negative bias voltage. The negative bias voltages V_A or V_B were gradually increased and desorption changes were measured. They are shown in Figure 6 and 7, respectively. In both cases, the ratio of $(Q-\Delta Q)/Q$ decreases with the increase in voltage. The maximum reduction of outgas is more than 50%. The reduction of outgas was probably caused by the orbits of some of the photoelectrons produced on the primary incident surface being changed or turned back by a negative electric field, and then by the photoelectrons hitting surfaces which had then been quickly cleaned up[7] and had low desorption rates. The difference between SS-S1 and SS-S3 in Figure 6 is the beam dose. In both bias voltages, V_A and V_B, $(Q-\Delta Q)/Q$ at a low beam dose is smaller than those at a high beam dose.

Figure 8 shows the desorption in a case of a positive bias voltage V_B being applied. When a positive bias gradually increases, the ratios, $(Q-\Delta Q)/Q$, of the outgas increase as in Figure 8(a). The increase in the desorption is mainly due to the

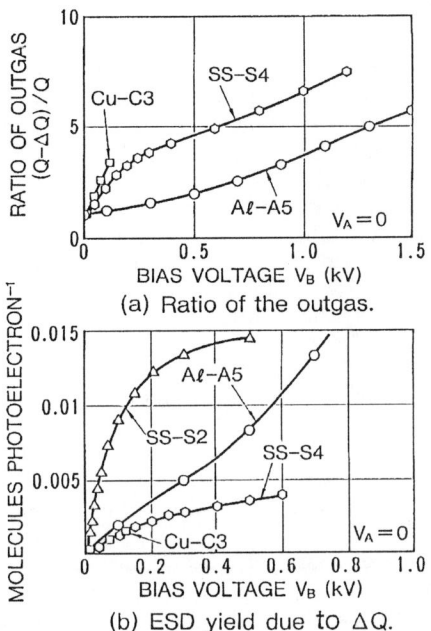

Fig. 8. Outgas and ESD yields as a function of the bias voltage V_B.

electron stimulated desorption(ESD) from grid B. It is because the number and energy of photoelectrons hitting grid B are increased by the electric field. However, it should be remembered that there are some electrons which were subject to orbit changes by the electric field and which hit the surface of the chamber. There are other factors inducing desorption. Photon incidence to the surface induces desorption and photoelectrons also induce desorption from the surface when they are emitted. The former phenomenon is not affected by the electric field. Therefore, the outgas includes both desorption from grid B electrode and from the test chamber surface.

Assuming that most desorption is due to the ESD from grid B, photoelectron desorption yields(molecules/electron) from grid B can be calculated, and they are shown in Figure 8(b). These are calculated from the deviations of the outgas, $|\Delta Q|$ and the currents of grid B. The high yield of SS-S2 is attributed to a low beam dose of 1 Ahr. The yields of Cu-C3 and that of SS-S4 at beam doses of more than 30 Ahr are low and almost equal. From this, the main gas source in SS-S4 and Cu-C3 are supposed to be the grid B which attracted the photoelectrons. Note that Aℓ-A5 shows a high photoelectron desorption yield despite a high beam dose. ESD from grid B is supposed to be dominant in SS-S4 and Cu-C3. However, ESD from the chamber can not be neglected in Aℓ-A5.

CONCLUSION

Effects of photoelectrons on photodesorption were investigated. The following conclusions were drawn from our investigation. (1) Photoelectron current caused by synchrotron radiation depends on the beam dose. The normal incidence photoelectron yield of the aluminum alloy test chamber is 0.013 under conditions of synchrotron radiation having a critical energy of 4.1 keV hitting the chamber with accumulated photons of about 10^{21}. This photoelectron yield is equivalent to that of a photon having a energy of 1.67 keV. (2) Photoelectrons contribute considerably to photodesorption, and the photodesorption can be reduced by electric fields.

REFERENCES

1. O.Gröbner, A.G.Mathewson, H.störi, P.Strubin, and R.Souchet, Vacuum, 33(7), 397 (1983).
2. T.Kobari and H.J.Halama, J. Vac. Sci. Technol., A5(4), 2355 (1987).
3. D.Bintinger, P.Limon, and R.A.Rosenberg, J. Vac. Sci. Technol., A7(1), 59 (1989)
4. H.C.Hseuh and Xiuhua Cui, J. Vac. Sci. Technol., A7(3), 2418 (1989).
5. S.Ueda, M.Matsumoto, T.Kobari, T.Ikeguchi, M.Kobayashi, and Y.Hori, 11th Int. Vacuum Congress, Köln, Sep.25th-Sep.29th (1989).
6. C.L.Foerster, H.Halama, and C.Lanni, J. Vac. Sci. Technol., A8(3), 2856(1990).
7. M.Kobayashi, M.Matsumoto, S.Ueda, J. Vac. Sci. Technol., A5(4), 2417 (1987).
8. O.Gröbner, A.G.Mathewson, P.Strubin, E.Alge, and R.Souchet, J. Vac. Sci. Technol., A7(2), 223 (1989).
9. M.Andritsuchky, O.Gröbner, A.G.Mathewson, F.Schumann, P.Strubin, and R.Souchet, Vacuum, 38(8-10), 933(1988).

GLOW DISCHARGE CLEANING EFFECTS ON ALUMINUM ALLOY BY TDS AND SIMS SURFACE ANALYSIS

G.Y.Hsiung, J.R.Chen*
Synchrotron Radiation Research Center, Hsinchu Science-Based Industrial Park, Hsinchu, Taiwan 30077, Republic of China
*also Institute of Nuclear Science, National Tsing-Hua University, Hsinchu, Taiwan 30043, Republic of China

Y.C.Liu
Department of Physics, National Tsing-Hua University, Hsinchu, Taiwan 30043, Republic of China

ABSTRACT

The glow discharge cleaning (GDC) effects on A6063 aluminum alloy were studied in-situ by the TDS and SIMS surface analysis methods. Four discharge gases, Ar, O_2, N_2 and H_2, were used. The positive and negative SIMS spectra showed that the surface contaminants, eg. C, Na, F and Cl etc., were effectively reduced after the GDC processes. However, the sputter coating effect was observed in the massive gaseous discharges. The mechanism of the H_2 glow discharge cleaning primarily involves chemical reactions, via the combination of hydrogen atoms with surface impurities. From the comparison of TDS spectra before and after the H_2 GDC, it is shown that the hydrogen atoms were retained on the sample surface with an desorption energy of 45.7 kcal/mole.

INTRODUCTION

In the vacuum system design of a synchrotron radiation source, photon-stimulated desorption effect is a dominant concern. The glow discharge cleaning (GDC) method was widely adopted to clean the surface of vacuum chamber wall.[1-4] For a stainless steel vacuum chamber, the contamination on the surface can be effectively reduced by the Ar + O_2 (10%) GDC method, and the deposited Ar gas can also be removed by a later bakeout at a high temperature of ~ 350°C.[5-7] For an aluminum alloy vacuum chamber, however, the maximum bakeout temperature is limited ($\leq$ 150°C) for the sake of mechanical strength. Therefore it is difficult to get rid of the deposited Ar atoms by bakeout. On the other hand, due to the fact that the interaction cross section of hydrogen with the circulating electron beam is much less than that of Ar, it seems that H_2 GDC is more promising than the Ar GDC method for the aluminum alloy vacuum chambers.

In our previous studies on the cleaning effects of GDC, by measuring the gases emitted from the surface of vacuum chamber during the discharge process and by analyzing the Auger electron spectra (AES) on the sample surfaces, there are two cleaning mechanisms on the surface, chemical cleaning and physical sputtering were verified.[8] The physical sputtering cleaning process is dominant for the heavy ions, such as Ar, and the chemical cleaning process is dominant for the reactive discharge gases, such as H_2.

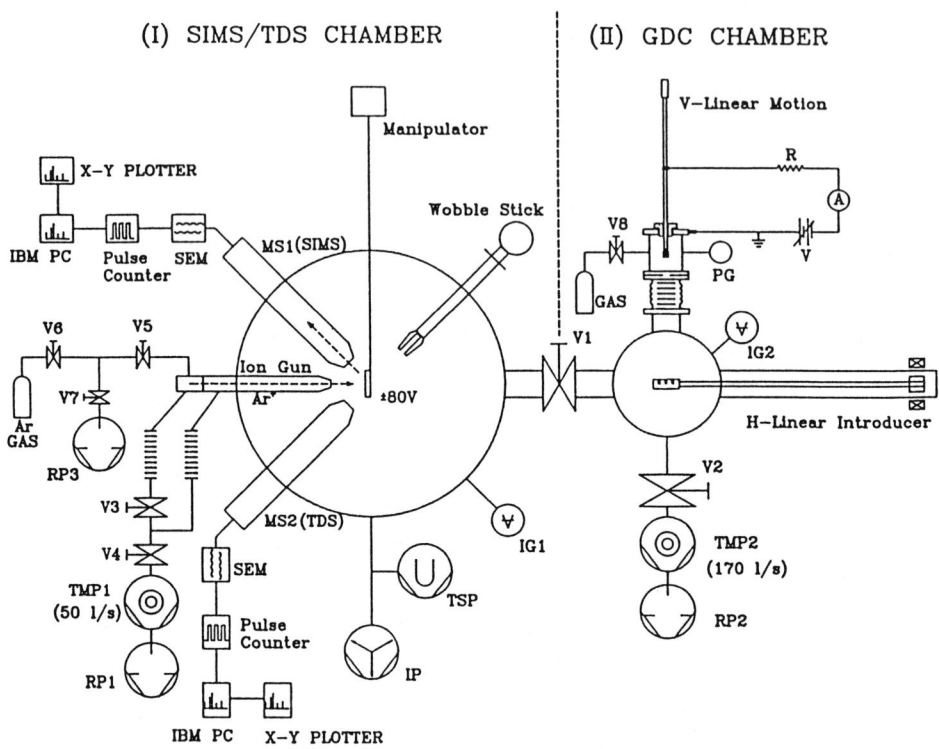

Fig.1 Schematic diagram of the experimental system. MS:quadrupole mass spectrometer; IG:ionization gauge; PG:pirani gauge; IP:ion pump; TMP:turbomolecular pump; RP:rotary pump; TSP:titanium sublimation pump; V(1-8):valve; SEM:secondary electron multiplier.

dominant for the reactive discharge gases, such as H_2.

In this experiment, two in-situ surface analyzing methods, the secondary ion mass spectroscopy (SIMS) and the thermal desorption spectroscopy (TDS) methods, are adopted to study the GDC effect. The experimental procedures and results are described in the following.

EXPERIMENTAL

The experimental system is shown in Fig.1. The system is divided into two parts, one of them is the GDC chamber in which the glow discharge process is performed, and the other is the surface analyzing chamber in which the SIMS and the TDS analyses are performed. The sample for this experiment was A6063 aluminum alloy originally mounted on a holder with vertical linear motion in the GDC chamber. The voltage applied to the sample was in the range of -500V to -900V, which depended on the gas species and the working pressure. The vertical holder was electrically isolated from the grounded GDC chamber - the anode. The discharge current density on he sample surface was ~ 50μA/cm² with a pressure of ~ 0.1mb. The discharge was stopped after an accumulated dosage of ~ 2×10^{18} ions/cm^2. Four kinds of discharge gases, Ar, O_2, N_2 and H_2, were used in this experiment. After the GDC process, the sample was introduced into the surface analyzing vacuum chamber by a horizontal linear introducer without breaking the vacuum. The re-contamination on the sample surface was thus avoided with this arrangement.

In the SIMS analysis, the sample was bombarded by Ar$^+$ ion beam with an energy of 3 keV. The beam current on a raster area of 2mm×2mm was ~ 160nA for the positive secondary ions and ~ 550nA for the negative secondary ions measurements. A bias of +80V (-80V) was applied to the sample for positive (negative) secondary ion detection, which increases the transmission efficiency of the secondary ions. In the TDS experiments, the heating range was from room temperature to 500°C with a heating rate of ~ 3°C/sec. The sample was preheated before the H_2 GDC in order to remove the interfering gases. After the H_2 GDC, the TDS spectrum was performed.

RESULTS AND DISCUSSIONS

The SIMS spectra of positive ions on A6063 aluminum alloy samples before and after the N_2 GDC treatments are shown in Fig.2(a) and (b), respectively. Figure 3(a) and (b) show the results of the

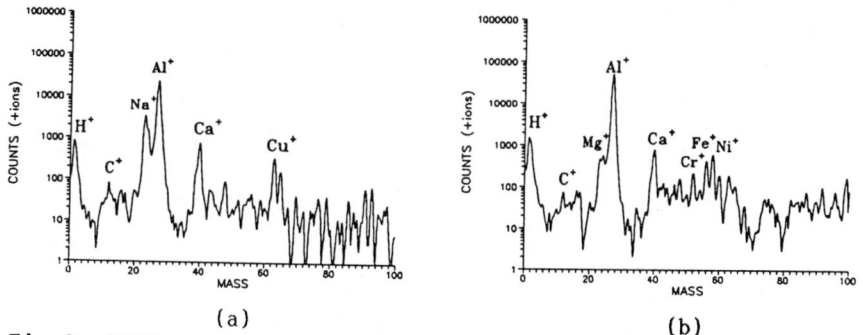

Fig.2 SIMS spectra of positive ions of A6063 aluminum alloy sample: (a) before the N_2 GDC; (b) after the N_2 GDC.

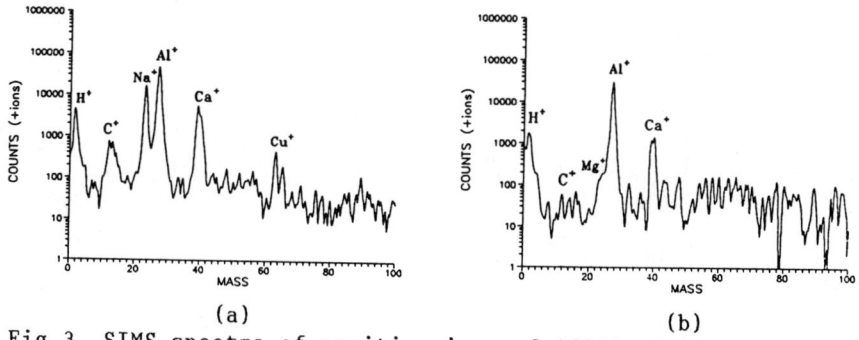

Fig.3 SIMS spectra of positive ions of A6063 aluminum alloy sample: (a) before the H_2 GDC; (b) after the H_2 GDC.

H₂ GDC. The results of Ar and O₂ GDC are similar to that of the N₂ GDC. The peaks of positive secondary ion from the surface of A6063 aluminum alloy before glow discharge treatments show mainly H⁺, C⁺, Na⁺, Al⁺, Ca⁺ and Cu⁺ ions. The Al⁺ signal is obtained from the composition of the Al sample. While the H⁺, Na⁺, Ca⁺ and Cu⁺ are probably the surface contaminants. The spectra in Fig.2(b) and 3(b) show that most of the surface impurities were reduced after the GDC processes. The Cr⁺, Fe⁺ and Ni⁺ peaks were observed in the spectra after the N₂ GDC. These elements were sputtered out of the stainless steel sample holder during the GDC processes. The degree of sputtering deposition is significant of the massive discharge gases in sequence of Ar > O₂ > N₂, while it is minor of the H₂ gas.

Figure 4(a) and (b) show the SIMS spectra of negative ions before and after the N₂ GDC treatment. Figure 5(a) and (b) show the results of the H₂ GDC. Before the GDC proceeded, the H⁻, C⁻, O⁻, F⁻, C₂⁻, O₂⁻, Cl⁻ and AlO⁻ peaks were observed. The O⁻, O₂⁻ and AlO⁻ peaks are comming from the aluminum oxide layer on the sample surface. However, the C⁻, F⁻, C₂⁻ and Cl⁻ peaks are due to the surface contamination. These peaks were reduced after the GDC processes.

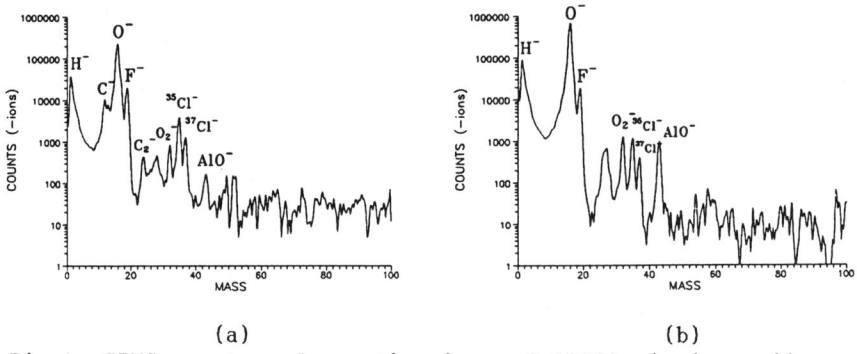

Fig.4 SIMS spectra of negative ions of A6063 aluminum alloy sample: (a) before the N₂ GDC; (b) after the N₂ GDC.

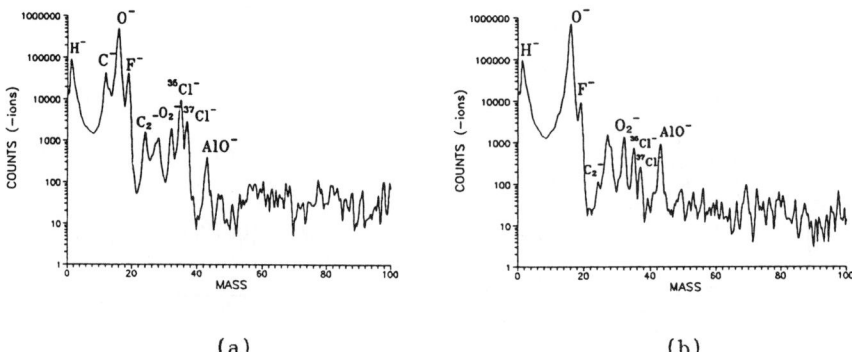

Fig.5 SIMS spectra of negative ions of A6063 aluminum alloy sample: (a) before the H₂ GDC; (b) after the H₂ GDC.

From the SIMS analysis, it can be noted that the contaminants on the sample surface were H, C, F, Na, Cl, Ca, Cu etc. and most of them can be removed effectively by the GDC method. These contaminations are probably from the handling process (hydrocarbons) or the chemical cleaning processes (HF, NaOH, HCl, Ca, etc.).

Due to the difficulty of investigating the discharge gases retained on the surface by the method of the SIMS, the TDS method was applied to measure the retained hydrogen atoms (or molecules) after the H_2 GDC. The TDS spectra are shown in Fig.6. The curves (a) and (b) show the desorbed H_2 gases from the sample before and after the H_2 GDC, respectively. It is observed that a large amount of H_2 molecules is desorbed after H_2 GDC. A broad peak located at ~ 450°C corresponds to an desorption energy of 45.7 kcal/mole.[9]

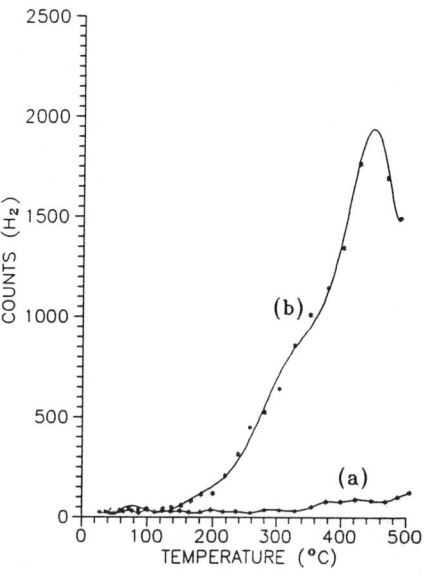

Fig.6 Thermal desorption spectra of H_2 gas: (a) before H_2 GDC; (b) after H_2 GDC. The heating rate is ~ 3°C/sec.

CONCLUSION

Due to the powerful trace-element analyzing cabability of the SIMS techniques on surfaces, the cleaning effects of the glow discharge treatment were clearly revealed. The results of the SIMS spectra, of positive and negative ions, showed that the contaminants on the A6063 aluminum alloy surface, H, C, Na, F, Cl, Ca and Cu, can be effectively removed by the GDC method. In addition, the deposition of Cr, Fe and Ni atoms on the sample surface after heavy ion GDC shows that the physical sputtering effect from the stainless steel components of the sample holder is serious. For the H_2 GDC process, however, the physical sputtering effect is insignificant; the surface contaminants are cleaned via chemical recombination with the hydrogen atoms.

In addition to the SIMS technique, the TDS method was applied to analyze the H_2 retention on the surface after the H_2 GDC. The TDS spectrum shows a large amount of hydrogen molecules desorbed from the surface of aluminum alloy sample after the H_2 GDC process. A broad peak located at ~450°C, corresponding to an desorption energy of 45.7 kcal/mole, was observed. Since the desorption eenergy is high, it seems that the H_2 gas retained on the surface is not easy to be removed by bakeout at the temperature below 150°C.

ACKNOWLEDGEMENTS

The authors wish to thank Mr. H.S.Tzeng for his kindly help in fabricating the vacuum components. A special thank also goes to Mr. W.J.Lin, Miss Y.C.Huang, Mr. K.L.Tsao, Mr. C.J.Nee, Mr. T.Y.Wu and Mr. S.H.Lin for their valuable contributions.

This work is supported by the National Science Council of R.O.C. under grand No. NSC80-0208-M213-01.

REFERENCES

1. R. S. Calder, Vacuum 24, 437 (1974).
2. H. Kitamura, Nucl. Inst. and Method 177, 107 (1980).
3. H. Stori, GERN-ISR-VA/81-28, July 1981.
4. H. C. Hseuh, T. S. Chou, and C. A. Christianson, J. Vac. Sci. Technol. A3(3), 518 (1985).
5. R. Calder, A. Grillot, F. Le Normand and A. Mathewson, CERN-ISR-VA/77-59, 1977.
6. A. G. Mathewson, Xth Italian National Congress on Vacuum Science and Technology, Stresa, Italy, 12-17 Oct. 1987.
7. H. F. Dylla, J. Vac. Sci. Technol. A6(3), 1276 (1988).
8. G. Y. Hsiung, J. R. Chen, C.Y. Tai and Y. C. Liu, American Institute Of Physics Conference Proceedings No.171, 250 (1988).
9. P. A. Redhead, Vacuum 12, 203(1962)

THE EFFECT OF "DIVERSEY" CLEANING ON THE SURFACE COMPOSITION OF 304 STAINLESS STEEL

T.W. Rusch, R.J. Liptak, and W.J. Eberle
Perkin-Elmer Corporation, Physical Electronics Division, Eden Prairie, MN 55344

ABSTRACT

Surface topography and chemical composition have been measured for 304 stainless steel coupons as received from the steel mill, after straight-lining, and during the Diversey DS-9 process used to prepare the surfaces of ultrahigh vacuum (UHV) vessels. Secondary electron images indicate that the acid brightening step of the Diversey process smooths the stainless steel surface by removing about 10 μm of material. Scanning tunneling microscope (STM) images with resolution better than 5 nm show no evidence of micro-porosity. Surface chemical analysis and composition depth profiles using Auger electron spectroscopy (AES) and x-ray photoelectron spectroscopy (XPS) are similar for the different surface conditions. A thin carbon layer is present over a 20-30 Å thick oxide in which Fe, Cr, and Ni are depleted from their bulk concentrations. The surface concentrations of Cr and Ni are higher in the straight-lined and processed samples. Secondary ion mass spectrometry (SIMS) depth profiles show that the Diversey process reduces the amount of C, Na, K, and Ca on the surface and that an oxide layer is formed with a lower concentration of OH.

INTRODUCTION

Vacuum system base pressure is determined by the pumping speed of the system and the rate of gas evolution within the system. A great deal of effort has been expended to develop and understand effective vacuum system surface preparation methods which reduce the outgassing rate so that lower base pressures can be attained with a given pumping speed. Because 300 series stainless steels have been used for the construction of ultrahigh vacuum systems for more than thirty years, a number of studies have been done to characterize the stainless steel surfaces produced by different preparation methods and to correlate these with outgassing rates.

The relationship between outgassing and stainless steel surface characteristics has been studied for surfaces treated by glass bead-blasting,[1] solvent and plasma cleaning,[2,3] electropolishing,[1,3,4,5] electrochemical buffing,[6] alkaline and acid etching.[7,8] Models to explain the beneficial effects of surface treatment involve three characteristics: hydrogen concentration in the material, surface roughness, and surface or near-surface chemical composition.

Assuming that the surface treatment removes gross contamination, the main outgassing constituent is hydrogen present in the material which diffuses to the surface and desorbs unless inhibited by a surface oxide barrier.[1,5] The roles of surface roughness and surface chemical composition are not clear because many of the processes which reduce the roughness also change the chemistry. Young[1] found that roughness changes in the range from 0.1-0.8 μm had no impact on outgassing rate. In addition, different outgassing rates have been observed from surfaces with similar chemical composition[4] and similar outgassing rates have been observed from surfaces with different chemical composition.[8] Existence of an oxide layer with pores ranging in size down to the nanometer scale has been invoked to explain a number of the observations.[5,7,9]

At Perkin-Elmer, Physical Electronics Division, the Diversey Process DS-9, a brightening acid etching process, is used to prepare the surfaces of stainless steel ultrahigh vacuum vessels. This was adopted about 20 years ago based on the beneficial results reported by Milleron who speculated that etching by the Diversey process reduced the outgassing rate by removing a "spongy surface layer" from the stainless steel surface and "leaving the real surface area much more nearly equal to the projected area of the metal".[7] Boschi and co-workers[8] have the only reported study where the outgassing rates of stainless steel treated with the Diversey DS-9 process were related to the surface chemistry. They found that the surface composition and depth distribution of Fe, Cr, Ni, O, and C as measured by Auger electron spectroscopy (AES) were similar for 304, 316L, and 316LN stainless steels. After baking to 350°C for 4 days, the outgassing rates were similar for each material.

This study extends the surface characterization of 304 stainless steel in previous studies in an attempt to address several questions. Does a porous or spongy surface layer exist? If so, is it removed by the Diversey process? Does the Diversey process reduce the surface area? At what point in the process does the surface smoothing occur? Do more subtle differences exist in the near-surface region which have not been observed to date?

STAINLESS STEEL CLEANING PROCEDURE

During the final steps at the steel mill, 304 stainless steel sheet is cold-rolled to near the desired thickness, heated in air, and then etched lightly or "pickled" with a mixture of hydrofluoric acid and nitric acid. After rinsing, the sheet is then rolled to its final thickness under hard chromium-coated rollers. Prior to receipt at Perkin-Elmer, the mill scale on stainless steel sheet is removed by belt sanding with a silicon nitride or aluminum oxide-coated abrasive, a process commonly called "straight-lining". Unfortunately, more details of these processes are not available.

After machining and welding, the stainless steel vacuum vessel is subjected to the fourteen step "Diversey process", which removes weld scale, reduces surface roughness, eliminates surface inclusions and forms a passive layer which has acceptable outgassing characteristics according to the manufacturer.[10]

The process steps are listed below.
1. Wash in detergent for 15-20 minutes.
2. Rinse in water
3. Immerse in oxidizing solution (Diversey Diverscale® 299*) for 45-60 minutes. The active ingredients are sodium hydroxide and potassium permanganate.
4. Rinse in water.
5. Immerse in hydrochloric acid de-scaler solution (Diversey Everite II*) for 10-15 seconds.
6. Rinse in water.

---

* This product is a proprietary formulation of the Diversey Wyandotte Corporation, Wyandotte, Michigan.

7. Immerse in chemical polish solution (Diversey DS 9-314*) for 5-20 minutes. The active ingredients are nitric, phosphoric, and hydrochloric acids.
8. Rinse in water.
9. Immerse in nitric acid bath for 30 minutes.
10. Rinse in water.
11. Immerse in ammonia neutralizer solution for 20 minutes.
12. Rinse in water.
13. Rinse in hot deionized water.
14. Dry in air.

EXPERIMENTAL RESULTS AND DISCUSSION

In order to more completely characterize 304 stainless steel cleaned by the Diversey process, about 125 coupons were processed in five batches. Coupons from each batch were removed after several of the rinse steps to monitor the evolution of the surface topography and composition. In addition, coupons from sheet as received from the mill and after straight-lining were characterized following degreasing in a heated vapor of trichloro-trifluoroethane and isopropyl alcohol. Preliminary micrographs and composition depth profiles showed that the surface topography and elemental depth distributions were the same on 3 mm thick sheet used to fabricate chambers as on thin stainless steel sheet, so 1 cm x 1 cm x 0.5 mm coupons were studied because they allowed more convenient handling in each of the analytical instruments.

SURFACE TOPOGRAPHY

Secondary electron micrographs were taken using a PHI Model 660 Scanning Auger Microprobe (SAM) on coupons as received from the mill, after straight-lining, and after process steps 4, 6, 8, 12, and 14. The secondary electron images indicated that major topographical changes occurred only during straight-lining and during chemical polishing (step 7). Because AES data indicated that significant surface chemical changes occurred only after straight-lining, after oxidation (step 3), and after chemical polishing (step 7), the study concentrated on the material as received from the mill, after straight-lining, and after the process was complete (step 14).

Micrographs at these three stages in the process are shown in Figure 1. The sheet as received from the mill has an "elephant hide" texture (Figure 1a) which is seen to be a surface scale with significant grain definition when viewed at 10x higher magnification (Figure 1b). This texture, produced by etching of grain boundaries during pickling, is very similar to that observed by Watanabe, et al [3] and Kato, et al [6] and could represent the "spongy surface layer" referred to by Milleron.[7]

Straight-lining produced a surface dominated by parallel abrasion marks but which still exhibited some areas with pits (Figures 1c and 1d). Identical structure has been observed after belt-polishing and buff-polishing.[5] The abrasion process did fold over metal to produce pockets and expose fissures which could trap contaminants so straight-lined material would not be suitable for UHV applications without further processing.

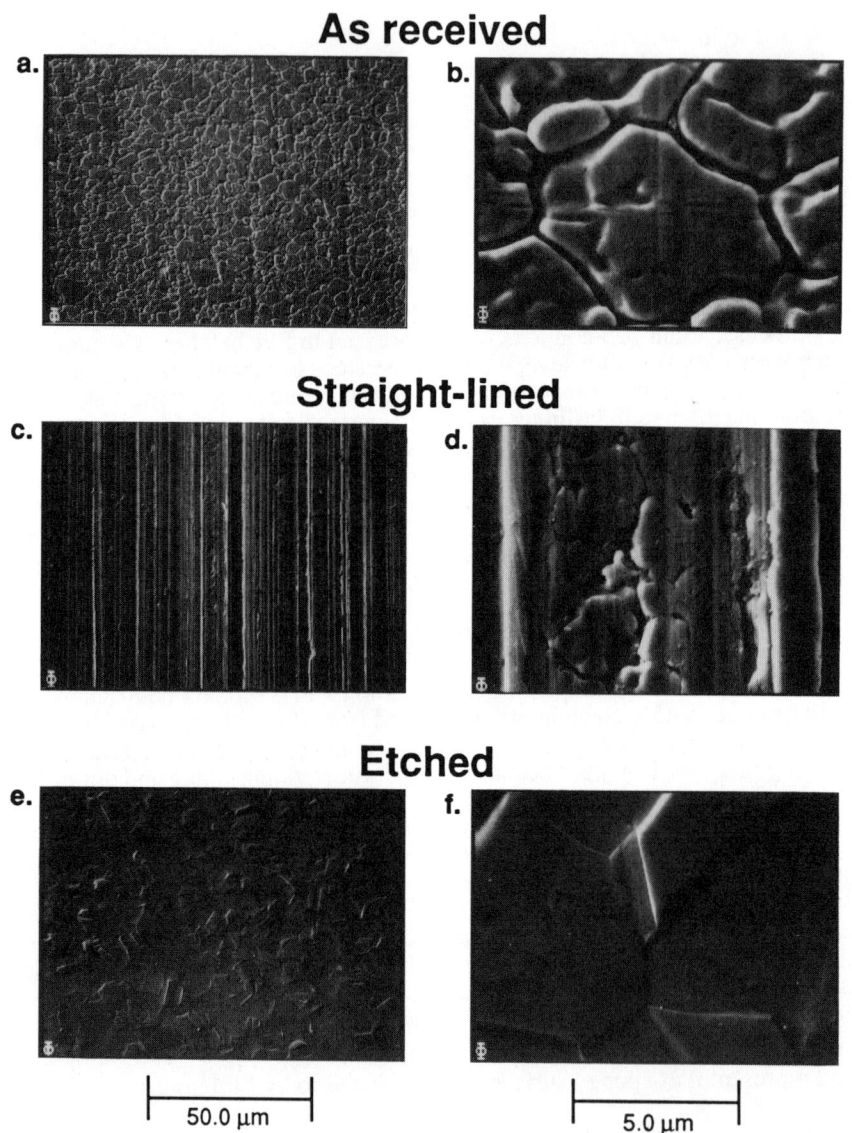

Figure 1. Secondary electron images of 304 stainless steel coupons as received from the mill (a and b), after straight-lining (c and d), and after Diversey processing (e and f). The original images were taken at magnifications of 500x and 5000x with a beam energy of 20 keV.

The chemical polishing in step 7 removed approximately 10 μm of material [10] and exposed grains of the base stainless steel. Subsequent immersion in nitric acid and ammonia with the corresponding rinses in steps 9 through 13 did not change the surface topography. Figures 1d and 1e are micrographs taken after step 14 which show no evidence of porosity or straight-lining texture. Micrographs were also taken after processing sheet as received from the mill with the omission of straight-lining. All evidence of the "elephant hide" texture was removed but a moderate density of pits existed which are thought to result from accelerated etching through pores in the pickled layer.

Micrographs at 50,000x, 10x higher magnification than Figures 1b,1d, and 1e, showed that the surfaces after each of the preparation steps were similar and relatively smooth on the scale of 0.1 μm. To look for differences in micro-porosity, higher spatial resolution images were obtained with a Digital Instruments Nanoscope II Scanning Tunneling Microscope (STM). The measurements were made in air using Pt-Ir tips fabricated by Microanalytical Services. Because of the thin oxide layer on the samples, images were difficult to obtain unless a tip-to-sample bias of at least 2 v was used.

STM images were taken on 10 μm x 10 μm areas of coupons as received from the mill, after straight-lining, and after step 14 (Figures 2a, 2c, and 2e, respectively) to establish a comparison with the secondary electron images in Figure 1. Then the magnification was increased by 100x to examine 100 nm x 100 nm areas on each coupon (Figure 2b,2d, and 2e). As is apparent, the micro-roughness is on the order of 5 nm, independent of the processing stage. No evidence of micro-pores was observed even when images were obtained with higher magnification. Although, as noted by Denley, on a rough surface, the tip contour can prevent the observation of features with dimensions comparable to the tip radius.[11] One additional possibility is that STM images could not be obtained on areas with a porous oxide due to low surface conductivity. To preclude this, atomic force microscopy should be used in future studies.

In summary, secondary electron images indicate that the acid brightening step of the Diversey process smooths the stainless steel surface by removing about 10 μm of material with distinct grain seperation caused by the milling process and ridges caused by straight-lining. STM images with resolution better than 5 nm show no evidence of micro-porosity.

SURFACE COMPOSITION

The surface composition and elemental depth profiles were characterized by three analytical techniques: Auger electron spectroscopy (AES) using a PHI Model 660 SAM, by x-ray photoelectron spectroscopy (XPS) and secondary ion mass spectrometry (SIMS) using a PHI Model 5500 Multi-technique System. AES spectra were measured on 100 μm x 80 μm areas of the coupons after being degreased and after step 14. The surface composition results obtained with a 10 keV electron beam with 1 μA of beam current are presented in Table 1.

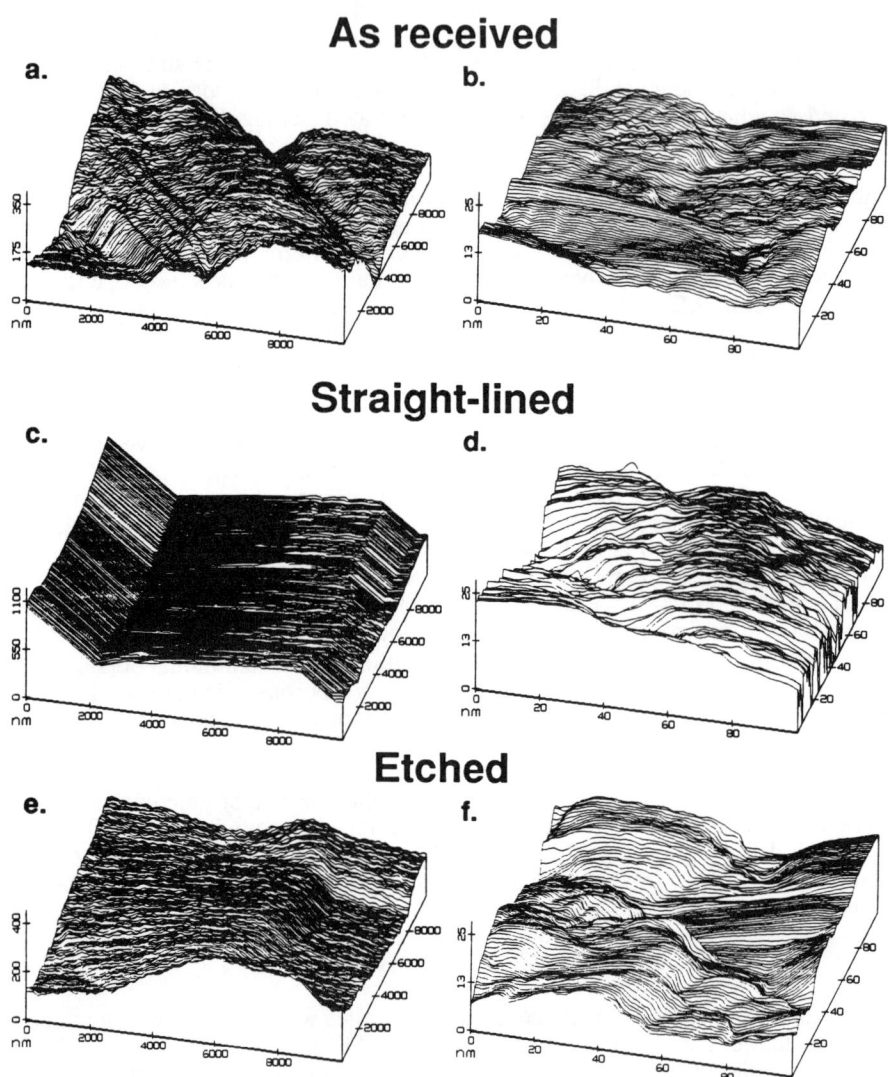

Figure 2. STM images of 304 stainless steel coupons as received from the mill (a and b), after straight-lining (c and d), and after Diversey processing (e and f). Images were taken with a tip-to-sample bias of 2 v.

Table 1. Surface composition in atomic % from AES after different treatments.

Treatment	C	O	Fe	Cr	Ni	Ca	P	N
As received, degrease ("elephant-hide")	36.9 ±1.6	28.9 ±1.7	13.1 ±1.0	1.2 ±0.1	1.4 ±0.2	13.4 ±0.5	-	0.7 ±0.1
Straight-lined, degrease	34.4	38.3	17.4	1.0	1.3	0.6	-	1.1
After step 14 ("etched")	13.6 ±1.2	50.2 ±0.5	16.8 ±0.5	11.1 ±0.3	4.3 ±0.2	1.2	1.3 ±0.4	1.3 ±0.1

From these results, it is clear that the Diversey process substantially increases the amount of chromium and nickel in the stainless steel surface region. Carbon has been reduced by a factor of 2.5, oxygen has increased, iron has remained relatively constant, but chromium has increased by a factor of 10 and nickel has increased by a factor of three. These trends are consistent with the observations of Yoshimura, et al for 304 stainless steel before and after electropolishing.[5] No comparison could be made with the results of Boschi, et al because they did not publish data on the surface composition prior to Diversey cleaning.[8]

COMPOSITION DEPTH PROFILES

Composition depth profiles were obtained in the near-surface region using AES with 4 keV argon ion sputtering. In order to have adequate depth resolution, AES data was taken after sputtering for 0.2 minute intervals with a nominal sputtering rate of 45 Å per minute (referred to the sputtering rate of tantalum pentoxide). The profile data was processed using the principle component analysis algorithm described by Watson in order to reduce the effects of spectral noise and the presence of overlapping peaks.[12] As shown in Figure 3, the depth profiles for major constituents are very similar. A thin carbon layer is present over a 20-30 Å thick oxide in which Fe, Cr, and Ni are depleted from their bulk concentrations. This is consistent with previous AES profiles on treated stainless steels.[6, 8,13,14] It should be noted that the profile in Figure 3c is the average of ten profiles taken on samples after steps 8, 12, and 14 which showed very few differences. This was done to improve the precision of the profile.

The composition scale is expanded in Figure 4 so the C, O, Cr and Ni profiles can be examined more carefully. Comparison of Figure 4a with Figures 4b and 4c, shows that the amount of surface carbon is greater for the material as received from the mill. In addition, the carbon concentration remains higher deeper into the sample. This could be ascribed to profile broadening due to surface roughness, except that the depth distribution of oxygen is comparable in all three figures. The surface concentrations of Cr and Ni are also higher in the straight-lined and processed samples. As noted in earlier work,[12,13] the Cr concentration drops slightly at the oxide-substrate interface and the Ni concentration is enhanced.

Depth profiles using x-ray photoelectron spectroscopy (XPS) were performed to further characterize the treated surfaces. Photoelectron line shifts were used to establish that Cr oxide was the predominant oxide constituent with very little Fe oxide

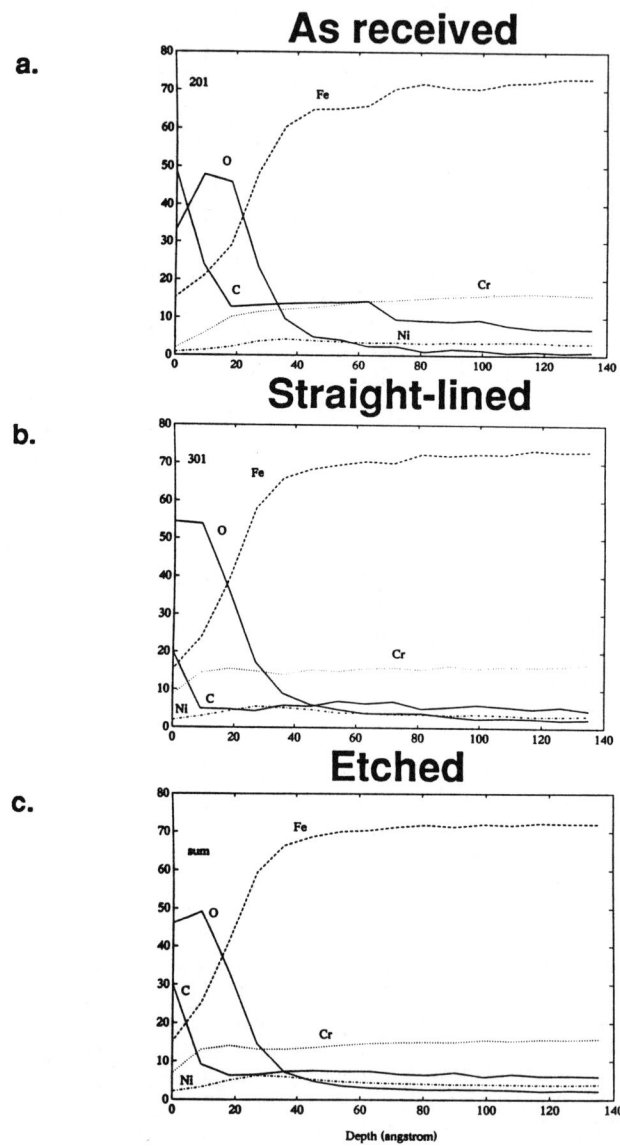

Figure 3. AES depth profiles of 304 stainless steel coupons as received from the mill (a), after straight-lining (b), and after Diversey processing (c).

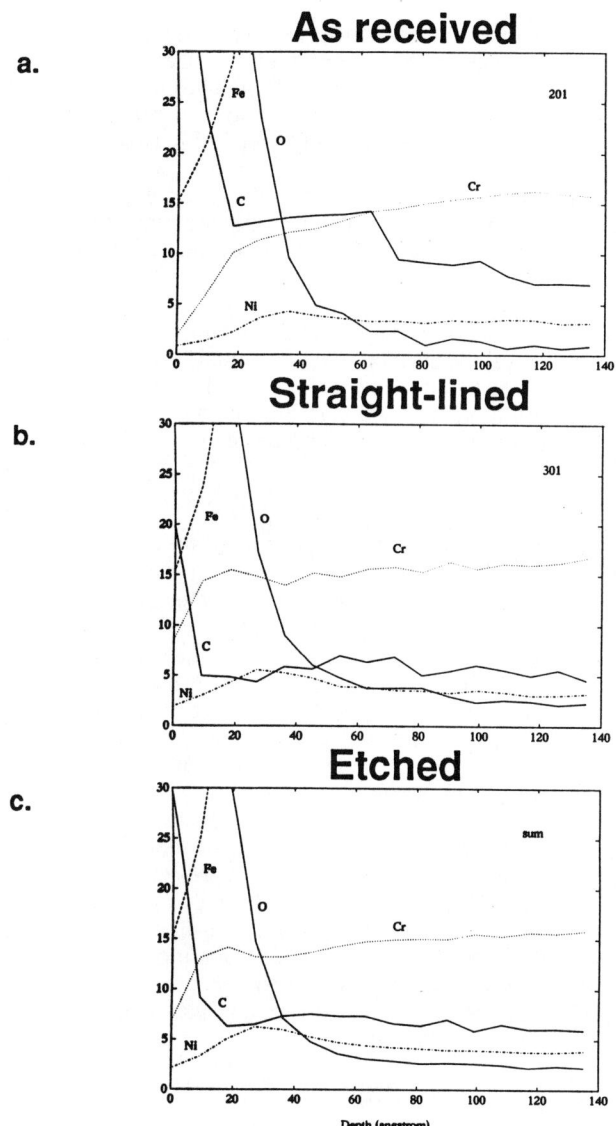

Figure 4. Expanded AES depth profiles of 304 stainless steel coupons as received from the mill (a), after straight-lining (b), and after Diversey processing (c).

and no observable Ni oxide. This latter observation is consistent with the work of Elfstrom and Olefjord.[12]

The AES and XPS depth profiles did not appear to have differences sufficient to explain the radically different outgassing behavior which has been observed for 304 stainless steel with this range of surface conditions. High depth resolution SIMS profiles were measured using 3 keV argon ions to search for other notable chemical differences. Figure 5 presents the raw data for C, O, Fe, and K (masses 12, 16, 56, and 39, respectively) as a function of sputtering time. Figure 6 presents the raw data for Cr, Ni, Na, Ca, and OH (masses 52, 58, 23, 40, and 17, respectively) as a function of sputtering time. Note that the signal amplitudes are plotted on a logarithmic scale. The nominal sputtering rate was 10Å per minute referred to the sputtering rate for silicon.

Comparison of Figure 5c with Figures 5a and 5b shows that the Diversey process reduces the amount of C and K on the surface. The Fe signal drops off with the oxygen signal due to the enhanced positive ion yields in oxides relative to metals. The apparent increased concentration of O and C in Figure 5b may be due to surface roughness effects if the straight-line direction was orthogonal to the ion beam direction.

Comparison of Figure 6c with Figures 6a and 6b shows that the Diversey process reduces the amount of Na and Ca on the surface. The higher Cr signal near the beginning of the profile in Figure 6c and the more rapid signal reduction with depth suggests that a greater fraction of the oxygen is reacted with chromium after processing or that the oxide may be more dense. Another very prominent difference in the profiles is that the OH signal drops off very rapidly in the sample after Diversey processing. This suggests that a mixed oxide-hydroxide layer may form on the stainless steel surface during processing at the mill or reaction in air during straight-lining. This layer may have a higher permeability to hydrogen or may be an additional source of hydrogen which could account for some of the increased outgassing of stainless steel prior to treatment.

In summary, AES and XPS depth profiles for major constituents are very similar for the different surface conditions. A thin carbon layer is present over a 20-30 Å thick oxide in which Fe, Cr, and Ni are depleted from their bulk concentrations. The surface concentrations of Cr and Ni are higher in the straight-lined and processed samples. At the oxide-substrate interface, the Cr concentration drops slightly and the Ni concentration is enhanced. SIMS depth profiles show that the Diversey process reduces the amount of C, Na, K, and Ca on the surface and that an oxide layer is formed with a lower concentration of OH.

CONCLUSIONS

Secondary electron images indicate that the acid brightening step of the Diversey process smooths the stainless steel surface by removing about 10 μm of material with distinct grain seperation caused by the milling process and ridges caused by straight-lining. STM images with resolution better than 5 nm show no evidence of micro-porosity. AES and XPS depth profiles for major constituents are very similar for the different surface conditions. A thin carbon layer is present over a 20-30 Å thick oxide in which Fe, Cr, and Ni are depleted from their bulk concentrations. The surface concentrations of Cr and Ni are higher in the straight-lined and processed

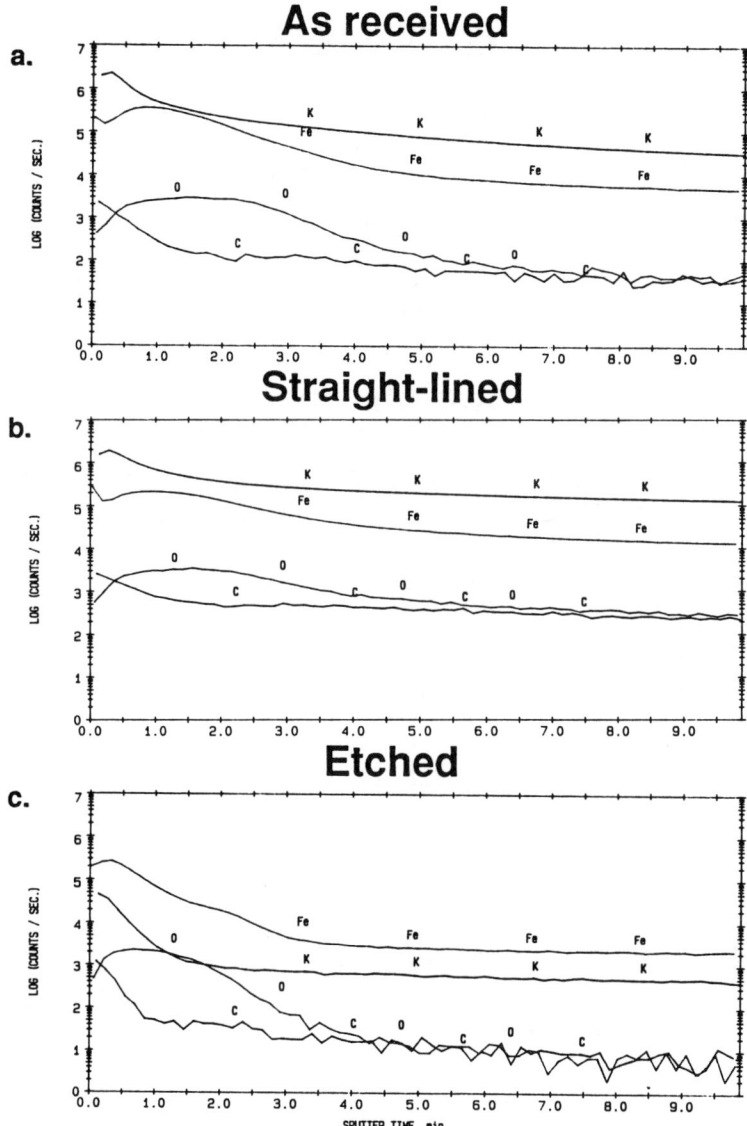

Figure 5. SIMS depth profiles of C, O, Fe, and K from 304 stainless steel coupons as received from the mill (a), after straight-lining (b), and after Diversey processing (c).

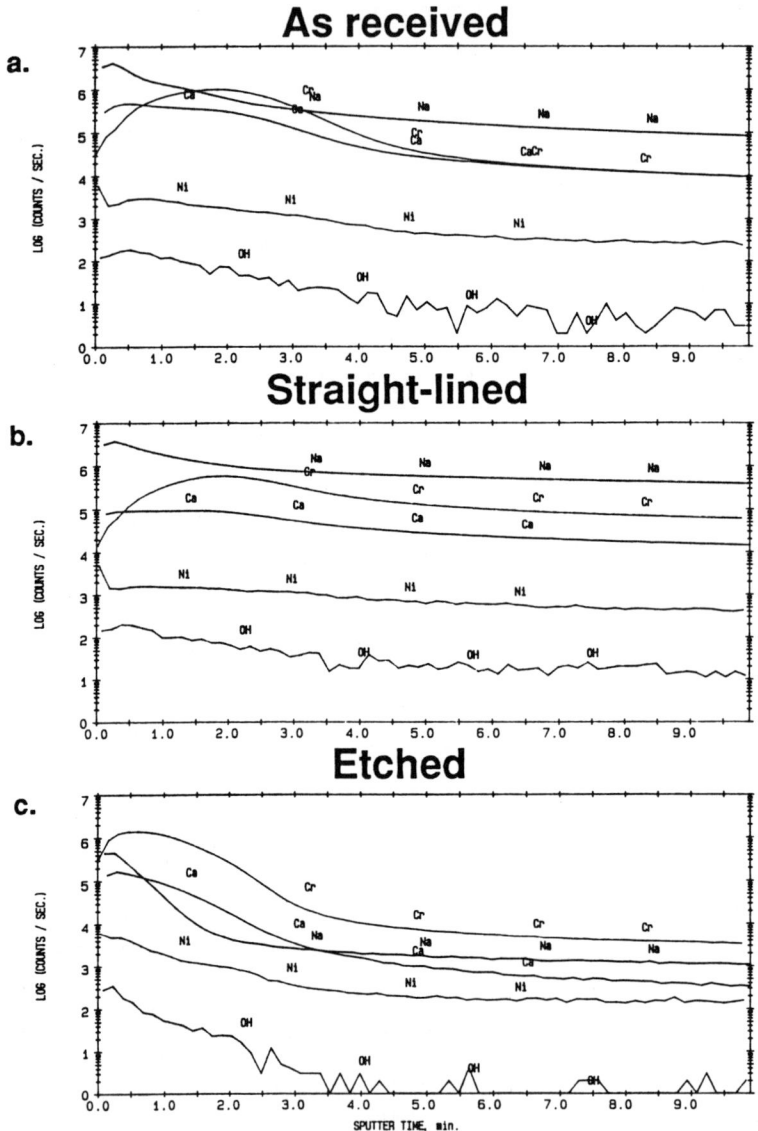

Figure 6. SIMS depth profiles of Cr, Ni, Na, Ca, and OH from 304 stainless steel coupons as received from the mill (a), after straight-lining (b), and after Diversey processing (c).

samples. At the oxide-substrate interface, the Cr concentration drops slightly and the Ni concentration is enhanced. SIMS depth profiles show that the Diversey process reduces the amount of C, Na, K, and Ca on the surface and that an oxide layer is formed with a lower concentration of OH.

ACKNOWLEDGEMENTS

The authors thank Brad Carlson, John Moulder, Peter Sobol, Masami Taguchi, and Robert Youngman of Perkin-Elmer for assistance with the sample preparation, XPS, STM and SIMS measurements.

REFERENCES

1. J.R. Young, J. Vac. Sci. Technol. 6, 398 (1969).
2. A.G. Mathewson, Vacuum 24, 505 (1974).
3. K. Watanabe, S. Maeda, T. Yamashina, and A.G. Mathewson, J. Nucl. Mater 93&94, 679 (1980).
4. K. Odaka, Y. Ishikawa, and M. Furuse, J. Vac. Sci. Technol. A 5, 2902 (1987).
5. N. Yoshimura, T. Sato, S. Adachi, and T. Kanazawa, J. Vac. Sci. Technol. A 8, 924 (1990).
6. S. Kato, M. Aono, K. Sato, and Y. Baba, J. Vac. Sci. Technol. A 8, 2860 (1990).
7. N. Milleron, IEEE Trans. Nucl. Sci. NS-14, 794 (1967).
8. A. Boschi, C. Ferro, G. Luzzi, and L. Papagno J. Vac. Sci. Technol. 16, 1037 (1979).
9. M-H. Achard, R. Calder, A. Mathewson, Vacuum 29, 53 (1979).
10. Diversey Wyandotte Technical and Operating Data Sheet, IM-34-1 (1985).
11. D.R. Denley, J. Vac. Sci. Technol. A 8, 603 (1990).
12. D.G. Watson, Surf. Interface Anal. 15, 516 (1990).
13. B.O. Elfstrom and I. Olefjord, Phys. Scr. 16, 436 (1977).
14. M. Seo and N. Sato, Trans. Jpn. Inst. Met. 21, 805 (1980).

PERFORMANCE CHARACTERISTICS OF LUMPED NEG PUMP

S.R.In*, S.Yokouchi, S.H.Be
RIKEN-JAERI SPring-8 Project Team
2-1,Hirosawa,Wako-shi,Saitama,351-01,Japan

T.Maruyama
Hachiohji Factory, Osaka Vacuum Ltd.,
1221,Kunugida-cho,Hachiohji 193,Japan

ABSTRACT

A prototype of the lumped NEG pump(LNP) which will be installed at the crotch of the SPring-8 storage ring was manufactured, and various vacuum performances and operating characteristics have been researched. Typical pumping speeds for H_2, CO, N_2, and CO_2, measured in a standard test chamber set on the top of the LNP, were 3000 l/s, 1700 l/s, 1000 l/s, and 1300 l/s respectively. We examined the impurity effect on the pumping speeds for H_2 and CO using the mixed gas and alternate injection scheme. The pumping speed of H_2 was considerably influenced by the CO molecules, while that of CO was almost unchanged even at high concentration of H_2. We made a general study of the equilibrium characteristics of the LNP in the activation period to provide the basic materials for the design of the supplement pumping system, and to make a pertinent operation schedule of the LNP. Some miscellaneous operating characteristics such as power consumption of the NEG modules and temperature rise of the adjacent chambers including the Al-alloy pump housing during the activation of NEG were also investigated.

INTRODUCTION

The vacuum system of SPring-8 is designed to utilize crotches and absorbers to intercept almost all synchrotron radiation produced by bending magnets. To protect other vacuum components from severe inrush of the photo-desorbed particles produced in the crotches and absorbers, and to maintain the ultimate pressure at a desired level, a proper pumping system which has sufficient pumping speed and capacity must be prepared. A Lumped NEG Pump(LNP) was manufactured and tested to examine such requirement.

The base pressure of the LNP side chamber was usually in the medium of 10^{-11} Torr range. However chamber pressure increased even up to 10^{-5} Torr range during acivation of NEG because of the large sorption-capacity of the LNP for hydrogen. Long activation period is anticipated to attain a low hydrogen sorption level considering a limited pumping speed in the practical machine. In order to design a supplement pumping system appropriately and make a reasonable operation schedule for the vacuum system employing non-evaporable getter(NEG) as a main pump, we need detailed information on the equilibrium characteristics of NEG material especially during the activation.

Three types of NEG ,i.e., strip, wafer module, and lumped mod-

* On leave from KAERI,Dajeon,Korea

ules, have distinct pumping charateristics even though they are made of the same material. If we know geometrical influences of the NEG cofigurations, we can estimate accurately the pumping speed of specified pump structure with the known sticking coefficient as an universal parameter. Distributed pump model and Monte Carlo simulation have been used to solve such problem.

In the following sections we present the results of the pumping speed measurements, and compare the observed values with those of Monte Carlo simulations of the LNP. Impurity effect on the pumping speed for H_2 and several operation characteristics of the LNP are also discussed.

PERFORMANCE TEST OF LNP

The proto-type LNP[1] consists of 14 St 707 NEG wafer modules installed concentrically in the cylindrical housing made of Al-alloy. Figure 1 is a schematic diagram of the exterior view and inner structure of the LNP. Inner virtual surface made by 14 wafer modules forms a strong adsorbing cylindrical channel of which diameter and length are all 250 mm. Diameter of the pumping port is 200 mm, and total weight of the LNP is about 45 Kg_f. Figure 2 shows the experimental setup for measuring the pumping speed and investigating the operational conditions of the LNP. The whole vacuum chamber is a composite system of Al-alloy LNP housing and the test chamber made of stainless steel assembled with an Al gasket.

Figure 3 shows the typical pumping speeds of the LNP for H_2, CO, N_2, and CO_2. CO is more reactive, and has the higher sticking coefficient for NEG than H_2. However the pumping speed of H_2 is much higher than that of CO in the LNP structure because NEG strip is folded into a highly integrated form in the LNP, and pumping speed depends strongly on the pumping channel conductance as well as the sticking coefficient. In the measurement of the pumping speed for CO_2 we have always experienced CO generation. The measured CO generation rates of the pressure gauge filaments considered as the main sources of spurious reactions[2] were max. 0.3 l/s for quadrupole mass analyser(QMA) and 0.5 l/s for BA gauge. Because the effective pumping speed for CO is a few 10 l/s in the LNP test chamber, CO concentration in the working gas is relatively small and measured pumping speed of CO_2 can be treated as a reasonable one. In Fig. 4 the calculated pumping speeds of the LNP for H_2 and CO using the data obtained with 1 NEG wafer module are compared with the observed pumping speeds of the LNP. In this cal-

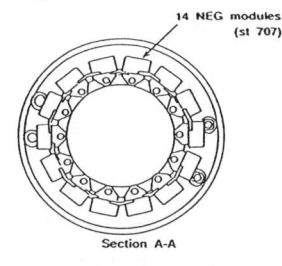

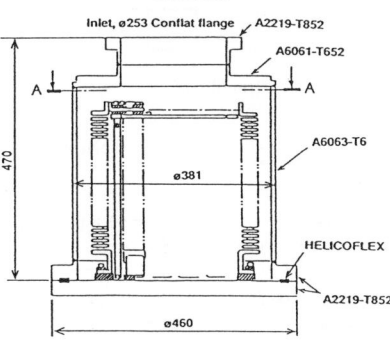

Fig.1. Schematic drawing of the LNP.

culation, we took into account the distributed characteristics of the LNP pumping channel, and the conductance to the pump entrance. Figure 4 shows slight but allowable discrepancies between the distributed pump model and the actual LNP structure.

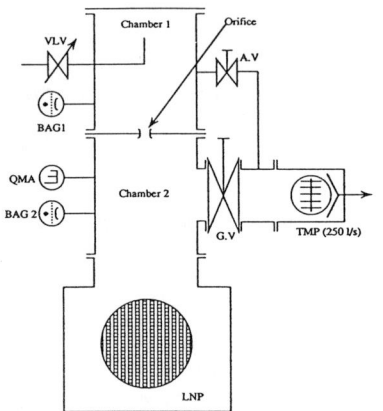

Fig.2. Experimental setup for the test of LNP.

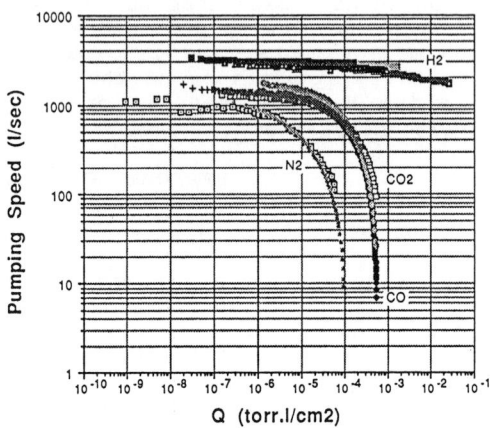

Fig.3. Pumping speed of LNP for H_2, CO, CO_2, and N_2.

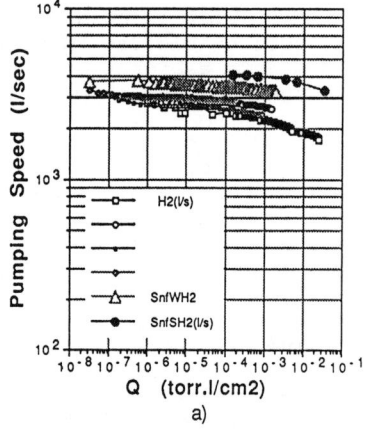

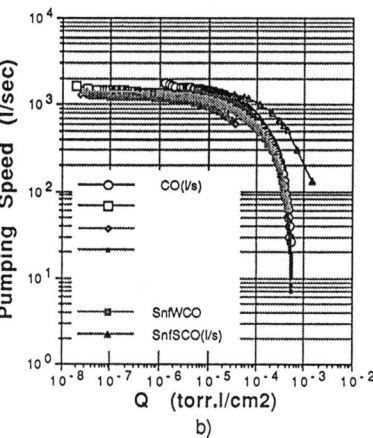

Fig.4. Comparison of a) calculated and b) observed pumping speeds of LNP.

Maximum temperature of the Al-alloy housing was controlled easily just below 150°C even without water cooling. This scheme of NEG activation had additional benefit that pump housing and adjacent chamber could be baked out during NEG activation. The influence of desorbed gas molecules from baked wall on the resultant pumping speed of the LNP after activation was tolerable. Figure 5 shows the temperature variations at several positions in the LNP and the test chamber during activation of NEG. In the typical control

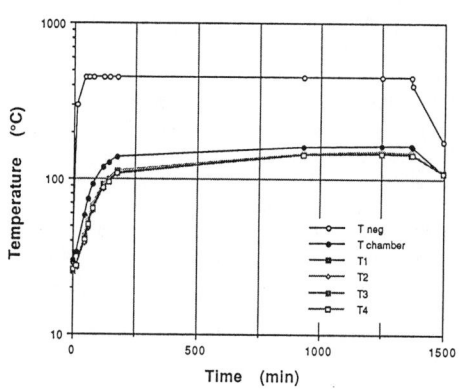

mode for NEG temperature, power consumption of the LNP was about 1.1 KW at the temperature rise stage, and about 700W at the flat-top temperature of 450°C.

Fig.5. Temperature variations in the LNP test chamber.

IMPURITY EFFECT ON THE PUMPING SPEED FOR H_2 AND CO

Most pumping speed data have been measured with high purity gas, which is somewhat different from the actual circumstances. To simulate practically expected atmosphere surrounding the NEG modules in the crotch, we made use of mixed gas in the measurement of the pumping speeds for H_2 and CO with 1 NEG module. The composition of the mixed gas is $80H_2 - 20CO$ in volume percent. We also examined the impurity effect by injecting H_2 and CO alternately and measuring the pumping speeds in the LNP. Figure 6 shows the pumping speed of H_2 and CO measured in the 1 NEG module experiment using mixed gas. In this figure the pumping speeds obtained with pure gas are also shown for comparison. The pumping speed of NEG for H_2 in mixed gas diminishes considerably comparing with that for pure H_2 as the sorption amount of CO increases. However the pumping speed curve

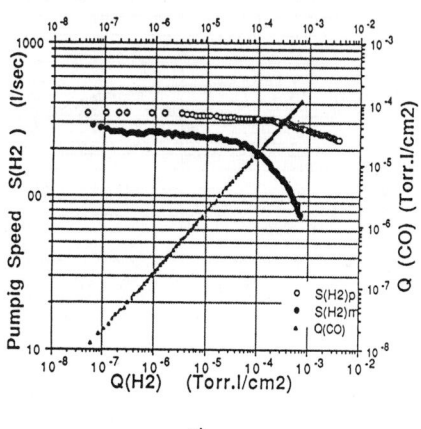
a)

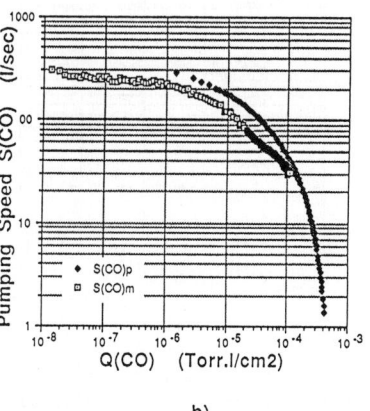
b)

Fig.6. Pumping speeds for a) H_2 and b) CO in mixed and pure gases.

for CO is almost unchanged regardless of the high H_2 concentration in the working gas. This indicates that the pumping capability of NEG for CO is attributed mainly to sticking of CO molecules on the surface of NEG, but that for H_2 is more sensitive to the diffusion of H atoms into the bulk of NEG. In Fig. 7 the pumping speeds of the LNP and 1 NEG module for H_2 in the mixed gas environment, normalized to those for pure H_2, as a function of the sorption amount of CO are shown. The pumping speed ratio for the LNP is calculated by using the mixed gas experiment data of 1 NEG module. Although this experiment was performed under an artificial condition which is still somewhat different from the actual one, with the profiles of two curves we can easily recognize and expect that the pumping capability of NEG for H_2 in the atmosphere of the practical vacuum system will be steeply depleted.

Figure 8 shows two graphs of the pumping speed curves for H_2 and CO obtained without NEG activations during the alternate injection of two gases in the LNP. In this figure upper curves are the normal pumping speed curves. Initial sorption density of CO was about 1×10^{-4} Torr.l/cm^2. In Fig. 8 a) the pumping speeds for H_2 without activation is about 30% of the normal value. This agrees well with that of Figure 7. As shown

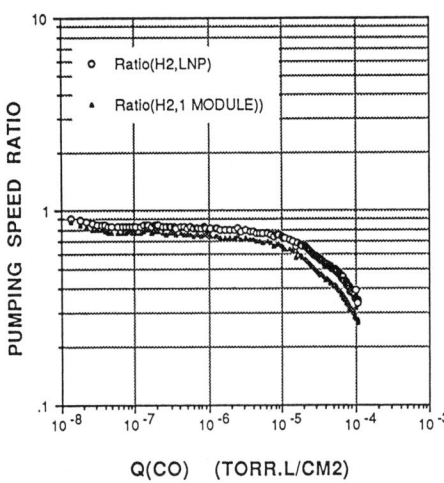

Fig.7. Ratios of H_2 pumping speeds in mixed gas to those in pure gas for LNP and 1 NEG module.

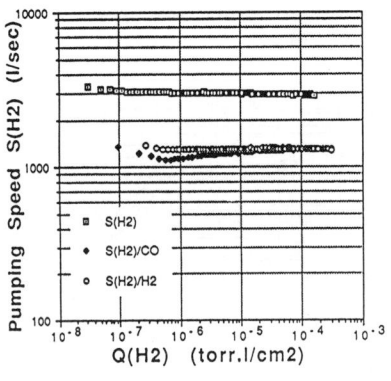

a)

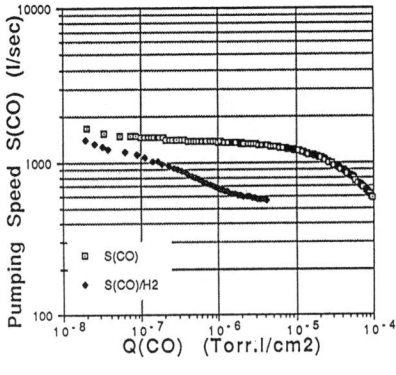

b)

Fig.8. Pumping speeds for a) H_2 and b) CO obtained by alternate injection, and comparison with the normal values.

in Fig. 8 b) the pumping speed for CO obtained without activation after H_2 injection becomes asymptotically the end value of previous CO experiment in spite of the additional sorption of H_2 between two CO experiments. This trend is consistent with that of Fig. 6.

MONTE CARLO CALCULATION OF THE PUMPING SPEED OF LNP

Pumping speed is expressed as a multiplication of the pumping probability at the top plane and the orifice conductance of the pump entrance port. Pumping probability depends on the sticking coefficient of NEG and the practical geometry of the whole pump system. Monte Carlo method is an effective tool of calculating the pumping probability. In the Monte Carlo program diffuse reflection and sticking probability are treated as the random processes. Pumping probability is obtained as a ratio of the total counts of absorption to the total counts of incidence of test particles.

Figure 9 shows the curves of the calculated pumping speed of the LNP as a function of sticking coefficient for H_2 and CO. By comparing with typical experiment data of pumping speed, sticking coefficients of NEG used in the LNP are estimated to be about 0.004 for H_2 and 0.03 for CO. Sticking coefficient for H_2 was directly measured in the experiment of equilibrium characteristics of the LNP, and was nearly 0.004. This means that calculating model and experimental conditions are consistent with each other. A few selected calculations of the pumping speed for different configurations of the NEG modules in the LNP for the purpose of considering the feasibility of pump size reduction are also shown in Fig.9. Because the conductances to the pumping port are almost the same for all configurations, pumping speeds of the LNP are not linearly proportional to the number of NEG modules used in the LNP especially for heavy molecules. The pumping speeds of the LNP with 6 NEG modules for H_2 and CO are about 50 % and 70 % of those with 14 NEG modules.

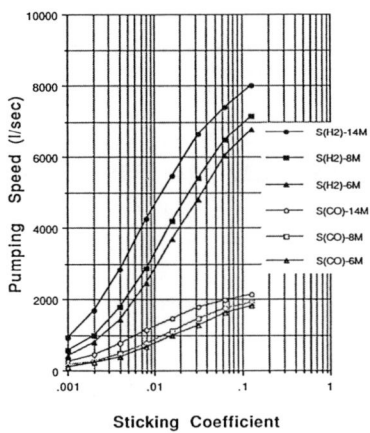 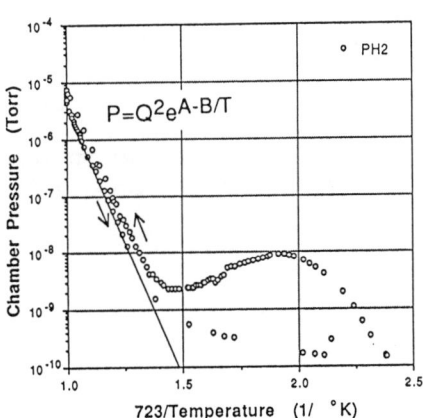

Fig.9. Pumping speeds calculated by Monte Carlo method in different LNP structures.

Fig.10. Chamber pressure change during NEG activation.

EQUILIBRIUM CHARACTERISTICS OF LNP

Figure 10 shows typical chamber pressure variation recorded in the temperature rise and fall stages during NEG activation. Measured pressure at the NEG temperature above about 300°C follows Sievert's law, and calculated parameter B from this figure is slightly larger than that recommended by maker. Figure 11 indicates the change of chamber pressure at the constant NEG temperature of 450°C. Vertical axis is the inverse square root of the measured chamber pressure. The gradient of the line is a function of initial pressure P_i and α, where α is the characteristic time taken to reduce the pressure to a quarter of the initial value. We should determine actual parameters of the vacuum system so that α becomes a resonable value.

As the effective pumping speed during NEG activation changes, chamber pressure and net outgassing rate are also changed; outgassing rate q with opening the gate valve in the LNP test chamber is expressed as $AsC_0(P_{eq}-P)$, and q' with closing the valve is given as $AsC_0(P_{eq}-P')$. Therefore sticking coefficient s is calculable with the known values of total NEG area A, unit orifice conductance C_0, measured pressure P and P' for two situations, and calculated q and q' in spite of that true P_{eq} is still unknown. Through this principle sticking coefficient for H_2 is calculated to be about 0.004. This value is almost the same as those estimated by comparing the data of pumping speed measurement and Monte Carlo calculation.

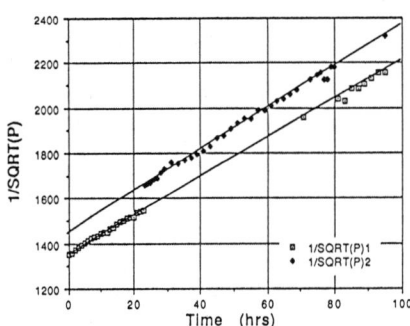

Fig.11. Variation of the chamber pressure at the NEG temperature of 450°C during activation.

The authors express their appreciation to Dr. H.Kamitsubo for his support of this work.

REFERENCES

1. S.Yokouchi et al., RIKEN Accel. Prog. Rep. 23, 138(1989)
2. P.E.Gear, Vacuum 26, 3(1976)

VACUUM PERFORMANCE OF VACUUM CHAMBER WITH A NEG STRIP OF 8 AND 16 m

S.Yokouchi, S.R.In,* T.Nishidono, H.Daibo, and S.H.Be

RIKEN-JAERI SPring-8 Project Team
RIKEN, Wako-shi, Saitama, 351-01 Japan

ABSTRACT

We manufactured 4m-long Al-alloy vacuum chambers, and investigated vacuum performance characteristics of the chambers with and without NEG strips.

In this paper we present a series of measurements of outgassing rate and the pumping speed of the NEG strips distributed in the chamber. The present experimental results are compared to those of Lee et al..[1] The ultimate pressures which can be achieved with a combination of NEG and other kinds of pumps such as a sputter ion pump (SIP) and a titanium sublimation pump (TSP) are discussed. Results of residual gas analysis are also discussed briefly.

INTRODUCTION

Super photon ring-8 GeV (SPring-8) is in the process of design as a high-brilliance synchrotron radiation source. In this storage ring, Zr-V-Fe (St 707) non-evaporable getter (NEG) strip is considered for use as a main pump. We are considering using an extruded 0.55%Mg-0.44%Si-Al alloy (A6063-T5 whose strength is equivalent to that of T6) as the chamber material to minimize synchrotron radiation (SR) induced desorption as well as thermal outgassing. This Al-alloy chamber is extruded in an atmosphere of Ar + O_2. Cross-sectional views of the straight section chamber (SSC) and the bending magnet chamber (BMC) are shown in Figs. 1 (a) and (b). These vacuum chambers consist of an electron beam chamber and a slot-isolated antechamber in which a NEG strip is installed. In the BMC, a distributed ion pump (DIP) will be installed in an additional slit-isolated chamber. SR is mostly intercepted by crotches and absorbers placed just downstream and upstream of a bending magnet, and avoids all other vacuum chamber walls around the storage ring.

To investigate vacuum performance characteristics of the chamber for the SPring-8 storage ring, we manufactured 4m-long Al-alloy vacuum chambers, and measured 1) outgassing rates with and without NEG strips, and 2) pumping speed of the NEG strips distributed in the chamber. Effects of combination of different pumps such as SIP and TSP on the ultimate pressure were also investigated.

OUTGASSING RATE OF THE CHAMBER

In the design of a pumping system, it is required to estimate

---

* On leave from KAERI, Daejon, Korea.

Vacuum Performance of Vacuum Chamber

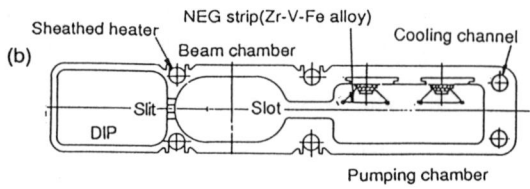
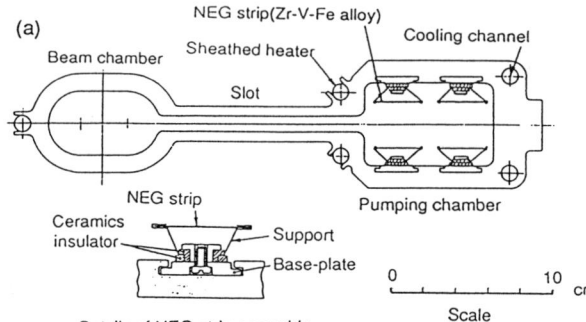

Fig. 1. Cross-sectional views of (a) the SSC and (b) the BMC.

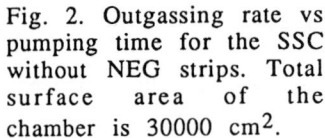

Details of NEG strip assembly

a thermal outgassing rate as well as a dynamic one due to photodesorption. In particular, in the design of a roughing pumping system we have to estimate the thermal outgassing rate during a bakeout of the chamber and an activation of the NEG strip. Therefore, we carried out experiments for measuring the outgassing rates of the chambers with and without NEG strips using the throughput method.

Results and discussion

Figure 2 shows the outgassing rate of the SSC itself without the NEG strips. We obtained an outgassing rate of the order of 10^{-10} Torr·l/sec/cm^2 at the chamber pressure of 4×10^{-6} Torr after ~20 hour pumping without a bakeout. After ~70 hour following a bakeout at 140°C for 40 hours, the outgassing rate of the chamber (Al-alloy) was measured to be 1×10^{-13} Torr·l/sec/cm^2 at the

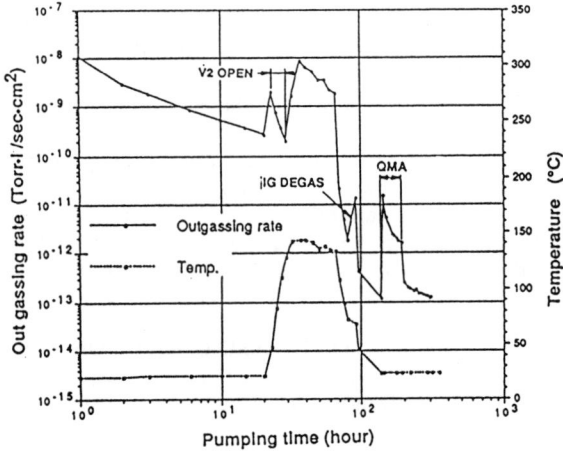

Fig. 2. Outgassing rate vs pumping time for the SSC without NEG strips. Total surface area of the chamber is 30000 cm^2.

chamber pressure of 2×10^{-9} Torr.

The outgassing rate of the BMC was also measured with virgin NEG strips in the chamber (Fig. 3). In this case, although we have additional sources of outgassing such as the surface of the NEG strips and their support (SUS 304 and ceramics), we got the overall outgassing rate of 10^{-10} Torr·l/sec/cm^2 after pumping as few as 10 hours without a bakeout. This turns out to be almost the same as that of the SSC. On the other hand, the outgassing rate after a ~70 hour evacuation following a bakeout similar to that given for the SSC was 2×10^{-12} Torr·l/sec/cm^2 at the chamber pressure of 4×10^{-8} Torr. We can calculate the outgassing rate of the NEG strips and the support from the data of the SSC and the BMC. The calculation result is 4×10^{-12} Torr·l/sec/cm^2, which is much higher than the outgassing rate of the Al-alloy chamber itself. This discrepancy was evidently due to an insufficient bakeout temperature for the NEG strips and the support.

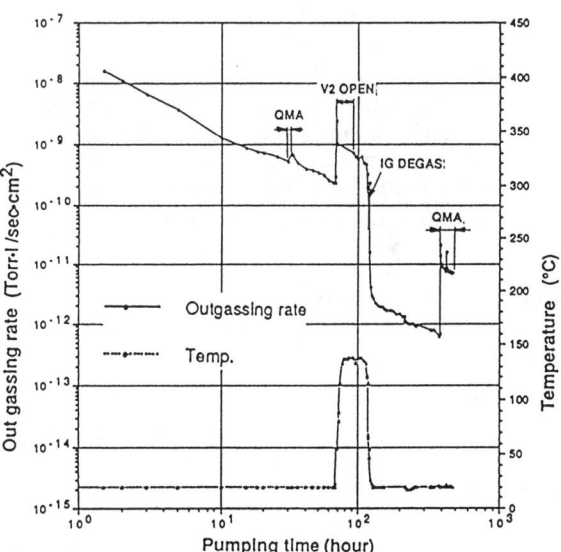

Fig. 3. Outgassing rate vs pumping time for the BMC with NEG strips. Total surface area of the sample, which is composed of the chamber itself, and the NEG strips and their support, is 56000 cm^2 (30700 cm^2 for the chamber itself, and 8700 and 16600 cm^2 for the NEG strips and the support, respectively). The NEG strips are extended to totally 1440 cm length.

PUMPING SPEED OF A NEG STRIP

Experimental

Figure 4 shows a schematic diagram of the measurement setup for the BMC. Four rows of NEG strip were installed in the SSC and two rows of NEG strip in the BMC. The effective length of the NEG strip per row is 360 cm for both the SSC and the BMC. The total effective surface areas of the NEG strips in the SSC and the BMC are 7776 and 3888 cm^2, respectively. The base pressure was in the range of 10^{-10} Torr after a bakeout at 140 °C for 40 hours. We measured a pumping speed of the NEG strips at room temperature for H_2 and CO after 25 hour 450°C and 50

Vacuum Performance of Vacuum Chamber

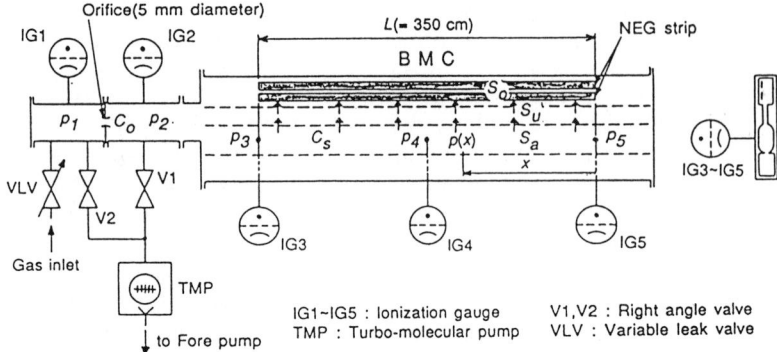

Fig. 4. Schematic diagram of measurement setup for the BMC. $P(x)$ is the pressure at x in Torr, S_a the average pumping speed at the beam chamber per unit length in l/sec/cm, S_u the pumping speed of the NEG strip per unit length in l/sec/cm, S_o the pumping speed of the NEG strip per unit area in l/sec/cm^2, C_s the conductance of the slot (See Fig.1) per unit length in l/sec/cm, L the length of the distributed NEG strip in cm, C_o the conductance of the orifice in l/sec, and P_1, P_2, P_3, P_4, and P_5 the pressure measured at IG1, IG2, IG3, IG4, and IG5 in Torr, respectively.

minute 450°C activations, respectively.

To evaluate the average pumping speed at the beam chamber per unit length S_a, we assumed such a simple model as Fig. 5. In this model, we get

$$Q'(x) = S(x)P(x) \tag{1}$$

and

$$P''(x) = S(x)P(x)/C \tag{2}$$

where $Q(x)$ is the throughput along the chamber axis in Torr·l/sec, $S(x)$ the pumping speed per unit length in l/sec/cm, $P(x)$ the pressure in Torr and C the conductance of the path per unit length in l·cm/sec. By integrating Eq.(1) from x=0 to x=L, we get $\int_0^L dQ = \int_0^L S(x)P(x)dx$. Since $\int_0^L dQ = Q_o$, that is given by $Q_o = C(P_1-P_2)$, S_a is defined by

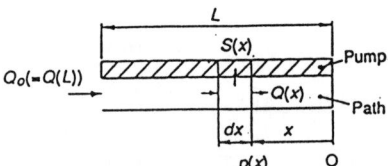

Fig. 5. Model for evaluation of the average pumping speed at the beam chamber.

$$S_a = Q_o/\int_0^L P(x)dx \tag{3}$$

Since it is difficult to know the exact solution of Eq.(2), we assume $P(x) = P_c(x)f(x)$, where $P_c(x)$ is the well known solution of Eq.(2) for constant $S(x)$ and $f(x)$ the correction factor of the quadratic. By using the measured pressure of P_3 and P_5, $P_c(x)$ is obtained from $P_c(x) = P_5\cosh(kx)$, where $k = (1/L)\cosh^{-1}(P_3/P_5)$. Since the calculated $P_c(L/2)$ does not agree with the measured P_4, $f(x)$ is introduced to compensate for this disagreement, and is determined by measured P_3, P_4, and P_5; $f(0) = f(L) = 1$ and $f(L/2) = P_4/P_c(L/2)$. Finally, we get the normalized pumping speed of the NEG strip per unit length by

$$S_u = (1/S_a - 1/C_s)^{-1}(350/360)/n \qquad (4)$$

where n is the number of NEG strips. The normalized pumping speed per unit area is given by

$$S_o = S_u/5.4 \qquad (5)$$

Results and discussion

Figures 6 (a) and (b) show the average pumping speeds for H_2 and CO at the SSC and BMC, respectively. By using these results and Eq.(5), we got the normalized pumping speeds of the NEG strips in the SSC and the BMC for H_2 and CO as shown in Figs 7 (a) and (b). The normalized pumping speed for H_2 is approximately constant because of bulk-diffusion occuring even at room temperature as well known.[2] On the other hand, that for CO decreases rapidly with an increase in the sorbed quantity.[3] The initial pumping speed is in the order of 10^{-1} l/sec/cm^2 for both H_2 and CO. The pumping speed variation with respect to the sorbed quantity is similar to that of a short piece of NEG strip for both H_2 and CO.

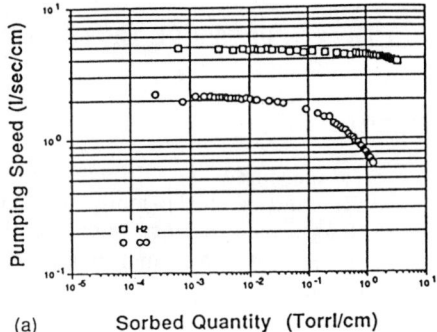

Fig. 6. Average pumping speed per unit length for H_2 and CO at the beam chamber of (a) the SSC and (b) the BMC.

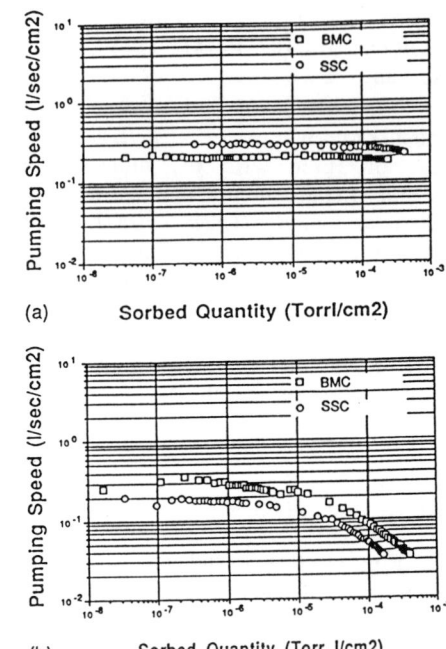

Fig. 7 Pumping speed per unit area of the NEG strip in the SSC and the BMC for (a) H_2 and (b) CO.

ULTIMATE PRESSURE OF THE CHAMBER WITH A COMBINATION OF PUMPS

Experimental

We measured the ultimate pressure of the SSC in which NEG strips, TSP, SIP and a turbo-molecular pump (TMP) were installed. The total active length and surface area of the NEG strips are 1440 cm and 7776 cm^2, respectively. Nominal pumping speed of SIP is 110 l/sec and that of TMP 300 l/sec.

Results and discussion

The pressure variation vs pumping time is shown in Fig.8. After 145 hour evacuation we carried out a bakeout of the chamber and an activation for the NEG strips simultaneously ((1) in Fig.8), and at the end we performed pre-conditioning for TSP and SIP ((2)). After that, we got the ultimate pressure of 5.6×10^{-11} Torr with NEG strips and TMP((3)). However, the pressure with only NEG strips increased to 1.2×10^{-8} Torr((4)) with a build-up of CH_4(m/e=16), as shown in Fig.9 (a), which could not be pumped by the NEG strips. We obtained the ultimate pressure of 4.7×10^{-11} Torr with the NEG strips and SIP((5)), and got a mass spectrum (Fig.9 (b)). Although TSP was flashed((6)), the pressure did not improve further ((7)). This means that additional TSP is not useful to achieve better ultimate pressure in the 10^{-11} Torr range.

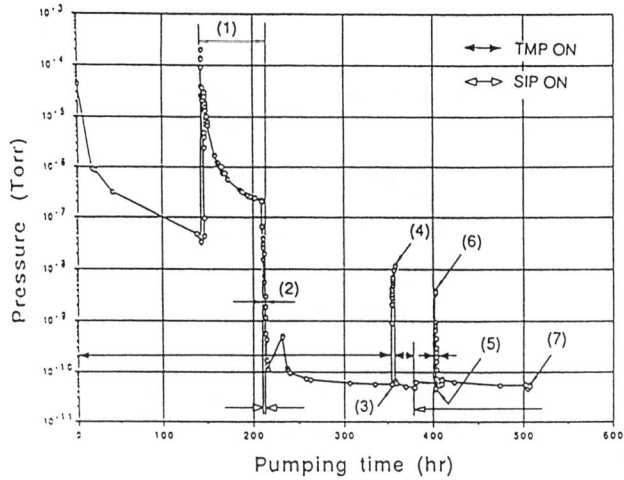

Fig.8. Pressure variation vs pumping time.

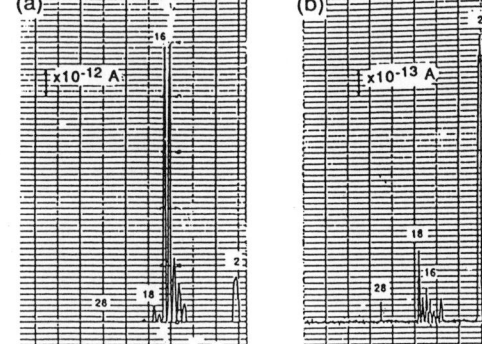

Fig.9. Mass spectrum of evacuated chamber (a) with only NEG strips, and (b) with the NEG strips and SIP.

CONCLUSIONS

In the experiments for investigating vacuum performance characteristics of the vacuum chambers, which were test-manufactured for the SPring-8 storage ring, we found the following :

1) We can get the outgassing rate of 1×10^{-13} Torr·l/sec/cm^2 for the chamber (Al-alloy) and of 2×10^{-12} Torr·l/sec/cm^2 for the chamber with NEG strips, after ~70 hour evacuation following a bakeout at 140°C for 40 hours.

2) The NEG strip (St 707) distributed in the chamber has an initial pumping speed of the order of 10^{-1} l/sec/cm^2 for H_2 and CO. Although its pumping speed for H_2 is relatively constant, that for CO decreases

drastically with an increase in the sorbed gas quantity. This variation in the pumping speed is similar to that of a NEG strip piece.

3) We can obtain the ultimate pressure of 4.7×10^{-11} Torr in the Al-alloy chamber with a combination of the NEG strips and SIP.

ACKNOWLEDGEMENTS

The authors are grateful to Dr. H. Kamitsubo for his support of this work.

REFERENCES

1. YoungPak Lee et al., J.Vac.Soc.Jpn. 33, 154 (1990).
2. R. J. Knize and J. L. Cecchi, J. Appl. Phys. 54, 3183 (1983).
3. C. Benvenuti and F. Francia, J. Vac. Sci. Technol. A6, 2528 (1988).

VACUUM SYSTEM DESIGN FOR A 1.2 GeV ELECTRON STORAGE RING WITH NON-EVAPORABLE GETTER PUMPING

H. F. Dylla,† D. M. Manos, J. C. Citrolo,
P. H. LaMarche, S. Raftopoulos, M. Ulrickson
Princeton Scientific Consultants, Inc.
P.O. Box 2181
Princeton, NJ 08543

A. G. Mathewson and A. Poncet
CERN, Geneva, Switzerland

F. Mazza
SAES Getters, Milano

ABSTRACT

We describe the design for the vacuum vessel and vacuum pumping system for a 1.2 GeV electron storage ring for light source applications. The hybrid vacuum vessel design consists of eight extruded aluminum (6063-T4) dipole chambers curved to the ring radius of 2.93 m, with interconnecting stainless (304L) steel straight vessel segments. The dipole chambers contain two 60 milliradian photon exit ports on the outer midplane, and an integral pumping chamber on the inner midplane which provides 500 l/s·m of active gas pumping with a circumferential array of a non-evaporable getter (NEG) material. Ion pumping stations are distributed along the straight vacuum vessel segments to provide for pumping of CH_4 and noble gases. The distributed pumping provided by the NEG offers a factor of five to ten times larger pumping speed in comparison to conventional designs based solely on the use of discrete or distributed ion pumps. Externally bonded cooling tubes on the dipole chambers offer a cost effective and fault tolerant means of removing the x-ray heat load. Estimates of the conditioning time for the vacuum vessel are presented based on the recent photodesorption data from aluminum surfaces.

1.0 INTRODUCTION

We describe the design for the vacuum vessel, vacuum pumping system and associated vacuum instrumentation for a 1.2 GeV, 400 mA electron storage ring compatible with conventional (room temperature) magnet technology. To meet typical electron storage time requirements ($\sim$ 10 hours), the vacuum system must maintain a total pressure of less than 10^{-9} torr during operation and must maintain base pressures of less than 10^{-10} torr during non-operational periods to prevent recontamination of the interior vessel surfaces by residual gas

---

† Present address; Continuous Electron Beam Accelerator Facility, 12000 Jefferson Avenue, Newport News, VA 23606

390 Vacuum System Design for 1.2 GeV Electron Storage Ring

adsorption. The primary operational gas load is due to photo-induced desorption by synchrotron radiation from the 1.2 GeV circulating electron beam. This photo-induced desorption gas load must be minimized by the proper choice of vessel fabrication materials and the use of vessel surface conditioning techniques. Since appendage pumping geometry for the storage ring is severely conductance limited, high speed distributed pumping is provided by a circumferential array of non-evaporable getter (NEG) material. A cost-effective bakeout capability provides the necessary in-situ vessel conditioning to minimize the system turn-around time after maintenance or modifications.

2.0 STORAGE RING VACUUM VESSEL

A schematic layout of the electron storage ring vacuum vessel is shown in Fig. 1. The 55.2 m circumference storage ring is specified for operation at 1.2 GeV and 400 mA[1] using a Chasman-Green magnetic lattice similar in design to the VUV ring at the National Synchrotron Light Source[2]. Synchrotron radiation at $\lambda_c = 9.5$ Å can be extracted from the 2.93 m radius dipole bending magnets. Dispersion free straight sections are available for interface of: 1) the injection system; 2) RF cavity; and 3) two insertion devices.

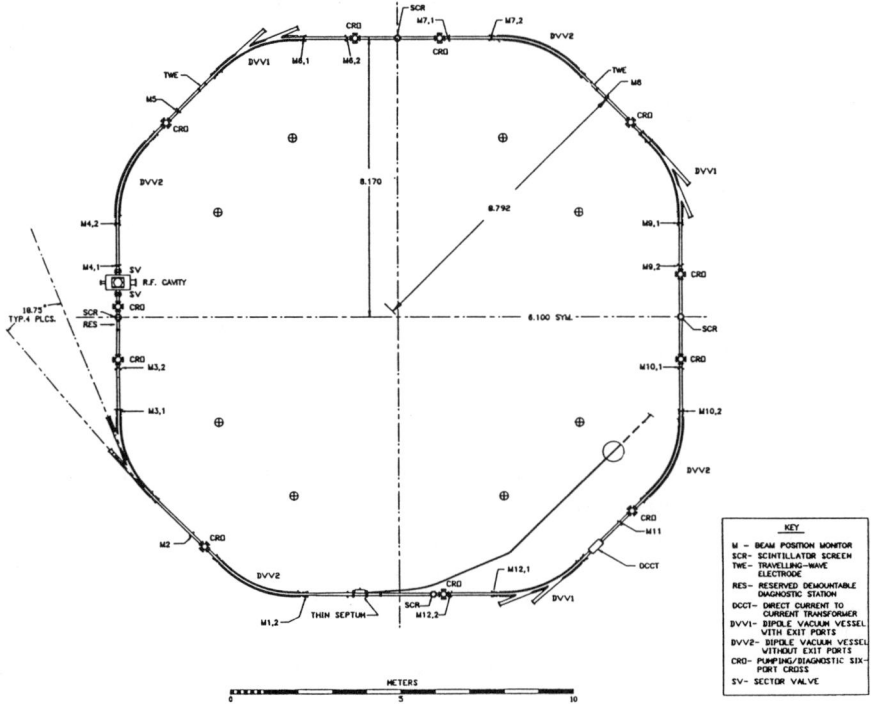

Fig. 1. Electron storage ring vacuum system.

The storage ring vacuum vessel is fabricated from three types of chambers: 1) eight, 2.3 m, 45° sections, curved to the 2.93 m major radius, to accommodate the dipole magnets; 2) four, 3 m straight sections; and 3) four, 6.1 m straight sections. The vessel straight sections are fabricated from 304 L stainless steel seamless tubing with an inner diameter of 70 mm and a wall thickness of 3 mm. The vessel segments are flanged with standard conflat-type UHV flanges with 15 cm outer diameters. The choice of stainless steel tubing for the straight section minimizes the system cost because of the ease of fabrication, welding and brazing. Low specific outgassing ($< 10^{-12}$ torr-liters/s·cm^2) and low photo-induced desorption yields ($< 10^{-3}$ molecules/photon) for stainless steel can be achieved with a combination of an alkaline detergent cleaning[3] and a high temperature vacuum pre-bake of vessel segments (300°C), followed by an in-situ 150°C vessel bakeout[4].

DIPOLE CHAMBERS

The dipole vacuum vessels are a multichambered, aluminum extrusion fabricated from type 6063-T4 aluminum. A view of the dipole chamber cross section is shown in Fig. 2. The beam chamber is nearly elliptical, with vertical and horizontal diameters of 51 mm and 66 mm, respectively. The minimum vessel wall thickness is 3 mm and the overall height of the vessel is 3 mm shorter than the 60 mm minimal dipole magnetic gap. An antechamber on the small major radius side of the dipole chamber contains slots for holding the non-evaporable getter (NEG) assembly used for distributed pumping (see Sec. 3). The NEG antechamber connects to the beam chamber through a chamfered 7 mm high by 7 mm wide pumping channel extending the length of the dipole chamber. The large major radius side of the dipole chambers is designed to accommodate the addition of two photon exit ports each emitting a 60 milliradian cone of x-ray radiation as shown in Fig. 3. The dipole chambers must be cooled during NEG activation and during storage ring operation due to the x-ray heat load (see below). The dipole chamber cooling is provided by externally bonded soft copper cooling tubes that can share the same facility cooling water as the magnet cooling loops. The cooling tubes are bonded to the dipole vessel with a high temperature, radiation-resistant high conductivity epoxy, which has a demonstrated longevity in adverse environments[5].

The dipole beam chamber/NEG antechamber is fabricated using standard extrusion techniques. Finite element modelling of the chamber (using NASTRAN) indicates that maximum stresses (2.8 ksi) are well below yield stresses (10–20 ksi). The extruded chambers are welded to special aluminum alloy (2219-T87), 15 cm conflat flanges[6,7] which mate to the stainless steel flanges on the bellows transition pieces described below. Following an alkaline detergent cleaning and a short (24 h) in-situ 150°C vacuum bakeout, the thermal outgassing rate of the aluminum dipole chambers can be reduced to below 1×10^{-12} torr-liters/s·cm^2.[8,9]

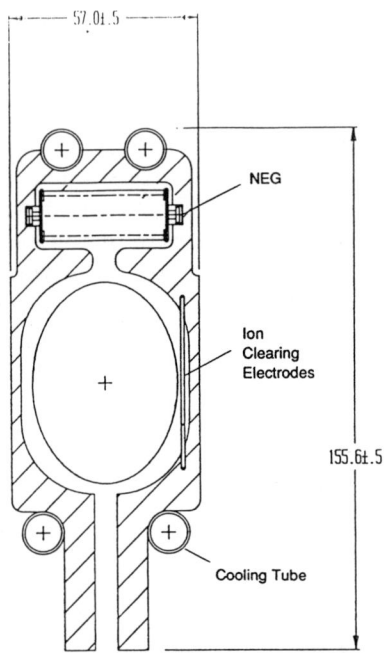

Fig. 2. Dipole vacuum chamber, cross-sectional view.
(dimensions in millimeters)

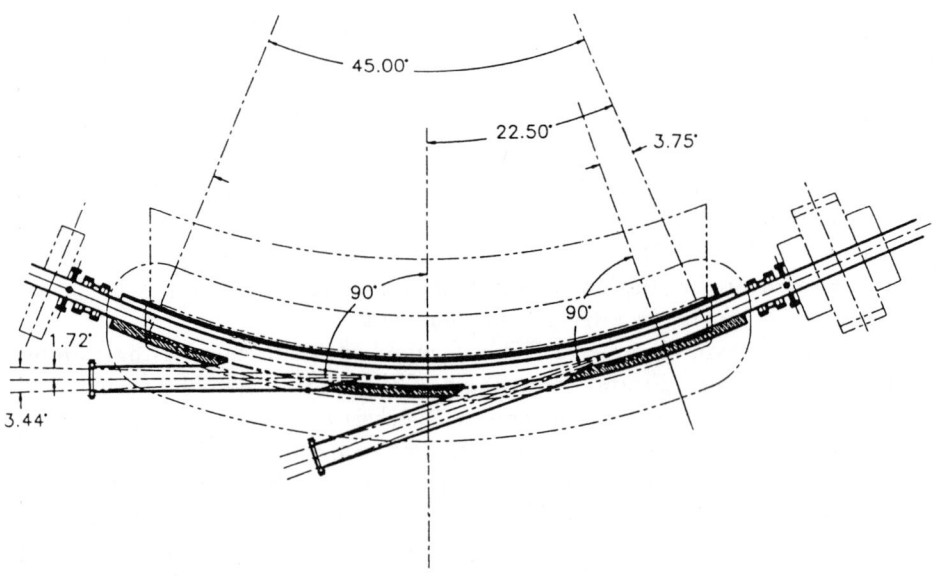

Fig. 3. Dipole vacuum chamber, plan-view showing photon exit ports.

HEAT LOAD TO THE DIPOLE CHAMBER DURING NORMAL OPERATION

The mass absorption coefficient for aluminum at wavelengths of 0.2 to 0.4 nm varies between 150 and 500 cm^2/g. Using 150 cm^2/g and the density of aluminum (2.7 g/cc), 90% of the photons are absorbed in the first 60 μm of the wall thickness, so the heat load can be considered as a surface heat flux for the purpose of thermal analysis. The beam energy loss is 1.2 kW/m (at 1.2 GeV, 400 mA and $R = 2.93$ m) with a vertical spread characterized by a Gaussian with a width of about 0.65 mm. This yields a peak heat load of about 70 W/cm^2. The angle of incidence of the heat flux is about 8.5°. Where the wall of the dipole chamber is penetrated by the exit ports, the angle of incidence becomes 90°, yielding a peak heat flux of about 700 W/cm^2 at the throat of the exit port. The temperature rise due to 70 W/cm^2 is approximately 30°C, assuming the cooling design shown in Fig. 2. In the region of the exit port thermal analysis indicates that the temperature rise will be about 150°C which is acceptable.

Eleven six-port crosses are provided on the straight vessel segments for pumping and diagnostic purposes. A slotted insert within the pumping cross minimizes beam impedance effects by preserving the vessel cross-section and also provides a pumping conductance of 10^3 l/s. The straight vacuum vessel segments are connected to the dipole chambers through a special transition piece (Fig. 4) which transforms the circular beam chamber in the straight segments to the elliptical beam chamber in the dipole chambers. This transition piece contains external bellows for ease of assembly and alignment, and to take up thermal expansion. A silver-plated sliding insert provides a smooth vessel cross-section and minimizes beam impedance effects.

To minimize the storage ring down-time during RF cavity maintenance two RF-shorted, in-line sector valves are included on the storage ring vessel on either side of the RF cavity. In addition, an isolation valve is placed at the end of the transfer line vacuum vessel to allow for injection system conditioning and maintenance activities independent of the storage ring vacuum system.

ION CLEARING ELECTRODES

A low profile ion clearing electrode is installed in the bottom of both the dipole and straight segment beam chambers. The electrode consists of alumina bonded to a 1 mm thick stainless steel mounting plate. The mounting plate is curved to fit loosely into a slot in the dipole sections so that the electrode position does not shift due to thermal stresses during bakeout or ring operation. The top surface of the alumina is covered with a thin film of a resistive ($\sim$ 1 kΩ·cm) glass to provide the ion collecting surface with minimum RF coupling to the electron beam[10]. The collector is as wide as possible to provide the maximum degree of ion collection for a given bias voltage. The structure provides adequate electrical isolation to hold off a minimum of 1 kV bias voltage.

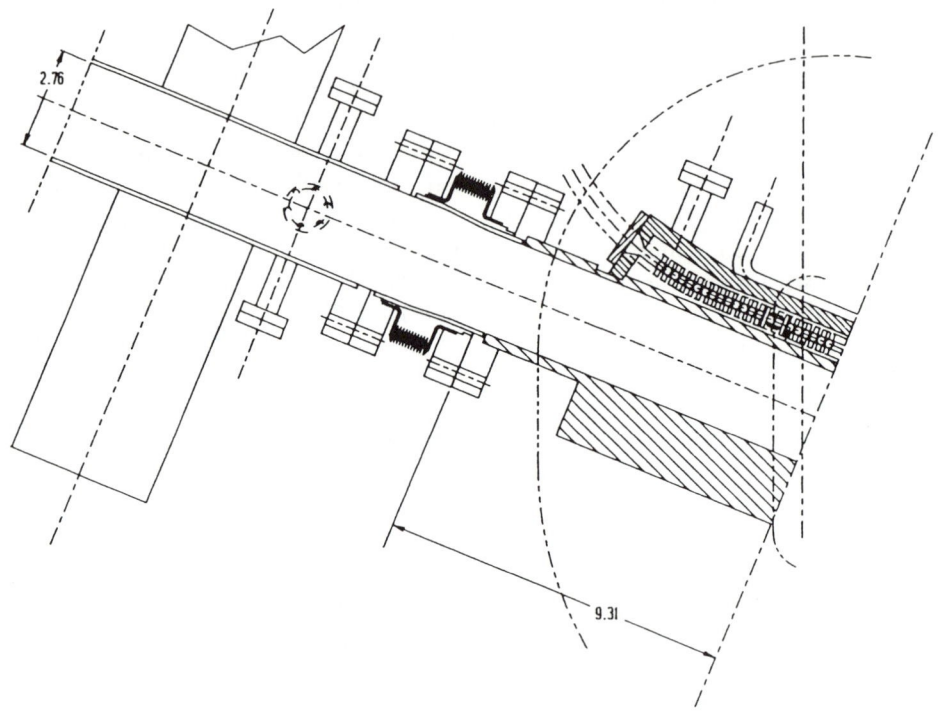

Fig. 4. Vacuum vessel transition piece: Dipole chamber to straight vessel segment.

3.0 VACUUM PUMPING SYSTEM

Complementary high vacuum pumping for the storage ring is provided by two different high vacuum pumping systems: 1) twelve discrete ion pumping stations, and 2) non-evaporable getter (NEG) strips positioned on the inner wall of the dipole chambers. Eleven, 45 l/s triode ion pumps are distributed around the ring on the bottom ports of the six-port pumping/diagnostic crosses. One, 400 l/s ion pump is located on the injection magnet chamber providing pumping for this chamber and the remaining lengths of straight storage ring vessel segments which are connected to the injection chamber. A dedicated power supply is specified for each ion pump to provide the necessary local pressure measurements and pressure interlocks for the vacuum control system. The outside ports of the six-port crosses are used to connect the roughing systems in two locations. The remaining midplane ports can be used for beam and vacuum diagnostics.

Vacuum pumping for the initial evacuation of the storage ring from atmospheric pressure and during vessel bakeout is provided by roughing systems, each consisting of a 50 l/s turbopump backed by a 2 cfm rotary-vane backing pump.

Although these pumps use oil for lubrication, there is considerable experience with these pumping systems in hydrocarbon-free vacuum environments. In addition, with the foreline of the turbopump properly configured, the turbopump acts as an accumulator for a helium leak detector attached to the foreline. This improves the response time of the helium signal and facilitates the leak-checking process.

The roughing system will pump the storage ring vessel from atmospheric pressure to 10^{-3} torr in less than an hour and to 10^{-5} torr in less than a day for an unconditioned vessel. A typical pump-down scenario is shown in Fig. 5. After proper conditioning of the vacuum vessel using the 150°C in-situ bakeout[3,8,9], the area specific outgassing rate of the vessel is expected to be $\leq 10^{-12}$ torr-liters/s·cm^2. Under these outgassing conditions, which can be achieved approximately three to four days after vessel pump-down, the base pressure in the storage ring can be maintained in the 10^{-10} torr range using the system ion pumps alone.

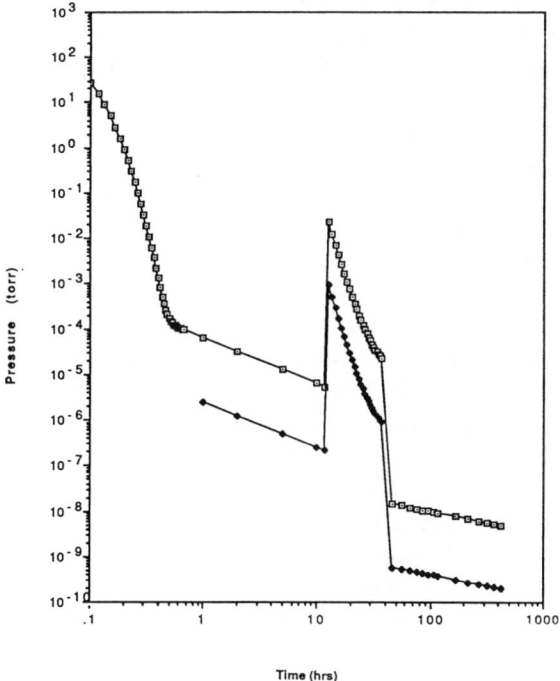

Fig. 5. A typical pump-down curve for the storage ring vacuum system assuming a 24 hour bakeout is initiated at ten hours after vessel evacuation with the turbomolecular roughing pumps (▫), and with the addition of the system ion pumps (◆).

NON-EVAPORABLE GETTER (NEG) PUMPS

In order to meet the pressure requirement of $< 10^{-9}$ torr during storage ring operation additional high vacuum pumping capacity beyond that provided by the ion pumps is required to pump the photo-induced desorption gas load. In large synchrotrons, linear or distributed pumps in the bending magnet sections offer the advantage of providing a large pumping speed where the dynamic gas load from photo-induced desorption is the greatest. Traditionally, the distributed pumping has been accomplished using in-line ion pumps employing a Penning discharge within the bending magnet field. A more cost-effective linear pumping scheme is the use of non-evaporable getters (NEG)[11] which have been tested and used in a number of accelerators including the Heavy Ion Transport Line at BNL[12] and the Large Electron Positron storage ring at CERN[13-16]. In addition, NEG pumping is planned for a number of next generation storage rings[17-19]. The NEG affords a number of advantages over integrated ion pumps including mechanical simplicity, much larger pumping speed, and reduced cost. In addition, the NEG does not require a magnetic field for operation. Power is required for activation; however, after activation, power is not required for operation. NEGs are not adversely affected by synchrotron radiation[20,21], and measured photo-induced desorption yields are lower than typical metallic surfaces[22].

The NEG array for the proposed storage ring consists of a pleated strip of getter material located in the antechamber of each dipole chamber. The assembly of the pleated strip is shown in Fig. 6. The getter material is an alloy of Zr and Al (SAES St 101) which is bonded to a 0.2 mm constantan substrate. The 15 mm wide strip is pleated in a way to form a structure of fins 38 mm high every 3 mm of length of the chamber. Its flexible structure allows an easy and fast insertion and substitution in the dipole antechamber. Contrary to the LEP case, which uses a straight strip, this design allows an increase of the specific pumping speed[23,24] and of the total capacity. The gain in net pumping speed is small because it is limited by the conductance of the slot between the beam chamber and the antechamber, but the pumping capacity is at least a factor of 5 larger than the LEP getter strip[14] (see Fig. 7). The pumping capability of the NEG is activated by passing current through the substrate and heating it to 700°C for approximately one hour. For this storage ring application, the NEG operates at room temperature and has a lifetime of about ten years (assuming two activations per year). The NEG array provides initial distributed pumping speeds of 500–1000 l/(s·m) for active gases such as CO, H_2O, and H_2, after correcting for the pumping slot conductance between the antechamber and beam chamber (Table I). The decrease in NEG pumping speed with surface loading of active gases can be significantly "regenerated" by heating the NEG to 400°C for $\sim$ 15 minutes. The NEG pumping speeds are approximately 100 times greater than the net speed provided by the ion pumps at the center of the dipole chambers. The ion pumps provide pumping for the much smaller gas load of chemically inert gases (CH_4, Ar, and He) that are not pumped by the NEG array.

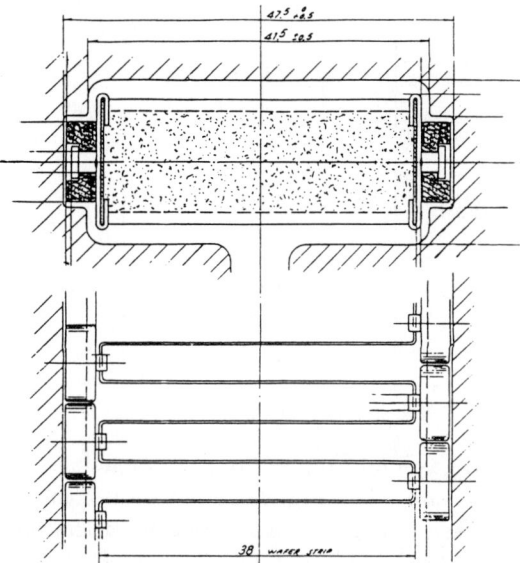

Fig. 6. Detail of the non-evaporable getter (NEG) pump assembly.

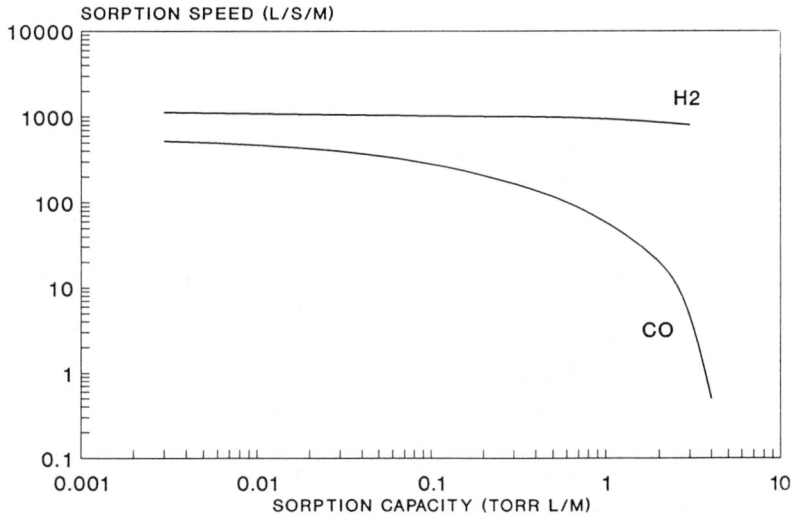

Fig. 7. Calculated net NEG pumping speed (at 25°C) as a function of loading provided to the dipole chamber for CO and H_2; with concurrent pumping of CO + H_2, the H_2 curve follows the CO curve as shown in Ref. 14.

Vacuum System Design for 1.2 GeV Electron Storage Ring

Table I   Net distributed pumping speed in the dipole chambers as a function of surface loading

Gas	Pump	Initial Pumping Speed (l/s/m)	Pumping Speed After 1 torr-liter/m of CO/CO_2 Loading
H_2	NEG	1000	400
H_2O	NEG/Wall	> 600	> 600
CO	NEG	600	100
CO_2	NEG	400	70
CH_4	Ion Pumps	10	10

VESSEL CONDITIONING

The photo-induced gas load from aluminum consists primarily of four residual gases: H_2, CO, CO_2, and CH_4. Fig. 8 shows a plot of the measured molecular desorption yields for these four gases as a function of the photon loading of an aluminum test chamber (baked at 150°C) exposed to $E_c = 2.95$ keV synchrotron radiation in the DCI storage ring[25-28]. Since the electron beam lifetime within the storage ring is determined primarily from bremsstrahlung on residual gas nuclei, the lifetime scales universally with the product of the mass and partial pressure of the gas. The molecular yield data of Fig. 8 shows that CO and CO_2 are the residual gases of concern with respect to beam lifetime determination. The initial CO and CO_2 desorption yields are in the range of 10^{-2} and then fall approximately as (dose)$^{-1}$ after a dose of 10^{19} photons/m.

During the initial operation of the storage ring, the usual "beam conditioning" process would be followed by slowly incrementing the beam current up to the design value (400 mA) as the photo-induced desorption yields fall with beam dose to the dipole vessels. For this storage ring an electron beam dose of 0.5 A·hr corresponds to a photon dose of 1×10^{23} photons/m in the dipole vessels. Assuming the DCI desorption yield curves (Fig. 8) are valid for the present machine, the total desorbed amount of CO and CO_2 is 1 torr-liter/m at the 0.5 A·hr dose, and the desorption yields will have fallen to 5×10^{-5} and 3×10^{-5} molecules/photon for CO and CO_2, respectively. At this point, the pumping speed of the NEG pumps will have fallen to 70 l/s according to the speed vs. loading data of Fig. 7, and the NEG pumps will require regeneration (i.e., heating to 400°C) to restore the pumping speed. According to this scenario, the subsequent beam dose can be increased by additional order of magnitude to 5 A·hr ($= 1 \times 10^{24}$ photons/m), before a second NEG regeneration is required. At this point the CO and CO_2 desorption yields are sufficiently low ($\sim 10^{-6}$) to support high current (400 mA) beam operation with long stored beam lifetimes (hrs). The total CO + CO_2 loading of the NEG (2 torr-liters/m) after a beam dose of 5 A·hr is still small compared to the maximum NEG capacity for CO + CO_2 which is ≈ 150 torr-liters/m.

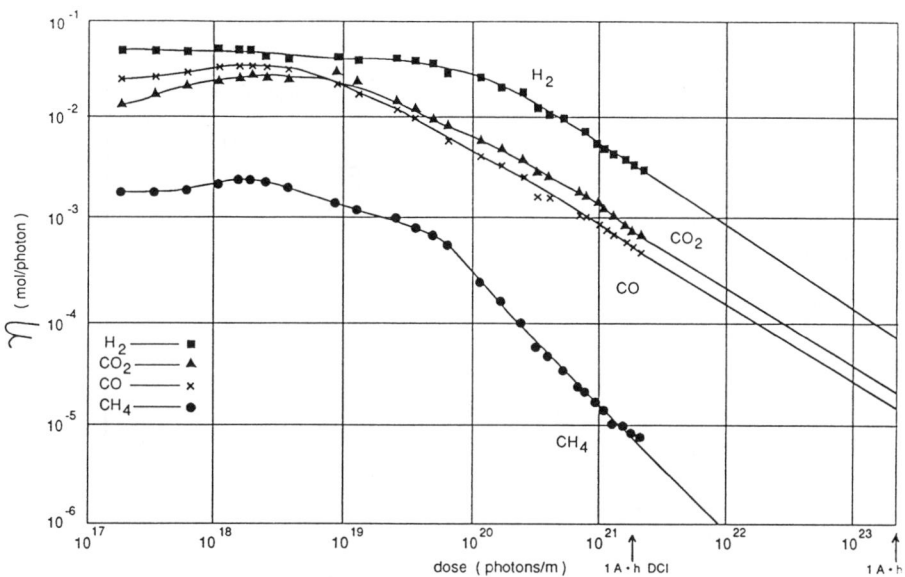

Fig. 8. Measured photodesorption molecular yields in the DCI storage ring (Ref. 28).

Basing the estimates of the beam conditioning time and NEG reconditioning interval on the DCI desorption yield data is very conservative. The photodesorption yield data from DCI (Fig. 8) was obtained with synchrotron radiation with $E_c = 2.95$ keV incident on the test chamber walls at an angle of 11 mrad. The photodesorption yield is believed to scale with electron beam energy E, as seen in the data also taken from DCI in Fig. 9, and to scale as $1/\sin\theta$, where θ is the angle of incidence of the synchrotron light onto the dipole chamber walls. In the present storage ring $E_c = 1.31$ keV and $\theta = 149$ millirad. Therefore, the photodesorption yields should be a factor of 5.6 lower due to the lower E_c, and a factor of 2.3 lower due to the higher angle of incidence. Taking the full factor of 13, (5.6×2.3) would extend the first NEG regeneration time to a dose 6.5 A·hr. The actual beam conditioning time and NEG regeneration interval will probably lie somewhere in between the original DCI-based estimate and the scaled estimate.

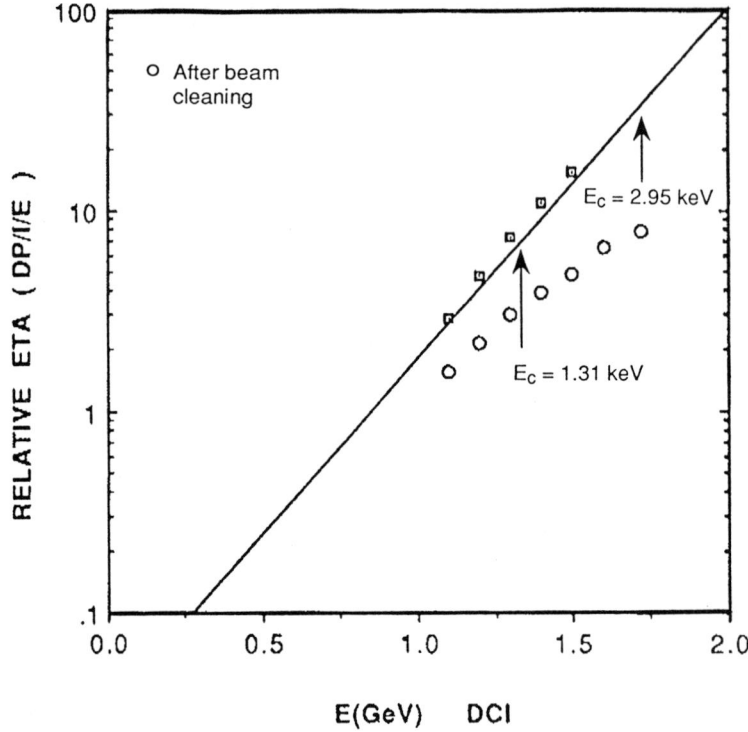

Fig. 9. Energy dependence of the photodesorption molecular yields as measured in DCI (from A. Mathewson, CERN).

NON-EVAPORABLE GETTER ACTIVATION

The NEG must be heated in vacuum to 700°C to dissolve the surface passivation layer and initiate pumping. This is accomplished using electrical heating of the getter strip. The power dissipated in the dipole chamber when the NEG is at 700°C is 5.1 kW per dipole chamber. A water flow of 24 gal/min (3 gpm per dipole) through the dipole cooling tubes keeps the temperature rise below 30°C during activation. This flow rate is similar to that required to remove the heat generated by the absorption of the synchrotron light in the chamber. Manufacturer's specifications[11] indicate that the NEG strips can be cycled through approximately twenty activation cycles (anticipated to be necessary only after atmospheric venting) before the material must be replaced. For routine maintenance activities, the dipole chambers can be vented to dry N_2, alleviating the need for a subsequent NEG activation cycle. Also, initial activation of the NEG strips is not required until beam alignment and beam optimization procedures are completed during the first phase of beam commissioning, thus eliminating the need for unproductive NEG activation cycles.

While the ring is operating, the NEG remains at ambient temperatures. The NEG operating modes are summarized in Table II.

Table II NEG operating modes

Mode	NEG Temperature	NEG Water Cooling
Bakeout	400°C	No
Regeneration	400°C	Yes
Activation	700°C	Yes
Operation	ambient	No

4.0 VACUUM VESSEL BAKEOUT

After initial assembly and after each atmospheric vent, the vacuum vessel can be baked at 150°C using electric heater tapes wrapped around the vessel. The straight sections of the vessel are covered with 12 mm of fiber glass insulation with a PVC overlayer. In the areas where there are quadrupole or sextupole magnets the insulation is removed in the areas of the pole tips. The heat lost through conduction in the thin ($\sim$ 1 mm) air gap to the pole tip is compensated by adding extra heating tape between the pole tips. The total power required to maintain the straight sections at 150°C is about 6 kW. In the dipole sections there is insulation only on the vertical surfaces of the dipole chamber. The heat loss through the thin (1 mm) air layer between the chamber and the magnet pole face is only 3.4 kW per chamber. This additional heat is supplied by heating the NEG strips to about 400°C and adding about 0.6 kW with an electrical heater strip along the exit port side of the chamber. The aluminum dipole chamber has sufficient thermal conductivity that the temperature difference between the hottest part and the coolest part of the chamber is only about 15 C. The total power required for bakeout is 35 kW. The magnet cooling water system must be operational during the bakeout to prevent overheating of the coils. The bakeout temperature is controlled by temperature controllers using thermocouples on the exterior of the chamber.

5.0 VACUUM INSTRUMENTATION

The vacuum system instrumentation includes a residual gas analyzer, redundant ion gauges, Pirani gauges; and capacitance manometers and a spinning rotor gauge for transfer calibrations diagnostics. A crude measurement of the pressure in the storage ring, accurate to ±50%, is obtained by monitoring the ion currents in the various ion pumps distributed around the ring. For more accurate pressure monitoring (±10%), a redundant set of Bayard-Alpert type ionization gauges is specified which spans the pressure range of 10^{-11}–10^{-4} torr.

ACKNOWLEDGMENTS

This work was supported in part by the Brobeck Division, Maxwell Laboratories, Inc.

The authors thank the following for helpful discussions of the vacuum system design: R. A. Jacobsen (DTI, Inc.), M. A. Green (Univ. of Wisconsin), B. Craft (LSU), H. Halama (BNL), A. Jackson (LBL), K. Kennedy (LBL), F. Pirota (SAES), V. Saile (LSU), H. Wiedemann (LBL), and M. S. Zisman (LBL).

REFERENCES

1. M. A. Green, H. F. Dylla, R. E. Hartline, R. A. Jacobsen, D. M. Manos, et. al., Nucl. Instr. Meth. Phys. Res. A291, 464 (1990).
2. N. Omur and S. White dePace (eds.), BNL Report No. 40156 (1986).
3. H. Halama, in Proc. AVS Topical Conference on Surface Conditioning of Vacuum Systems, Los Angeles, April 1989; R. A. Langley et. al., eds., AIP Conf. Proc. No. 199 (AIP, NY, 1990), p. 93.
4. A. G. Mathewson, in Proc. AVS Topical Conference on Surface Conditioning of Vacuum Systems, Los Angeles, April 1989; R. A. Langley et. al., eds. AIP Conf. Proc. No. 199 (AIP, NY, 1990), p. 110.
5. Omegabond 200™ high temperature epoxy, Omega Engineering, Inc., Omega Engineering Catalog, p. J6 (1987).
6. H. Ishimaru, J. Vac. Sci. Technol. A2, 1170 (1984).
7. Manufactured by ULVAC or Hakuda Corporation.
8. A. G. Mathewson, J.-P. Bacher, K. Booth, R. S. Calder, G. Dominichini, et. al., J. Vac. Sci. Technol. A7, 77 (1989).
9. J. R. Chen, K. Narushima, H. Ishimaru, J. Vac. Sci. Technol. A3, 2188 (1985).
10. F. Caspers, J. P. Delahay, J. C. Godot, K. Hübner, A. Poncet, Proc. 1988 European Particle Accelerator Conference, Rome, 1988.
11. Manufactured by SAES, SPA (Milan).
12. H. C. Hseuh in Proc. AVS Topical Conference on Vacuum Systems for Compact and Advanced Synchrotron Light Sources, Brookhaven, May 1988; H. Halama et. al., eds., AIP Conf. Proc. No. 171 (AIP, NY 1988), p. 108.
13. C. Benvenuti, Nucl. Inst. Meth. 205, 391 (1983).
14. C. Benvenuti and F. Francia, J. Vac. Sci. Technol. A8, 3864 (1990).
15. O. Gröbner, Proc. Second European Particle Accelerator Conference, (Nice, 1990) in press (also CERN Report AT-VA-90-09).
16. H.-P. Reinhard, et. al., Proc. XI International Vacuum Congress, (Köln, 1989), Vacuum, in press.
17. R. Niemann, this conference.
18. S. H. Be, this conference.
19. J. C. Schuchman, J. Aloia, H. Hsieh, T. Kim, S. Pjerov, Proc. 1989 Particle Accelerator Conference, IEEE, New York (1989).

20. J. J. Welch and C. K. Sinclair, SLAC Report No. SLAC-PUB-3975, May 1986.
21. T. Kobari, et. al. in <u>Proc. AVS Topical Conference on Vacuum Systems for Compact and Advanced Synchrotron Light Sources</u>, Brookhaven, May 1988; H. Halama et. al., eds., AIP Conf. Proc. No. 171 (AIP, NY 1988), p. 100.
22. H.-P. Reinhard et. al., in <u>Proc. 9th Intern. Vacuum Congress</u>, Madrid, 1983 (A.S.E.V.A, Madrid 1983), p. 131.
23. P. della Porta, B. Ferrario, M. Borghi, J. Vac. Sci. Technol. 7, 300 (1970).
24. L. Rosai, B. Ferrario, P. della Porta, J. Vac. Sci. Technol. 15, 746 (1978).
25. O. Gröbner et. al., Vacuum <u>33</u>, 397 (1983).
26. M. Andritschky, CERN Report No. LEP-VA/MA, January 1989.
27. M. Andritschky, O. Gröbner, A. G. Mathewson, F. Schumann, P. Strubin, and R. Souchet, Vacuum <u>38</u>, 8 (1988).
28. A. G. Mathewson, O. Gröbner, P. Strubin, P. Marin, and R. Souchet, this conference.

Modern Residual Gas Analyser (RGA)

Applications and Developments in the Operation of Particle Accelerators and Experimental Facilities

S. P. Shannon and A. P. James
VG Quadrupoles Ltd., Cheshire, England, CW10 0HS

ABSTRACT

When designing a particle accelerator or storage ring the ultimate vacuum pressure is a prime consideration. This is because the better the vacuum, the fewer the number of collisions between the beam and the residual gas molecules, resulting in a longer beam lifetime — a significant criterion. The total pressure readings from BA gauges are useful for pressure distribution data, but an RGA is required to monitor the gas species. The data which RGA's provide is useful, not only to the vacuum scientist, but everyone connected with the construction and use of a synchrotron.

It is usual to employ RGA diagnostics in the early stages of machine construction. A typical sequence of tests would be for leak checking, glow-discharge and bakeout monitoring, re-checking for leaks and for contamination (either hydrocarbons or cleaning solvents). When a machine is being commissioned, the RGA can monitor for water to vacuum leaks (water cooled absorbers). Later, when the machine is operational, the RGA can show, via the local surface outgassing, when radiation is hitting new surfaces or obstructions.

This paper details how RGA diagnostics can be employed, the criteria used, and the extent of the value of the RGA. With today's new generation of microprocessor controlled RGA's it is possible to have simultaneous operation and data acquisition from a number of analysers without multiplexing. Multiplexing can, however, allow up to 64 analysers to be controlled from one RGA controller.

General Principles of Operation

A residual gas analyser system[1] consists of three main components; an analyser, a head amplifier/RF generator and a control unit. The analyser fits to the vacuum system, and the RF generator connects between the analyser and control unit (FIGURE 1). The source area of the analyser ionises the local gas molecules, extracting the ions to the mass filter in the control unit. The ions with a selected mass-to-charge ratio (m/q) are then transmitted to a detector. The detector can be a Faraday Cup or Secondary Electron Multiplier (SEM), generating a small output signal proportional to the partial pressure. The electronics in the control unit 'sweep' to select m/q values for the mass filter to pass — thus generating a mass scan or spectrum.

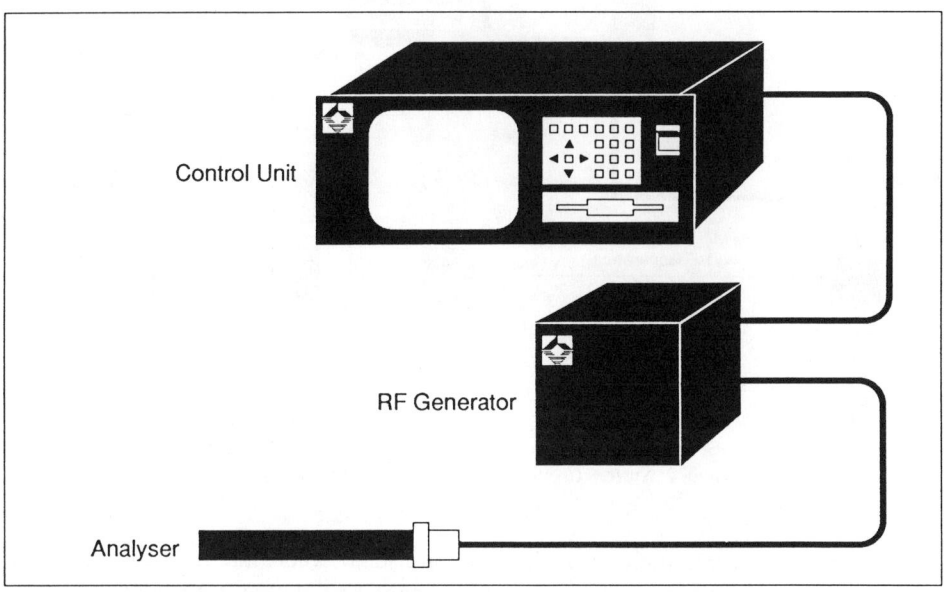

FIGURE 1: *Typical Configuration of an RGA System*

Multiple Head Operation

Particle accelerators and electron storage rings are now being used in the fabrication of IC's requiring sub 0.5 micron lithography[2] as well as for research machines. The accelerators and associated beam lines tend to be on a relatively large scale and need good UHV conditions for long operating cycles.

Large machines (*e.g.* SRS, NSLS, LEP) have many tens of analysers, often linked to a main control point and a host computer. Data from the analyser heads is switched or multiplexed on the central control screen.

Simultaneous RGA Monitoring

A single RGA control unit can analyse and display data from a number of operational analysers (FIGURE 2). A master control unit communicates to remote devices via an RS232 link and displays data from the selected analyser(s) which are all simultaneously operating using their own microprocessor controlled electronics. This configuration has none of the switching or settling delays sometimes associated with switching between analysers on a multiplexed system.

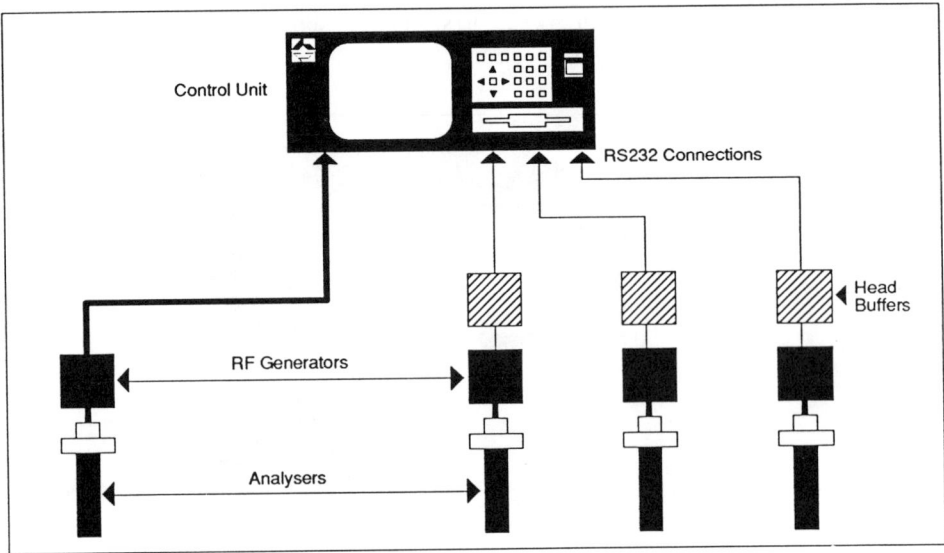

FIGURE 2: *System Configuration for Simultaneous Operation*

Hardware configurations of this type are ideal where the analysis points are separated by long distances. The cost of RS232 cables is relatively low compared to the multi-way cables normally used to carry the signals and filament current. Optical fibre links may also be used for noise immunity or if the analyser system is being 'floated' above ground potential.

Multiplexed Operation

On large machines it is appropriate to install a multiplexed system where a single RGA controller can be switched to one of many analysers (FIGURE 3). This can be a very cost effective way of monitoring high numbers of analysers (typically 16 to 64). Automatic and timed cycling of the enabled analysers can take place under computer control. The RGA controller in this application can have a number of control points, each with a keyboard and monitor, to enable remote operation at different locations.

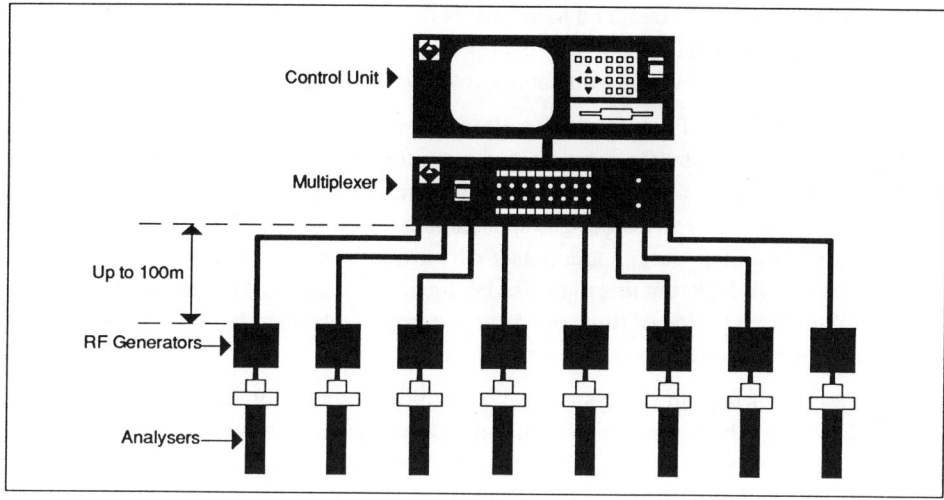

FIGURE 3: *Multiplexed System*

All the normal functions of the controller are unaffected by the installation of the multiplexer. In manual mode, channel selection is via push buttons on the front panel. Channels may be disabled to allow maintenance work while other heads are maintained in operational mode. Each channel has a trip facility, connected to a pressure gauge, to prevent a channel being brought into operation at high pressure. When the heads are in standby mode, each filament is powered to keep the source at operating temperature, with other potentials to the head biased to ensure rapid equilibrium conditions when brought 'on line'.

Benefits of Usage

Leak Detection

A leak in the system causes an additional gas load. This is undesirable as it limits the base pressure or introduces unwanted gas species (*e.g.* oxygen) into the vacuum system. A fundamental mode of operation for an RGA is 'leak detect', a standard option on all instruments.

Helium leak testing using an RGA is performed more easily than using a standard leak detector with the following added benefits:-

(i) Liquid nitrogen is not required to prevent 'backstreaming'.

(ii) The equipment can be left running continuously and unattended, as relay contacts are provided to switch external units in the event of a leak during 'silent hours'.

(iii) With the RGA head permanently installed on the system there is no need for coupling up any external pumping equipment, thereby avoiding a break in the vacuum envelope before starting the leak detection procedure.

408 Modern Residual Gas Analyser RGA

Since the RGA can be tuned to any mass (a leak detector is optimised for a single mass) any trace gas can be used. This is of particular interest on cryo pumped or super-cooled magnet systems where you cannot readily use helium; an alternative is argon.

Leak checking a large system is often a very time consuming process. By first taking a scan of the system, a semi-trained operator can quickly decide whether a leak is present by the 'peaks' present in the mass scan. If there is a leak in the system, the atmospheric gases (and any other gas species local to the leak) will be introduced into the vacuum system in larger amounts than normally seen. High pressure is often indicative of an air leak, but this may also be due to outgassing. If the chamber has been at atmosphere for a period of time, or has new components installed, then water vapour could have been absorbed on their surfaces.

The type of leak in the system is also important. By monitoring an appropriate partial pressure with respect to time you can differentiate between real and virtual leaks. Whilst the high total pressure of a leaking system is readily obvious, only the RGA can deduce the type of leak — for example:-

(i)   Atmosphere to vacuum leak.
(ii) Water to vacuum (from water cooled surfaces or an un-baked system).
(iii) Cleaning solvent from a blind hole.
(iv) Backstreaming from a faulty mechanical pump.
(v)   Surface outgassing due to beam cleaning of new areas which have not previously been exposed to 'beam' (if the increase in pressure is proportional to the beam parameters).

Gate valves are often used in large numbers on beam lines to isolate various vacuum chambers. By comparing gas scans from either side, you can check for seat leaks without removing the valve.

Mass Spectrum

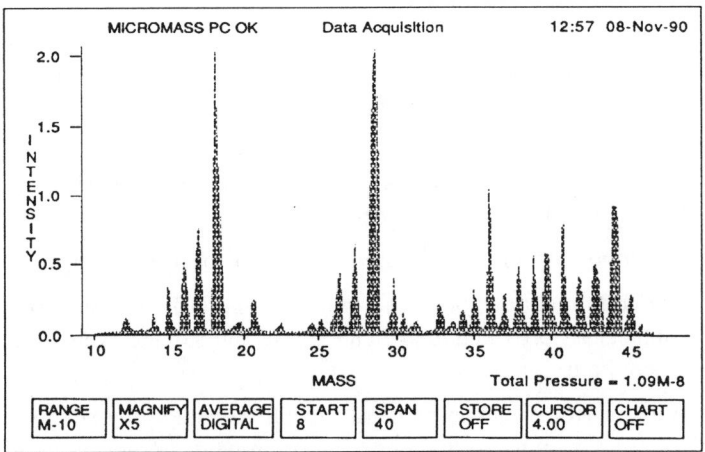

FIGURE 4: *Mass spectrum reveals components of total pressure*

An analogue scan allows the user to see the peak shape (*e.g.* resolution) and its relative position (mass alignment), it is therefore ideal for setting up and calibrating the instrument. On a digitally controlled RGA, an analogue scan is taken at small increments of mass value and the results are placed alongside each other on the display to re-generate the analogue scan. As each peak might require around 50 readings there is an associated time penalty, albeit very small, in taking such a scan.

In most instances, the user is only really interested in knowing which masses are there and their relative amounts. This is best displayed as a histogram or bar chart where only a single reading per mass is taken relating to the peak height. This mode also has a significant decrease in scanning time since fewer readings are taken.

Contamination Monitoring & Automated Searches

The experienced RGA user can easily recognise various types of spectrum (*e.g.* an air leak). As an aid to identification, the latest generation of microprocessor controlled RGA's often have a built-in Library Search mode to interpret the spectra.

Relating the peaks in the spectra to particular gas species can be aided by the use of 'cracking patterns'. A cracking pattern or 'fragmentation table' is a set of data for a gas species, for which the principal m/q is allocated 100% and then the secondary and tertiary peaks etc. are allocated percentage values proportional to their contributions. Water (H_2O), for example, can be ionised to a number of fragments:

Species	Mass	%
H_2O^+	18	100
HO^+	17	21
O^+	16	2

TABLE I - *Cracking Pattern for Water*

Following an automated library search the user is presented with a table of gas species and their relative proportions. It is normal for such a search routine to produce a 'confidence factor' enabling the user to monitor how well the spectra has been interpreted. A high confidence factor means that all the gas species were identified. A low confidence factor means that the library search was unable to produce a match for all the gas species in the system.

The more elegant RGA software packages allow spectra to be deconvoluted utilising a commercial mass spectral database *e.g.* NIST[3] with over 50,000 entries.

Computerised searches can have their limitations. If the gas species in your system is not in the library data, then obviously it cannot be matched. Cracking patterns have been calculated for both magnetic and quadrupole type mass spectrometers; if the cracking patterns are not appropriate for your instrument the results will be inconsistent.

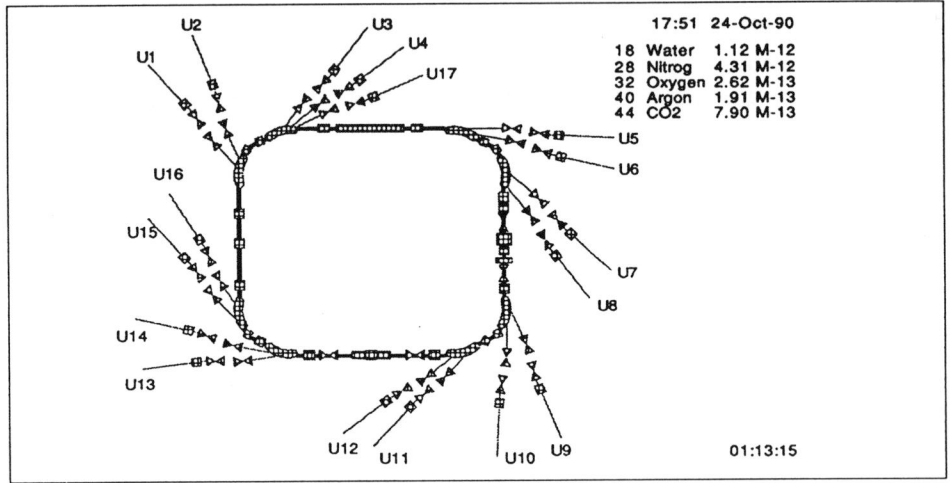

FIGURE 5: *All RGA heads can be shown on one display*

System Diagnostics

Vacuum requirements vary in different parts of an accelerator complex. At the junction of the beam line to the main ring, sometimes referred to as the 'beam port', there is normally a vacuum valve and radiation shutter. The operating crew will often need to ensure that the beam line is suitably 'clean', by taking an RGA scan and checking for contamination, before opening to UHV of the main ring.

Practical Considerations

Lines of Sight

If a 'line of sight' exists between the beam and the ion source of the analyser then interactions are likely to occur, giving false readings in the RGA. This is because electrons/ions can enter the quadrupole from the beam, altering the ionisation process in the ion source.

UHV total pressure measurement is normally carried out using a BA (Bayard Alpert) gauge. Construction is similar to the ion source region of an RGA and generates electrons to ionise the gas molecules. Precautions must be taken when using these two gauges to avoid interference with each others operation.

The general solution is to mount the analyser in an elbow configuration, requiring at least three reflections to enter the ion source area.

Radiation

On a circular machine, the synchrotron radiation is transmitted tangentially outwards, and so by positioning the analysers on the inside of the ring a number of problems can be avoided:-

(i) The plastic sheaths of the connecting cables can become hardened and brittle due to radiation effects. This can be eliminated by keeping the cables on the inside or in metal ducting.

(ii) Secondary electrons can be generated if radiation strikes the tube used to mount the analyser. This can also affect the results from the RGA by creating extra ions in the ion source.

(iii) Some electronics (such as the junction of transistors) can be affected by radiation. Again careful location, and the use of metal enclosures, can avoid problems.

If necessary lead blocks can be used when the ideal locating position is not available.

The collector signal from the detector in the analyser can be as low as 1×10^{-14} A. If a strong RF source, such as a klystron or RF cavity, is in the local vicinity then, under some conditions, it is possible to get RF superimposed on the collector signal. The use of copper knit mesh fastened around the RGA cables, with one end earthed, is often a simple and effective cure.

Bakeout

Bakeout is normally carried out at 250°C with Conflat flanges, and 150°C with O–rings. This initially increases the desorption of water and then produces a much lower background at room temperature.

The RGA can be used during bakeout, *e.g.* monitoring for mass 18, to detect the end point of bakeout (assuming the pressure is satisfactory for the sensor to operate), and also to detect whether any leaks have occured. It may be necessary to use a 'thermal extender' on the analyser for operation at bakeout temperatures. This moves the connector away from the analyser, avoiding destruction of the connector or local electronics by the heat.

In situ bakeout of a large system is expensive in terms of personnel and resources, yet it must be carefully controlled and monitored to avoid a disaster. Potential pitfalls are:-

(i) Too high a temperature can cause the braze of feedthroughs to become porous.

(ii) Junctions made of materials with dissimilar values of thermal conductivity/ expansivity can cause a leak to occur both in the heating and cooling parts of the cycle due to temperature differential.

(iii) If water cooled surfaces have not been drained, distortion, porosity or bursting may result as, at 100°C, the water will turn to steam and expand, creating a high pressure compared to that of the vacuum system.

Magnetic Fields

The trajectory of ions in the quadrupole mass filter will be influenced by external strong magnetic fields (>100 gauss) by altering the spiral path. This undesired side effect can occur, for example, if the RGA is situated too close to a Dipole magnet. A simple cure is to 'wrap' the analyser elbow in a jacket fabricated from mu-metal foil.

UHV Compatibility

As the vacuum pressure will have an influence on the beam lifetime due to particle collisions (particularly on an electron ring as the electrons are so 'light'), operation in the UHV region is required. All components within the vacuum envelope must, therefore, be UHV compatible.

At HV and UHV pressure levels, molecules desorbing from surfaces will limit the lowest pressure obtainable. It is therefore interesting to monitor which gas species come off the chamber walls. The gases from the RGA head on a small volume system can produce a dominant contribution, thus the need to be able to degas the analyser. Partial pressure analysis is essential in understanding residual gases, desorbed gases and very small leaks at pressures below 1×10^{-8} torr.

The analyser is normally of stainless steel and ceramic construction with copper connecting leads making it UHV compatible, *i.e.* hydrocarbon free, with a low outgassing rate and bakeable.

Ultra High Vacuum Source

At UHV pressures the RGA itself can become a source of outgassing, spurious peaks at 16, 19, 35 & 37 amu have been observed[4]. This effect has been termed ESD (Electron Stimulated Desorption) and is due to electrons from the hot filament striking the metal surfaces around the source area and desorbing ions. This contribution to the spectrum can be minimised by careful design. FIGURE 6 shows a UHV spectrum taken using a UHV source constructed of platinum mesh, thus reducing the bulk of metal, and specially treated to minimise desorption of ions. This method of manufacture leads to a 'UHV' source with noticeably superior performance at pressures below 10^{-9} torr.

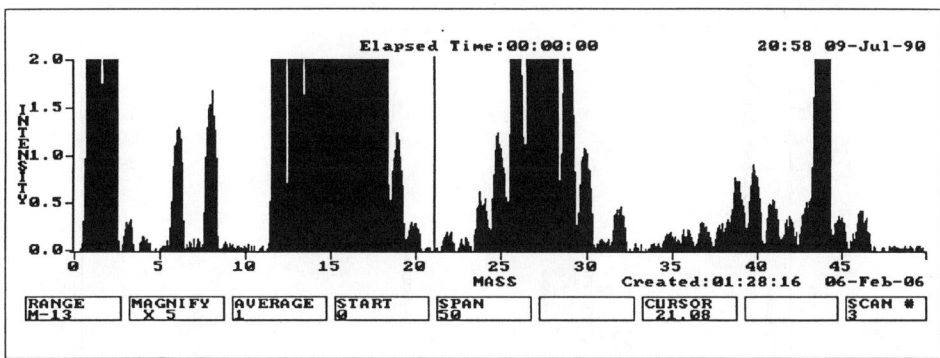

Total Pressure = 1 x 10⁻¹⁰ mbar

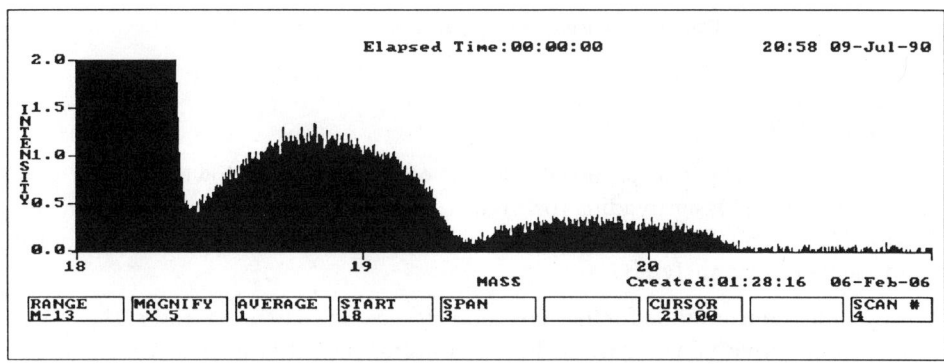

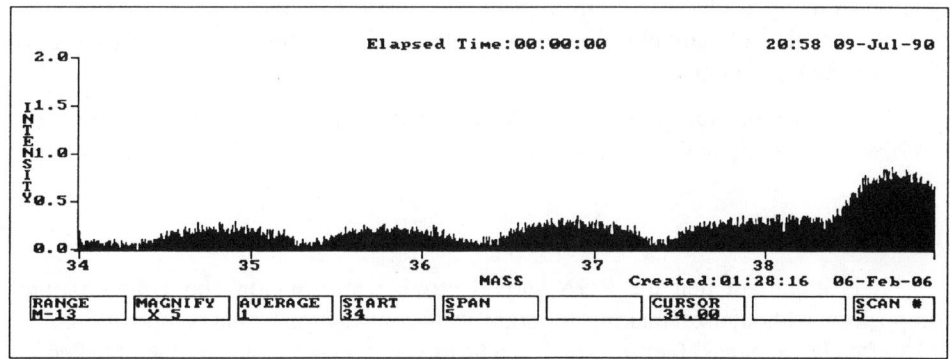

Very low ESD

FIGURE 6: *Performance of an RGA fitted with a UHV source*

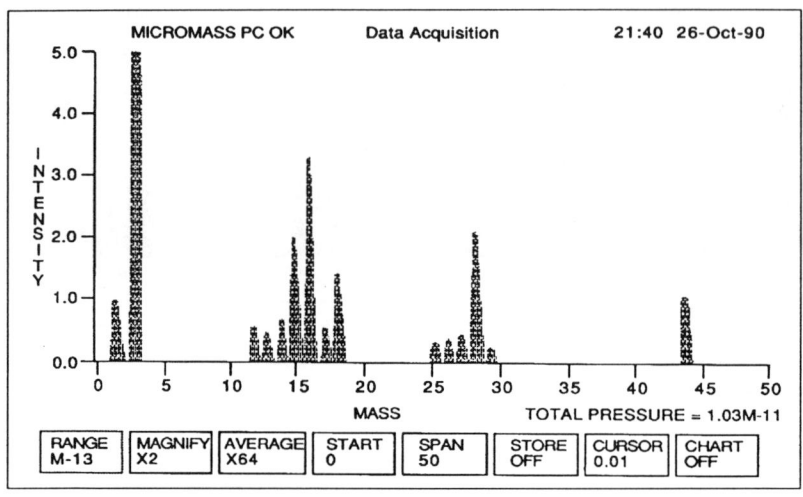

FIGURE 7: *Typical RGA Spectrum from UHV system*

Detection Limits

The Faraday cup detector is only capable of measurements down to the 1×10^{-10} torr region. So, for pressure readings below this (down to 1×10^{-14} torr levels), Secondary Electron Multipliers (SEM's) are used. Analysers which have both types of detector are called 'dual detector analysers'.

An important criterion for data collection is the 'signal to noise' ratio. The RGA, on its more sensitive gain ranges, can have a collector current as small as 1×10^{-14} A. If the cable is knocked or subject to vibration, spurious signals are produced which swamp the detector current, causing a noisy output. This effect, microphony, is due to a small emf being generated when the cables are disturbed. The effect can be minimised by electronics using a 'chopper' amplifier which subtracts background signal from the pressure signal. Alternatively, the signal cable can be fastened to a solid fixture to eliminate the microphony.

The typical residual gases in a UHV system at low pressures and temperature are H_2 (90%), H_2O, CO and CO_2. FIGURE 7 shows an example of a good clean UHV system.

Degas

It is important that the RGA should provide information about the vacuum composition without itself making any significant contribution to that composition. The RGA often has a 'degas' facility which can be used on its own or at the time of bakeout. The 'degas' facility electron bombards the source area accelerating the desorption of water. The resulting power is ramped over a period of minutes to avoid any gas bursts (which may stall a pump) and allows time for the gases to be pumped away.

Connecting the RGA to Other Devices

Operations such as leak checking should not be seen as a stand-alone manual operator test. It is more likely to be one of many steps in an experimental environment which is increasingly becoming more automated[5]. Here the control system will be required to switch RGA emission on and off, take measurements, check if alarm levels are breached, etc.

Multiple User Operation

There is often a requirement to control an RGA from two or more locations and to view it on external monitor displays. For example, a beam line might be 100m long, and the user will not want to repeatedly walk its full length just to check the vacuum scan.

One method is to take the output from the graphics card, fitted in the RGA, to a splitter box. This is a passive device that sends the same signal to a remote monitor that is sent to the RGA monitor. If the length is greater than 100m then the graphics card in the RGA does not have enough power to send the video signal down such distances without signal degradation. A commercially available video buffer box will overcome this difficulty.

An RS232 keyboard can act as a second 'remote' controller centre when linked into the RGA via an RS232 port with a suitable software device driver. In this configuration you can drive up to 100 feet; if the distance required is longer, then commercial RS232 line drivers will be required.

Host Computer

If the RGA has an RS232 port for integrating the unit to a host computer, this can be used for running remotely in undesirable environments, (*e.g.* radiation areas or over long distances). The data link is normally RS232 (IEEE-488 optional) and can be supplied in a variety of configurations with:

(i)   Line drivers to give a remote operation capability of many kilometres.

(ii) Optical fibre links, allowing the experimenter to 'float' the mass spectrometer electronics above ground potential.

(iii) Data line protectors to guard against electrical pick-up in 'noisy' environments.

Remote Operation with User Programs

Although the software supplied with the modern mass spectrometers is very comprehensive, the user may want to treat the RGA as a black box and write a computer program to control the specific functions required. Microprocessor type RGA's are generally capable of receiving instructions from a host computer, collecting the relevant data and transferring it back to the host on request via a data link. The benefits of this are that you utilise the sophisticated data measuring routines built into the mass spectrometer control unit, and you only need to write the code for the data handling (*e.g.* graphics or number crunching) for your own application.

Commands and responses take the form of a simple ASCII string. The acquired data is stored in a large buffer until requested to be sent as an ASCII string to the host, thus ensuring that no data is lost.

Remote Operation With VG 'Super-Serial' Software

A typical application of this facility is when the mass spectrometer is installed on a system some distance from the office or laboratory, but you still wish to control it using a standard PC and the VG software normally supplied with the instrument. On your computer screen you will see all the graphics normally displayed on the mass spectrometer, and you will be able to control the instrument using your keyboard.

Graphics Plotter

RGA controllers can now produce colour output on HPGL (Hewlett Packard Graphics Language) compatible plotters. This allows the user to optimise the use of existing equipment often found in a laboratory environment. The standard of output is ideal for technical reports and publications with the software adding the time and date for archiving data in chronological order.

Printer

With a printer, the software freezes the screen until it has performed a raster type scan of the page, sending data corresponding to each point on the screen.

If required, both a printer and plotter can be attached to a single instrument as they would use different ports and output software drivers.

Alarms

For safety reasons, not all interlocks operate via software. It is common practice to 'daisy chain' relay contacts, such that a pair of 'closed' floating contacts are available for 'good' conditions and are 'open' circuit for a 'fault' condition. In this way, if there is a power failure, there is no supply to energise a relay coil and 'open' the contacts to a fail safe condition.

Alarms can also take a number of other forms:
(i) TTL outputs for high and low breaches of set levels.
(ii) Text over data line to a host computer, (*e.g.* RS232).
(iii) Text messages on the display.

A microprocessor controlled RGA system can sum the partial pressures over a range of masses (*e.g.* hydrocarbons) and can continuously monitor species generating alarm outputs as necessary.

Analogue Outputs and Inputs

In a multiple channel type operating mode, where the partial pressures for selected masses are displayed, it is often desirable to have an analogue output proportional to the signal for external monitoring or feedback purposes. This is normally in the form of a 0 to 10V signal per channel enabled and is automatically carried out with no need for user interaction.

The mass spectrometer can read up to 8 external DC voltages (-10V to +10V). This information can then be re-scaled and titled, allowing it to be plotted alongside spectral data on the screen and saved to disk if desired. Analogue input data can also be reviewed (if previously stored) in a manner similar to that for spectral data. Two examples that can be displayed as physical parameters are temperature and ion gauge readings.

It is also possible to read in analogue parameters via CAMAC for example. A control system would normally 'read in' analogue outputs from ion gauges and have the reading in digital format. It is possible for an RGA controller to have a CAMAC software driver and interface so that it can read in these values.

Conclusions

The RGA is like a microscope inside the vacuum system and allows the user to identify which gases are in the system, enabling him to take steps to reach an ultimate vacuum. The quadrupole type RGA is now a powerful and versatile instrument in the field of mass spectrometry, and is an invaluable tool for the vacuum scientist. It is capable of being integrated into the most sophisticated system.

REFERENCES

1. J. H. BATEY, *Vacuum*, 37, 659-68, (1987).
2. S. P. SHANNON, *Microelectronics Manufacturing and Testing, Vol 10, No 10*, 16-18, (1987).
3. NIST, NATIONAL INSTITUTE OF STANDARDS AND TECHNOLOGY, Gaithersburg, MD 20899.
4. M. J. DRINKWINE and D. LICHTMANN, *Partial Pressure Analyser and Analysis*, American Vacuum Society Monograph Series.
5. A. P. JAMES, *Vacuum*, 37, 677-80, (1987).

VACUUM VALVES FOR SYNCHROTRONS AND STORAGE RINGS

John A. Freeman
VAT, Inc., Woburn, MA. 01801

ABSTRACT

Key issues in selecting valves for synchrotrons, the sealing technologies used in the valves, and the types of valves commonly chosen are reviewed. Valves with metal static seals and viton plate seals are often used for isolating vacuum pumps or used on experimental lines when the environment permits them. Other elastomers can extend radiation resistance or bake limits. For radiation environments, the seals in the valve actuator are considered. In high radiation or deep UHV environments all-metal valves are necessary. Sector valves with RF contacts isolate sections of the storage ring when closed and provide good RF continuity along the walls of the ring when open. For some experiments, gate valves with a membrane or window are needed. If an experimental line is accidently vented, a fast closing system isolates it to prevent venting of the ring and other experiments. To reduce the risk of equipment damage or injury a beam stopper closes and absorbs the beam line energy.

INTRODUCTION

Both the static seals to the outside world and the internal gate or plate seal influence the attainable vacuum, outgassing, bakeability, radiation limit, cycle life, and serviceability of the valve. One should also consider the actuator type, position indicators, allowable differential pressure when closed and during opening, and the response to power or compressed air failure. Valves used in synchrotrons can be grouped as sector valves, pump isolation valves, beamline isolation valves, fast closing systems, and window valves. Figure 1 shows the typical locations of valves in UHV beam lines and storage rings. Beam stoppers share construction techniques with valves and function to protect the vacuum valves and personnel.

METAL STATIC SEALS

Beam lifetimes are enhanced by using metal flange seals to minimize extraneous gas in the system from leaks, outgassing, or permeation. Most valves have a removable

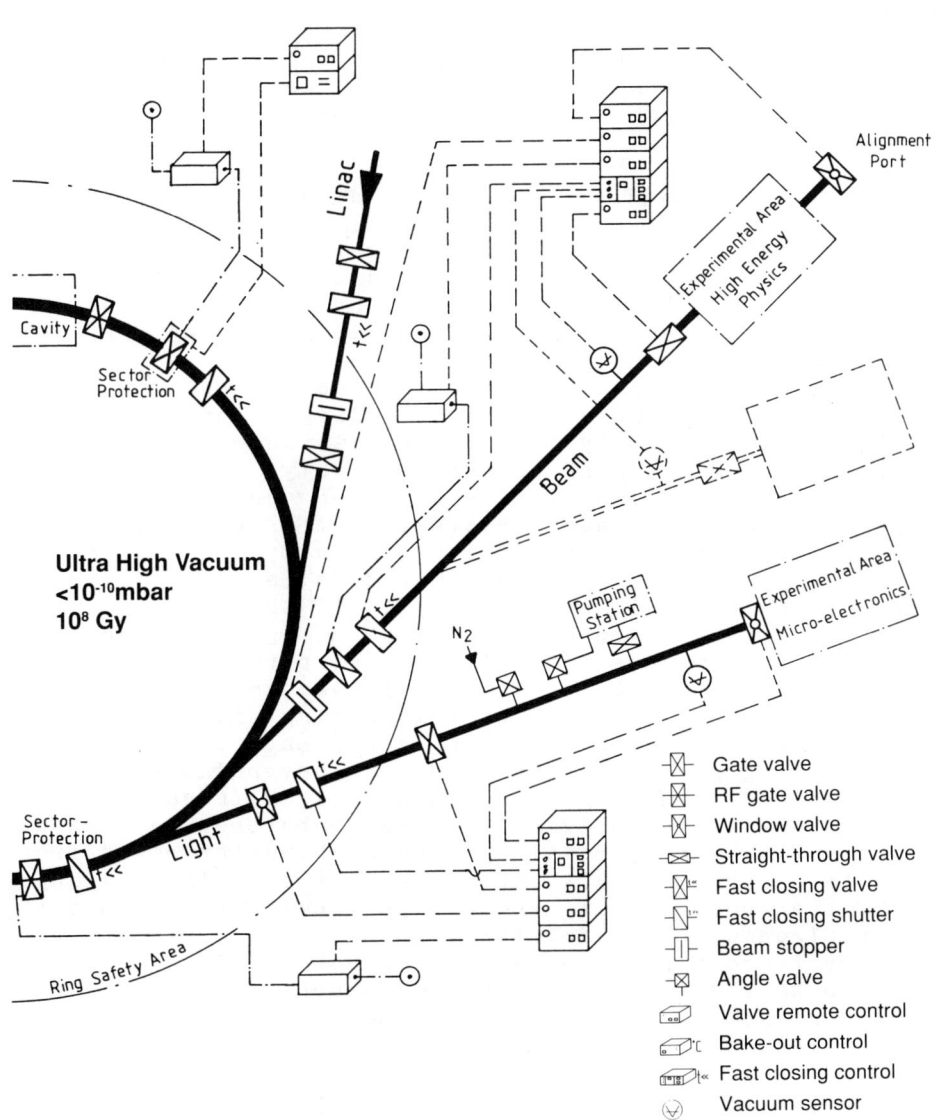

Figure #1

bonnet flange to access the internal mechanism for maintenance; to be consistent with the vacuum objectives of the synchrotron, this bonnet flange also should be metal sealed. On some valves welded bonnet flanges are a lower cost alternative to removable metal sealed bonnets, but welded bonnets make the valve either disposable or costly to repair: This is of no concern if the expected cycle life of the valve is longer than the expected life of the synchrotron.

ELASTOMER SEALED VALVE PLATES

The most common internal seal material on the valve plate is viton. Viton sealed valves have a relatively low cost compared to all-metal valves, and require a sealing force of only a few kg per cm of sealing line which allows the use of compact actuators. Some valves with viton plate seals can be baked for short periods as high as 250 Celsius open (seal not compressed) or 200 Celsius closed. Viton tolerates approximately 10E7 rad radiation exposure before it deteriorates too much to seal reliably. However, the relatively high degassing rate of viton contributes extraneous gas to a vacuum system, and for a standard helium leak test the leak rate across a typical viton sealed plate is about 1x10E-9 mbar xl/s, at least five times higher than all-metal valves. If there is a substantial total area of viton sealing against atmosphere or in the system, the total leak rate, permeation rate, or outgassing rate can limit achievable vacuum.

The recent innovation of vulcanizing viton directly to a sealing plate instead of using a separate o-ring enhances viton sealing for ultrahigh vacuum in several ways. There is no trapped dead space under the seal, eliminating a source of gas. Vulcanizing the seal onto the plate results in Viton with very consistent vacuum characteristics from batch to batch. Finally, the risk of the o-ring pulling out of the groove, using an improper seal, or damage during seal replacement is eliminated.

Perfluoro compounds such as Dupont's Kalrez have been shown to have degassing properties superior to viton under certain conditions, such as after methanol cleaning (1), and can increase bakeability limits of valves to 250 - 300 Celsius. However, perfluoro o-rings have a very high compression set and high sticking coefficient. For example, if baked at 200 Celsius and cooled while compressed, a perfluoro seal will remain flat, creating sealing problems afterwards, and if the valve is opened, the o-ring may pull out of the groove.

ALL-METAL VALVES

Extreme UHV environments require the low outgassing, leak rates, and permeation of baked metal seals. In Synchrotrons operating at less than $1 \times 10E-9$ mbar, all metal seals (including the valve seat seal) are essential. With few exceptions, all-metal valves are used as sector valves, as the primary valves isolating beam lines from the ring, and even on beam line pump out ports where the seal is exposed on one side to atmosphere for long periods of time.

Two basic categories of metal seat seals exist: soft metal seals such as copper or gold which are relatively inelastic and permanently deformed when the seal is made, and hard metal seals, typically silver coated stainless steel. Soft metal seals require relatively low closing force, and if closed carefully, can be resealed dozens or even hundreds of times. However, each closure cuts or compresses the seal further, so the actuator position and closing torque must be higher each time. One over-tightening can result in the need for a seal replacement. Because each closure needs a different force, pneumatic soft metal sealed valves use relatively complex pneumatics or are not available at all, depending on the size and manufacturer. Nonetheless, soft-seal manual angle valves are relatively inexpensive and reliable, and thousands are in use. Soft-seal all metal valves can provide acceptable service life if handled properly and records of the closing position are kept.

Elastic hard metal seat seals of silver coated stainless steel require high closing forces (a few hundred kg per cm of sealing line), resulting in larger handwheels or pneumatic operators. However, they have some advantages. Since the seal is not permanently deformed, it is typically good for tens of thousands of closures without replacement and requires the same closing torque each time. Therefore, service life is high and not as operator sensitive as soft-metal sealed valves. Finally, pneumatic versions are available in a wide variety of sizes and configurations.

ACTUATORS

On synchrotrons, manual valves with hand cranks or hex-heads (used with a wrench or hand-wheel) are common on pump-out ports. Various manual actuators such as levers, quick-turn, torque-limiting handwheels, etc. are available. For remotely driven valves, double-acting pneumatic valves are most common . Other types of actuators such as electromagnetic drives, AC or DC motors, spring open/air close, or air open/spring close, are available on some sizes and types of valves.

Actuators transfer motion to the valve plate through either a dynamic shaft seal or expandable metal bellows. The leak rate of dynamic shaft seals is too high for UHV environments and they can't be baked, so most valves on synchrotrons use bellows seals.

POSITION INDICATORS

Position indicators (limit switches) provide a confirmation that the valve has reached its fully open or closed position. They are standard on many (but not all) pneumatic valves and an option on some manual ones. On some valves double limit switches are available in case one switch fails. Ceramic limit switches for bake zones or radiation zones are also available. Valves with mechanical visual position indicators on them are useful to provide a local indication and verification of the switch reading.

BAKEABILITY

Bake-out limits of a valve are determined by its seals, actuator, hardware such as solenoids and position indicators, and the valve's construction. Valves with metal static seals but viton seat seals are available bakeable to 200 Celsius closed or 250 Celsius open. All metal valves are available for baking at 300 Celsius and even 450 Celsius. Bakeability of the actuator and hardware frequently are different than the valve itself. Pneumatic actuators bakeable to 200 Celsius are common, and special ones for to 300 Celsius are available on some valves, but such temperatures require a remoted solenoid and either no position indicators or bakeable ones. Finally, the heating and cooling rates are also important, especially on all metal valves which tend to have critical tolerances. Different expansion rates between the valve body and internal mechanism can result in leaks across the seat or valve damage.

RADIATION LIMITS

Valves with viton seals are limited to about 10E7 rad exposure. Bodies and mechanisms of all-metal valves are radiation resistent to 10E10 rad, but the pneumatic actuators are limited to 10E7 rad by the viton pneumatic seals in the actuator cylinder: This limit can be increased to 10E8 rad by using Buna-N, but the temperature limit is reduced to 80 Celsius. Also, lead shielding of the actuator could be used. Special pneumatics without elastomers are available on some valves to extend the exposure limit to at least 10E10 rad, which is also the limit for most manual all-metal valves.

MAGNETIC FIELDS

Synchrotron valves consist primarily of stainless steel 304, 304L or 316L, but there can be magnetic parts in the actuator or position indicators. For highly sensitive applications, special valves are available, made from carefully selected materials with low magnetic permeability. All-metal aluminum valves are also available and may be suitable for applications extremely sensitive to magnetic materials.

DIFFERENTIAL PRESSURE

Different valve types and sizes have different limits of pressure they can withstand across the seat while closed or during opening. The manufacturer's specified limits should be followed carefully. Most angle valves can be opened with 1 bar pressure differential on the valve plate. Some viton sealed gate valves can also be opened with 1 bar differential under certain mounting conditions, but most cannot. All-metal gate valves can sustain expensive damage by opening against excess differential pressure, so care should be taken through procedures or interlocks to assure pressure is reduced on both sides of the plate to within the manufacturers limits before opening the valve.

POWER OR AIR FAILURE

Double acting pneumatic valves typically close during a power failure (currentless closed) but can be configured as currentless open, or by use of an impulse solenoid, to remain in position. Spring open valves open at power failure, spring close valves close.

Most valves, but not all, lock closed, so if the compressed air fails the valve remains sealed. On valves that spring open or do not mechanically lock shut, an air failure means the valve would open fully or partly; to avoid this an airline check valve is needed. If air fails while the valve is open, the gate may slide partially closed depending on its orientation in the system: such closures are avoided by mounting the valve so gravity holds it open, or by using a check valve to maintain local air pressure.

SECTOR VALVES

Sector valves allow a ring to be installed, leak-checked, and vacuum conditioned in sections, and sections can be serviced without venting the entire ring. To optimize the UHV environment and because radiation is present, all-metal valves are preferred. RF current on the walls of the beam tube follows the beam around the ring, but if it comes to a gate valve in the beam tube, it will follow the internal surfaces of the valve body to get to across the valve. Therefore, the RF current is delayed relative to the beam, de-stabilizing and de-energizing the beam. Sector valves with low resistance contacts or "RF Fingers" which effectively simulate the beam tube profile when the valve is open are used to allow the RF current to stay in phase with the beam, and flow unimpeded around the ring.

Sector valves should be configured with impulse solenoids to hold their position during power interruptions, and should be locked both open and closed to remain unaffected by air pressure problems.

PUMP ISOLATION VALVES

To evacuate the system initially, mobile pumpstands are connected to isolation valves on the ring or beam line. On systems where conductance is highly limited and the initial evacuation time is not critical, manual angle valves can be an effective, economical choice. If the pump is coupled through a short high conductance line, a gate may be preferred. To avoid having an elastomer seal between vacuum and atmosphere for long periods, manual all-metal angle valves are used. If a hex-head actuator is used and the handwheels
to operate them reside with the pump cart, accidental valve opening risk is minimized. In addition, if the pump cart is to be left unattended, it should include a viton-sealed automatic isolation valve in case of pump failure.

BEAM STOPPERS

Experimental lines normally are isolated from the storage ring by a gate valve near where the line enters the ring. Since this gate would be damaged by the beam, a water-cooled beam stopper is lowered between the isolation valve and and ring whenever the experimental line is not in use. Beam stoppers use metal static seals and bellows, but they are non-sealing. Most beam stoppers are custom designed by the synchrotron facility, but commercial units are available for dissipating up to 30W per sq. mm or 7kW total.

EMERGENCY BEAM LINE ISOLATION, FAST CLOSING SYSTEMS

Breaking a port, gauge or feedthrough can cause a high pressure shockwave moving at least 330 meters per second to propagate down the experimental line, rapidly venting the storage ring, and all the other lines. To avoid such a catastrophe, fast closing valves, fast closing shutters, and sometimes acoustic delay lines are used.

Where an experimental line branches into smaller lines (1" - 1 1/2"), simple yet effective fast closing devices can be utilized. All-metal valves closing in about 35 ms are available, suitable for about 50 closures before maintenance. Fast closing gates up to 1 1/2" I.D. are available which close and seal in under 8 ms, using a viton gate seal good for thousands of closures. Fast closing valves over 1 1/2" I.D. have a viton plate seal and close from 13 ms to 20 ms., depending on their size. For a faster response or for all-metal sealing systems, a fast, non-sealing all metal shutter is used together with a vacuum isolation valve. Shutters close in under 10ms with a leak rate under
2 mbar x l/s, stopping the shock wave and allowing several seconds for a sealing valve to close behind it.

If the time for a sudden air in-rush to go from its source to the fast valve/shutter is too short, acoustic delay can be designed into the experimental line to delay the shock wave and provide additional time (2).

WINDOW VALVES

Valves with a transparent window may be required at one or more points to allow an experimental line to be installed, evacuated, and aligned with lasers, or occassionally re-aligned while under vacuum. Valves with windows of glass, sapphire, beryllium or other materials, or thin foils, are often used at the end of beam lines to separate an experimental chamber at the end of the line. The window must withstand pressure differentials that depend on its use. All-metal valves use permanently bonded windows. Valves with viton seat seals can use easily exchanged viton-sealed windows. In some cases the window needs no seal at all, providing a non-sealing separation of the experiment from the rest of the beam line.

INSULATING VACUUMS

One additional use of vacuum valves in synchrotrons is for sealing the insulating vacuum on super cooled magnets. This function is typically performed with inexpensive viton-sealed angle valves or special portable pump-out valves that can be moved from location to location leaving behind only a vacuum tight cap. The use of a portable valve assembly and multiple vacuum tight caps reduces costs and reduces the chance of accidental opening of the ports.

SUMMARY

Selection of the correct valve for the application is important for optimizing a synchrotron's vacuum performance and therefore its beam quality and lifetime. The valves play an important role in the synchrotron's up time, safety, serviceability and convenience. Understanding the issues involved in the valve selection and the valve types helps users and designers in their discussions with the valve manufacturers, and leads to better valve designs, more reliable and efficient synchrotrons, and more satisfied users.

REFERENCES

1. Dr. G. Rao, CEBAF paper TN-0185, unpublished study
2. K. Sakai et al Symposium of 4th Proceeding Accelerator Science and Technology, Japan

Author Index

A

Anashin, V. V., 30

B

Be, S. H., 102, 110, 374, 381
Benaroya, R., 84
Bernardini, M., 188
Biasci, J. C., 266
Bulygin, A. N., 30

C

Chen, J. R., 173, 355
Chin, J., 219
Choi, M., 84
Chung, S. M., 202
Citrolo, J. C., 389
Crank, P. A., 183

D

Daibo, H., 102, 381
Dortwegt, R. J., 84
Dylla, H. F., 389

E

Eberle, W. J., 361

F

Foerster, C. L., 325
Freeman, J. A., 418

G

Gavrilov, N. G., 30
Geiler, D. E., 210
Goeppner, G., 84, 124
Gomes, P. A. P., 118
Gonczy, J., 84
Gröbner, O., 18, 313

H

Halama, H., 39
Halbach, K., 219
Hassenzahl, W., 219
Hill, S. F., 183
Hisamatsu, H., 1
Hori, Y., 347
Hoshi, Y., 142
Howell, J., 84
Hoyer, E., 219
Hsiung, G. Y., 355
Humphries, D., 219

I

Ikeguchi, T., 347
In, S. R., 102, 374, 381
Ishimaru, H., 1

J

James, A. P., 404

K

Kanazawa K., 1
Karmarkar, M. G., 155
Kennedy, K., 197
Kil, K. H., 202
Kim, C. K., 202
Kincaid, B., 219
Kobari, T., 347
Kobayashi, M., 332, 347
Kollerov, E. P., 30
Konishi, K., 110
Korn, G., 325
Korchuganov, V. N., 30
Krieger, C., 84

L

LaMarche, P. H., 389
Lancaster, H., 219

Li, G., 168
Liptak, R. J., 361
Liu, Y. C., 173, 355

M

Mairs, T., 266
Mandai, S., 142
Manos, D. M., 389
Marin, P., 52, 313
Maruyama, T., 374
Mathewson, A. G., 313, 389
Mazza, F., 389
Matsumoto, M., 347
Momose, T., 1
Morimoto, Y., 102, 110

N

Nakashizu, T., 142
Newnam, B. E., 278
Nielsen, R., 84
Niemann, R., 84
Nikitin, A. I., 30
Nishidono, T., 102, 142, 381

O

Oikawa, Y., 102
Oishi, M., 142
Osipov, V. N., 30

P

Park, C. D., 202
Patel, R. J., 155
Plate, D., 219
Poncet, A., 389

R

Raftopoulos, S., 389
Ramamurthi, S. S., 155
Rehn, V., 235

Reid, R. J., 183
Renier, M., 71
Roop, B., 84
Rusch, T. W., 361

S

Schmied, D., 71
Schuchman, J. C., 300
Shannon, S. P., 404
Shirakura, T., 110
Souchet, R., 313
Strubin, P., 18, 313
Suetsugu, Y., 1

T

Takahashi, S., 102, 110
Trakhtenberg, E. M., 30
Trickett, B., 71

U

Ueda, S., 347
Uesaka, M., 142
Ulrickson, M., 389

W

Watanabe, K., 102
Wherle, R., 84

X

Xu, W., 152
Xu, X., 152

Y

Yao, C., 152
Yokouchi, S., 102, 110, 374, 381

AIP Conference Proceedings

		L.C. Number	ISBN
No. 108	The Time Projection Chamber (TRIUMF, Vancouver, 1983)	83-83445	0-88318-307-2
No. 109	Random Walks and Their Applications in the Physical and Biological Sciences (NBS/La Jolla Institute, 1982)	84-70208	0-88318-308-0
No. 110	Hadron Substructure in Nuclear Physics (Indiana University, 1983)	84-70165	0-88318-309-9
No. 111	Production and Neutralization of Negative Ions and Beams (3rd Int'l Symposium) (Brookhaven, NY, 1983)	84-70379	0-88318-310-2
No. 112	Particles and Fields – 1983 (APS/DPF, Blacksburg, VA)	84-70378	0-88318-311-0
No. 113	Experimental Meson Spectroscopy – 1983 (7th International Conference, Brookhaven, NY)	84-70910	0-88318-312-9
No. 114	Low Energy Tests of Conservation Laws in Particle Physics (Blacksburg, VA, 1983)	84-71157	0-88318-313-7
No. 115	High Energy Transients in Astrophysics (Santa Cruz, CA, 1983)	84-71205	0-88318-314-5
No. 116	Problems in Unification and Supergravity (La Jolla Institute, 1983)	84-71246	0-88318-315-3
No. 117	Polarized Proton Ion Sources (TRIUMF, Vancouver, 1983)	84-71235	0-88318-316-1
No. 118	Free Electron Generation of Extreme Ultraviolet Coherent Radiation (Brookhaven/OSA, 1983)	84-71539	0-88318-317-X
No. 119	Laser Techniques in the Extreme Ultraviolet (OSA, Boulder, CO, 1984)	84-72128	0-88318-318-8
No. 120	Optical Effects in Amorphous Semiconductors (Snowbird, UT, 1984)	84-72419	0-88318-319-6
No. 121	High Energy e^+e^- Interactions (Vanderbilt, 1984)	84-72632	0-88318-320-X
No. 122	The Physics of VLSI (Xerox, Palo Alto, CA, 1984)	84-72729	0-88318-321-8
No. 123	Intersections Between Particle and Nuclear Physics (Steamboat Springs, CO, 1984)	84-72790	0-88318-322-6
No. 124	Neutron-Nucleus Collisions: A Probe of Nuclear Structure (Burr Oak State Park, 1984)	84-73216	0-88318-323-4
No. 125	Capture Gamma-Ray Spectroscopy and Related Topics – 1984 (Int'l Symposium, Knoxville, TN)	84-73303	0-88318-324-2
No. 126	Solar Neutrinos and Neutrino Astronomy (Homestake, 1984)	84-63143	0-88318-325-0
No. 127	Physics of High Energy Particle Accelerators (BNL/SUNY Summer School, 1983)	85-70057	0-88318-326-9
No. 128	Nuclear Physics with Stored, Cooled Beams (McCormick's Creek State Park, IN, 1984)	85-71167	0-88318-327-7

No.	Title	LCCN	ISBN
No. 129	Radiofrequency Plasma Heating (Sixth Topical Conference) (Callaway Gardens, GA, 1985)	85-48027	0-88318-328-5
No. 130	Laser Acceleration of Particles (Malibu, CA, 1985)	85-48028	0-88318-329-3
No. 131	Workshop on Polarized ^{3}He Beams and Targets (Princeton, NJ, 1984)	85-48026	0-88318-330-7
No. 132	Hadron Spectroscopy–1985 (International Conference, Univ. of Maryland)	85-72537	0-88318-331-5
No. 133	Hadronic Probes and Nuclear Interactions (Arizona State University, 1985)	85-72638	0-88318-332-3
No. 134	The State of High Energy Physics (BNL/SUNY Summer School, 1983)	85-73170	0-88318-333-1
No. 135	Energy Sources: Conservation and Renewables (APS, Washington, DC, 1985)	85-73019	0-88318-334-X
No. 136	Atomic Theory Workshop on Relativistic and QED Effects in Heavy Atoms (Gaithersburg, MD, 1985)	85-73790	0-88318-335-8
No. 137	Polymer-Flow Interaction (La Jolla Institute, 1985)	85-73915	0-88318-336-6
No. 138	Frontiers in Electronic Materials and Processing (Houston, TX, 1985)	86-70108	0-88318-337-4
No. 139	High-Current, High-Brightness, and High-Duty Factor Ion Injectors (La Jolla Institute, 1985)	86-70245	0-88318-338-2
No. 140	Boron-Rich Solids (Albuquerque, NM, 1985)	86-70246	0-88318-339-0
No. 141	Gamma-Ray Bursts (Stanford, CA, 1984)	86-70761	0-88318-340-4
No. 142	Nuclear Structure at High Spin, Excitation, and Momentum Transfer (Indiana University, 1985)	86-70837	0-88318-341-2
No. 143	Mexican School of Particles and Fields (Oaxtepec, México, 1984)	86-81187	0-88318-342-0
No. 144	Magnetospheric Phenomena in Astrophysics (Los Alamos, NM, 1984)	86-71149	0-88318-343-9
No. 145	Polarized Beams at SSC & Polarized Antiprotons (Ann Arbor, MI & Bodega Bay, CA, 1985)	86-71343	0-88318-344-7
No. 146	Advances in Laser Science–I (Dallas, TX, 1985)	86-71536	0-88318-345-5
No. 147	Short Wavelength Coherent Radiation: Generation and Applications (Monterey, CA, 1986)	86-71674	0-88318-346-3
No. 148	Space Colonization: Technology and The Liberal Arts (Geneva, NY, 1985)	86-71675	0-88318-347-1
No. 149	Physics and Chemistry of Protective Coatings (Universal City, CA, 1985)	86-72019	0-88318-348-X
No. 150	Intersections Between Particle and Nuclear Physics (Lake Louise, Canada, 1986)	86-72018	0-88318-349-8

No. 151	Neural Networks for Computing (Snowbird, UT, 1986)	86-72481	0-88318-351-X
No. 152	Heavy Ion Inertial Fusion (Washington, DC, 1986)	86-73185	0-88318-352-8
No. 153	Physics of Particle Accelerators (SLAC Summer School, 1985) (Fermilab Summer School, 1984)	87-70103	0-88318-353-6
No. 154	Physics and Chemistry of Porous Media—II (Ridge Field, CT, 1986)	83-73640	0-88318-354-4
No. 155	The Galactic Center: Proceedings of the Symposium Honoring C. H. Townes (Berkeley, CA, 1986)	86-73186	0-88318-355-2
No. 156	Advanced Accelerator Concepts (Madison, WI, 1986)	87-70635	0-88318-358-0
No. 157	Stability of Amorphous Silicon Alloy Materials and Devices (Palo Alto, CA, 1987)	87-70990	0-88318-359-9
No. 158	Production and Neutralization of Negative Ions and Beams (Brookhaven, NY, 1986)	87-71695	0-88318-358-7
No. 159	Applications of Radio-Frequency Power to Plasma: Seventh Topical Conference (Kissimmee, FL, 1987)	87-71812	0-88318-359-5
No. 160	Advances in Laser Science–II (Seattle, WA, 1986)	87-71962	0-88318-360-9
No. 161	Electron Scattering in Nuclear and Particle Science: In Commemoration of the 35th Anniversary of the Lyman-Hanson-Scott Experiment (Urbana, IL, 1986)	87-72403	0-88318-361-7
No. 162	Few-Body Systems and Multiparticle Dynamics (Crystal City, VA, 1987)	87-72594	0-88318-362-5
No. 163	Pion–Nucleus Physics: Future Directions and New Facilities at LAMPF (Los Alamos, NM, 1987)	87-72961	0-88318-363-3
No. 164	Nuclei Far from Stability: Fifth International Conference (Rosseau Lake, ON, 1987)	87-73214	0-88318-364-1
No. 165	Thin Film Processing and Characterization of High-Temperature Superconductors (Anaheim, CA, 1987)	87-73420	0-88318-365-X
No. 166	Photovoltaic Safety (Denver, CO, 1988)	88-42854	0-88318-366-8
No. 167	Deposition and Growth: Limits for Microelectronics (Anaheim, CA, 1987)	88-71432	0-88318-367-6
No. 168	Atomic Processes in Plasmas (Santa Fe, NM, 1987)	88-71273	0-88318-368-4
No. 169	Modern Physics in America: A Michelson-Morley Centennial Symposium (Cleveland, OH, 1987)	88-71348	0-88318-369-2
No. 170	Nuclear Spectroscopy of Astrophysical Sources (Washington, DC, 1987)	88-71625	0-88318-370-6
No. 171	Vacuum Design of Advanced and Compact Synchrotron Light Sources (Upton, NY, 1988)	88-71824	0-88318-371-4

No. 172	Advances in Laser Science–III: Proceedings of the International Laser Science Conference (Atlantic City, NJ, 1987)	88-71879	0-88318-372-2
No. 173	Cooperative Networks in Physics Education (Oaxtepec, Mexico, 1987)	88-72091	0-88318-373-0
No. 174	Radio Wave Scattering in the Interstellar Medium (San Diego, CA, 1988)	88-72092	0-88318-374-9
No. 175	Non-neutral Plasma Physics (Washington, DC, 1988)	88-72275	0-88318-375-7
No. 176	Intersections Between Particle and Nuclear Physics (Third International Conference) (Rockport, ME, 1988)	88-62535	0-88318-376-5
No. 177	Linear Accelerator and Beam Optics Codes (La Jolla, CA, 1988)	88-46074	0-88318-377-3
No. 178	Nuclear Arms Technologies in the 1990s (Washington, DC, 1988)	88-83262	0-88318-378-1
No. 179	The Michelson Era in American Science: 1870–1930 (Cleveland, OH, 1987)	88-83369	0-88318-379-X
No. 180	Frontiers in Science: International Symposium (Urbana, IL, 1987)	88-83526	0-88318-380-3
No. 181	Muon-Catalyzed Fusion (Sanibel Island, FL, 1988)	88-83636	0-88318-381-1
No. 182	High T_c Superconducting Thin Films, Devices, and Application (Atlanta, GA, 1988)	88-03947	0-88318-382-X
No. 183	Cosmic Abundances of Matter (Minneapolis, MN, 1988)	89-80147	0-88318-383-8
No. 184	Physics of Particle Accelerators (Ithaca, NY, 1988)	89-83575	0-88318-384-6
No. 185	Glueballs, Hybrids, and Exotic Hadrons (Upton, NY, 1988)	89-83513	0-88318-385-4
No. 186	High-Energy Radiation Background in Space (Sanibel Island, FL, 1987)	89-83833	0-88318-386-2
No. 187	High-Energy Spin Physics (Minneapolis, MN, 1988)	89-83948	0-88318-387-0
No. 188	International Symposium on Electron Beam Ion Sources and their Applications (Upton, NY, 1988)	89-84343	0-88318-388-9
No. 189	Relativistic, Quantum Electrodynamic, and Weak Interaction Effects in Atoms (Santa Barbara, CA, 1988)	89-84431	0-88318-389-7
No. 190	Radio-frequency Power in Plasmas (Irvine, CA, 1989)	89-45805	0-88318-397-8
No. 191	Advances in Laser Science–IV (Atlanta, GA, 1988)	89-85595	0-88318-391-9
No. 192	Vacuum Mechatronics (First International Workshop) (Santa Barbara, CA, 1989)	89-45905	0-88318-394-3
No. 193	Advanced Accelerator Concepts (Lake Arrowhead, CA, 1989)	89-45914	0-88318-393-5

No. 194	Quantum Fluids and Solids—1989 (Gainesville, FL, 1989)	89-81079	0-88318-395-1
No. 195	Dense Z-Pinches (Laguna Beach, CA, 1989)	89-46212	0-88318-396-X
No. 196	Heavy Quark Physics (Ithaca, NY, 1989)	89-81583	0-88318-644-6
No. 197	Drops and Bubbles (Monterey, CA, 1988)	89-46360	0-88318-392-7
No. 198	Astrophysics in Antarctica (Newark, DE, 1989)	89-46421	0-88318-398-6
No. 199	Surface Conditioning of Vacuum Systems (Los Angeles, CA, 1989)	89-82542	0-88318-756-6
No. 200	High T_c Superconducting Thin Films: Processing, Characterization, and Applications (Boston, MA, 1989)	90-80006	0-88318-759-0
No. 201	QED Stucture Functions (Ann Arbor, MI, 1989)	90-80229	0-88318-671-3
No. 202	NASA Workshop on Physics From a Lunar Base (Stanford, CA, 1989)	90-55073	0-88318-646-2
No. 203	Particle Astrophysics: The NASA Cosmic Ray Program for the 1990s and Beyond (Greenbelt, MD, 1989)	90-55077	0-88318-763-9
No. 204	Aspects of Electron–Molecule Scattering and Photoionization (New Haven, CT, 1989)	90-55175	0-88318-764-7
No. 205	The Physics of Electronic and Atomic Collisions (XVI International Conference) (New York, NY, 1989)	90-53183	0-88318-390-0
No. 206	Atomic Processes in Plasmas (Gaithersburg, MD, 1989)	90-55265	0-88318-769-8
No. 207	Astrophysics from the Moon (Annapolis, MD, 1990)	90-55582	0-88318-770-1
No. 208	Current Topics in Shock Waves (Bethlehem, PA, 1989)	90-55617	0-88318-776-0
No. 209	Computing for High Luminosity and High Intensity Facilities (Santa Fe, NM, 1990)	90-55634	0-88318-786-8
No. 210	Production and Neutralization of Negative Ions and Beams (Brookhaven, NY, 1990)	90-55316	0-88318-786-8
No. 211	High-Energy Astrophysics in the 21st Century (Taos, NM, 1989)	90-55644	0-88318-803-1
No. 212	Accelerator Instrumentation (Brookhaven, NY, 1989)	90-55838	0-88318-645-4
No. 213	Frontiers in Condensed Matter Theory (New York, NY, 1989)	90-6421	0-88318-771-X 0-88318-772-8 (pbk.)
No. 214	Beam Dynamics Issues of High-Luminosity Asymmetric Collider Rings (Berkeley, CA, 1990)	90-55857	0-88318-767-1
No. 215	X-Ray and Inner-Shell Processes (Knoxville, TN, 1990)	90-84700	0-88318-790-6

No. 216	Spectral Line Shapes, Vol. 6 (Austin, TX, 1990)	90-06278	0-88318-791-4
No. 217	Space Nuclear Power Systems (Albuquerque, NM, 1991)	90-56220	0-88318-838-4
No. 218	Positron Beams for Solids and Surfaces (London, Canada, 1990)	90-56407	0-88318-842-2
No. 219	Superconductivity and Its Applications (Buffalo, NY, 1990)	91-55020	0-88318-835-X
No. 220	High Energy Gamma-Ray Astronomy (Ann Arbor, MI, 1990)	91-70876	0-88318-812-0
No. 221	Particle Production Near Threshold (Nashville, IN, 1990)	91-55134	0-88318-829-5
No. 222	After the First Three Minutes (College Park, MD, 1990)	91-55214	0-88318-828-7
No. 223	Polarized Collider Workshop (University Park, PA, 1990)	91-71303	0-88318-826-0
No. 224	LAMPF Workshop on (π, K) Physics (Los Alamos, NM, 1990)	91-71304	0-88318-825-2
No. 225	Half Collision Resonance Phenomena in Molecules (Caracus, Venezuela, 1990)	91-55210	0-88318-840-6
No. 226	The Living Cell in Four Dimensions (Gif sur Yvette, France, 1990)	91-55209	0-88318-794-9
No. 227	Advanced Processing and Characterization Technologies (Clearwater, FL, 1991)	91-55194	0-88318-910-0
No. 228	Anomalous Nuclear Effects in Deuterium/Solid Systems (Provo, UT, 1990)	91-55245	0-88318-833-3
No. 229	Accelerator Instrumentation (Batavia, IL, 1990)	91-55347	0-88318-832-1
No. 230	Nonlinear Dynamics and Particle Acceleration (Tsukuba, Japan, 1990)	91-55348	0-88318-824-4
No. 231	Boron-Rich Solids (Albuquerque, NM, 1990)	91-53024	0-88318-793-4
No. 232	Gamma-Ray Line Astrophysics (Paris–Saclay, France, 1990)	91-55492	0-88318-875-9
No. 233	Atomic Physics 12 (Ann Arbor, MI, 1990)	91-55595	088318-811-2
No. 234	Amorphous Silicon Materials and Solar Cells (Denver, CO, 1991)	91-55575	088318-831-7
No. 235	Physics and Chemistry of MCT and Novel IR Detector Materials (San Francisco, CA, 1990)	91-55493	0-88318-931-3
No. 236	Vacuum Design of Synchrotron Light Sources (Argonne, IL, 1990)	91-55527	0-88318-873-2
No. 237	Kent M. Terwilliger Memorial Symposium (Ann Arbor, MI, 1989)	91-55576	0-88318-788-4